Simon Beecham and Julia Piantadosi (Eds.)

Water Resources in a Variable and Changing Climate

MDPI

This book is a reprint of the special issue that appeared in the online open access journal *Water* (ISSN 2073-4441) in 2014 and 2015 (available at: http://www.mdpi.com/journal/water/special_issues/water_climate).

Guest Editors
Simon Beecham
Division of Information Technology
Engineering and the Environment
School of Natural and Built Environments, University of South Australia
South Australia 5001
Australia

Julia Piantadosi
Division of Information Technology
Engineering and the Environment
School of Information Technology and Mathematical Sciences
Centre for Industrial and Applied Mathematics, University of South Australia
South Australia 5001
Australia

Editorial Office
MDPI AG
Klybeckstrasse 64
Basel, Switzerland

Publisher
Shu-Kun Lin

Managing Editor
Cherry Gong

1. Edition 2015

MDPI • Basel • Beijing • Wuhan

ISBN 978-3-03842-082-8 (PDF)
ISBN 978-3-03842-083-5 (Hbk)

Table of Contents

List of Contributors

Asma Abahussain: Department of Natural Resources and Environment, College of Graduate Studies, Arabian Gulf University, Manama 26671, Bahrain

Euloge K. Agbossou: Faculty of Agricultural Sciences, University of Abomey-Calavi, Cotonou 01 BP. 526, Benin

Mushtaque Ahmed: Department of Soils, Water & Agricultural Engineering, College of Agricultural and Marine Sciences, Sultan Qaboos University, PO Box 34, Al-Khod 123, Oman; Department of Civil Engineering, KU Leuven, Kasteelpark Arenberg 40, bus 2448, Leuven 3001, Belgium

Mary Akurut: Department of Civil Engineering, KU Leuven, Kasteelpark Arenberg 40, bus 2448, Leuven 3001, Belgium

Mohammed Saif Al-Kalbani: Centre for Mountain Studies, Perth College, University of the Highlands and Islands, Crieff Road, Perth PH1 2NX, UK

Ricardo Arias: Facultad de Ciencias and Centro de Investigaciones Científicas Avanzadas (CICA), University of A Coruña, Campus de A Coruña, A Coruña 15071, Spain

Rupak Aryal: Centre for Water Management and Reuse, School of Natural and Built Environments, University of South Australia, Mawson Lakes, SA 5095, Australia

Simon Beecham: Centre for Water Management and Reuse (CWMR), School of Natural and Built Environments, University of South Australia, Mawson Lakes, SA 5095, Australia

Morgan Bida: Laboratory Science Technology Program, Department of Science & Mathematics, Rochester Institute of Technology, Rochester, NY 14623, USA

Henning Bjornlund: School of Commerce, University of South Australia, Adelaide 5000, Australia

John Boland: Centre for Industrial and Applied Mathematics, University of South Australia, Mawson Lakes Boulevard, Mawson Lakes, SA 5095, Australia

Aymar Y. Bossa: West African Science Service Center on Climate Change and Adapted Land Use—WASCAL, Ouagadougou 06 P.O. Box 9507, Burkina Faso; Department of Geography, University of Bonn, Meckenheimer Allee 166, Bonn 53115, Germany

Robert Brooks: Faculty of Business and Economics, Monash University, Caulfield East 3145, Australia

Mike D. Burch: South Australian Water Corporation, Adelaide, SA 5000, Australia

Robert I. Daly: South Australian Water Corporation, Adelaide, SA 5000, Australia

Amalia Davies: Pakistan Strategy Support Program (PSSP), the International Food Policy Research Institute (IFPRI), IFPRI-PSSP Office #006, Islamabad, Pakistan

Stephen Davies: Pakistan Strategy Support Program (PSSP), the International Food Policy Research Institute (IFPRI), IFPRI-PSSP Office #006, Islamabad, Pakistan

Bernd Diekkrüger: Department of Geography, University of Bonn, Meckenheimer Allee 166, Bonn 53115, Germany

Eihab Fathelrahman: Department of Agribusiness and Consumer Sciences, College of Food and Agriculture, United Arab Emirates University, Al-Magam Campus, P.O. Box 15551, Al-Ain, United Arab Emirates

Alistair Grinham: School of Civil Engineering, The University of Queensland, St Lucia, QLD 4072, Australia

Janel Hanrahan: Atmospheric Sciences Department, Lyndon State College, PO Box 919, Lyndonville, VT 05851, USA

Edwyna Harris: Department of Economics, Monash University, Clayton 3168, Australia

Wen-Cheng Huang: Department of Harbor and River Engineering, National Taiwan Ocean University, Keelung 20224, Taiwan

Mohammad Kamruzzaman: Centre for Water Management and Reuse (CWMR), School of Natural and Built Environments, University of South Australia, Mawson Lakes, SA 5095, Australia

Jan Jacob Keizer: Centre for Environmental and Marine Studies (CESAM) and Department Environment & Planning, University of Aveiro, Aveiro 3810-193, Portugal

Jonathan E. Kenny: Department of Chemistry, Tufts University, Medford, MA 02155, USA

Sergey Kravtsov: Department of Mathematical Sciences, Atmospheric Sciences Group, University of Wisconsin-Milwaukee, P.O. Box 413, Milwaukee, WI 53201, USA

Jyun-Long Lee: Department of Harbor and River Engineering, National Taiwan Ocean University, Keelung 20224, Taiwan

Zongli Li: General Institute of Water Resources and Hydropower Planning and Design, Ministry of Water Resources of China, Beijing 100120, China

Hong-Ming Liu: Department of Civil and Disaster Prevention Engineering, National United University, Miaoli 36003, Taiwan

Wen-Cheng Liu: Department of Civil and Disaster Prevention Engineering, National United University, Miaoli 36003, Taiwan; Taiwan Typhoon and Flood Research Institute, National Applied Research Laboratories, Taipei 10093, Taiwan

Charles B. Niwagaba: Department of Civil and Environmental Engineering, Makerere University Kampala, P. O. Box 7062, Kampala 00256, Uganda

Joao Pedro Nunes: Centre for Environmental and Marine Studies (CESAM) and Department Environment & Planning, University of Aveiro, Aveiro 3810-193, Portugal

Timothy O'Higgins: Scottish Association for Marine Sciences, University of the Highlands and Islands, Scottish Marine Institute, Dunstaffnage, Argyll PA37 1QA, UK

Todd Pagano: Laboratory Science Technology Program, Department of Science & Mathematics, Rochester Institute of Technology, Rochester, NY 14623, USA

Martin F. Price: Centre for Mountain Studies, Perth College, University of the Highlands and Islands, Crieff Road, Perth PH1 2NX, UK

James Pritchett: Department of Agricultural and Resource Economics, Colorado State University, Campus Mail 1172, Fort Collins, CO 80523, USA

M. Luz Rodríguez-Blanco: Facultad de Ciencias and Centro de Investigaciones Científicas Avanzadas (CICA), University of A Coruña, Campus de A Coruña, A Coruña 15071, Spain
Paul Roebber: Department of Mathematical Sciences, Atmospheric Sciences Group, University of Wisconsin-Milwaukee, P.O. Box 413, Milwaukee, WI 53201, USA
Md Sumon Shahriar: CSIRO Computational Informatics (CCI), Hobart, TAS 7001, Australia
Leszek Sobkowiak: Institute of Physical Geography and Environmental Planning, Adam Mickiewicz University, Poznan 61-680, Poland
M. Mercedes Taboada-Castro: Facultad de Ciencias and Centro de Investigaciones Científicas Avanzadas (CICA), University of A Coruña, Campus de A Coruña, A Coruña 15071, Spain
M. Teresa Taboada-Castro: Facultad de Ciencias and Centro de Investigaciones Científicas Avanzadas (CICA), University of A Coruña, Campus de A Coruña, A Coruña 15071, Spain
Leon van der Linden: South Australian Water Corporation, Adelaide, SA 5000, Australia
Ann Wheeler: School of Commerce, University of South Australia, Adelaide 5000, Australia
Patrick Willems: Department of Civil Engineering, KU Leuven, Kasteelpark Arenberg 40, bus 2448, Leuven 3001, Belgium; Department of Hydrology and Hydraulic Engineering, Vrije Universiteit Brussel, Pleinlaan 2, Brussels 1050, Belgium
Jun Xia: Key Laboratory of Water Cycle & Related Land Surface Processes, Institute of Geographic Sciences and Natural Resources Research, Chinese Academy of Sciences, Beijing 100101, China; State Key Laboratory of Water Resources & Hydropower Engineering Science, Wuhan University Wuhan 430072, China
Lingling Zhao: Guangzhou Institute of Geography, No.100 Xianliezhong Road, Guangzhou 510070, China; Key Laboratory of Water Cycle & Related Land Surface Processes, Institute of Geographic Sciences and Natural Resources Research, Chinese Academy of Sciences, Beijing 100101, China
Alec Zuo: School of Commerce, University of South Australia, Adelaide 5000, Australia

About the Guest Editors

Simon Beecham, PhD, is currently Pro Vice-Chancellor of the Division of Information Technology, Engineering and the Environment at the University of South Australia. His research interests centre around the effects of climate change on integrated water management and he is currently Chair of the International Water Association's International Group on Urban Rainfall. He is also a Fellow of Engineers Australia and a Fellow of the Australian Institute of Company Directors.

Julia Piantadosi, PhD. Julia is a member of the Centre for Industrial and Applied Mathematics at the University of South Australia. Her research focus lies in the application of advanced mathematical tools to the solution of unsolved problems relating to the management of water supply and distribution and stochastic rainfall modelling on various time scales for a variety of applications. Other research interests include proactive responses to climate change, risk analysis, integrated challenges around the water-energy nexus and waste management.

Preface

Climate change will bring about significant changes to the capacity of, and the demand on, water resources. The resulting changes include increasing climate variability that is expected to affect hydrologic conditions. The effects of climate variability on various meteorological variables have been extensively observed in many regions around the world. Atmospheric circulation, topography, land use and other regional features modify global changes to produce unique patterns of change at the regional scale. As the future changes to these water resources cannot be measured in the present, hydrological models are critical in the planning required to adapt our water resource management strategies to future climate conditions. Such models include catchment runoff models, reservoir management models, flood prediction models, groundwater recharge and flow models, and crop water balance models. In water-scarce regions such as Australia, urban water systems are particularly vulnerable to rapid population growth and climate change. In the presence of climate change induced uncertainty, urban water systems need to be more resilient and multi-sourced. Decreasing volumetric rainfall trends have an effect on reservoir yield and operation practices. Severe intensity rainfall events can cause failure of drainage system capacity and subsequent urban flood inundation problems. Policy makers, end users and leading researchers need to work together to develop a consistent approach to interpreting the effects of climate variability and change on water resources.

This Special Edition includes papers by international experts who have investigated climate change impacts on a variety of systems including irrigation and water markets, land use changes and vegetation growth, lake water levels and quality and sea level rises. These investigations have been conducted in many regions of the world including the USA, China, East Africa, Australia, Taiwan and the Sultanate of Oman.

Simon Beecham and Julia Piantadosi
Guest Editors

Assessment of Short Term Rainfall and Stream Flows in South Australia

Mohammad Kamruzzaman, Md Sumon Shahriar and Simon Beecham

Abstract: The aim of this study is to assess the relationship between rainfall and stream flow at Broughton River in Mooroola, Torrance River in Mount Pleasant, and Wakefield River near Rhyine, in South Australia, from 1990 to 2010. Initially, we present a short term relationship between rainfall and stream flow, in terms of correlations, lagged correlations, and estimated variability between wavelet coefficients at each level. A deterministic regression based response model is used to detect linear, quadratic and polynomial trends, while allowing for seasonality effects. Antecedent rainfall data were considered to predict stream flow. The best fitting model was selected based on maximum adjusted R^2 values (R^2_{adj}), minimum sigma square (σ^2), and a minimum Akaike Information Criterion (AIC). The best performance in the response model is lag rainfall, which indicates at least one day and up to 7 days (past) difference in rainfall, including offset cross products of lag rainfall. With the inclusion of antecedent stream flow as an input with one day time lag, the result shows a significant improvement of the R^2_{adj} values from 0.18, 0.26 and 0.14 to 0.35, 0.42 and 0.21 at Broughton River, Torrance River and Wakefield River, respectively. A benchmark comparison was made with an Artificial Neural Network analysis. The optimization strategy involved adopting a minimum mean absolute error (MAE).

Reprinted from *Water*. Cite as: Kamruzzaman, M.; Shahriar, M.S.; Beecham, S. Assessment of Short Term Rainfall and Stream Flows in South Australia. *Water* **2014**, *6*, 3528-3544.

1. Introduction

A review of rainfall-runoff modeling has been given by [1]. Rainfall and stream flow models can be applied to a diverse range of purposes including daily control of reservoirs, projecting future stream flows and flood management. Rainfall and stream-flow models can be classified as physically based, conceptual and empirical. Physically-based models include the Système Hydrologique Européan with sediment and solute transport [2] and Gridded Surface Subsurface Hydrologic Analysis [3] both of which require extensive spatial and temporal data and typically are used for small catchments. An example of a conceptual based model is the Modèle du Génie Rural à 4 paramètres Journalier (GR4J), which has been developed for understanding catchment hydrological behavior [4]. Other examples of conceptual rainfall-runoff models are the Sacramento Soil Moisture Accounting Model [5] and the SIMulation and HYDrologic model (SIMHYD) [6], which can be applied either as a lumped or gridded application. SIMHYD estimates daily stream flows from daily rainfall and areal potential evapotranspiration data. The class of empirical models includes time series models [7–13]. An advantage of an empirical model is that it can be fitted to situations where the hydrological data are restricted to rainfall and stream flow time series. A further advantage is that in a parametric test, a distribution can be fitted for assessing the hydrological behavior for any time period in any region. In addition, they can represent either linear or non-linear relationships. Time series models perform as well as physically-based alternatives [14]. Combined a conceptual model with an artificial neural network (ANN) for forecasting inflow into the Daecheong Dam in Korea [15]. Compared the wavelet decompositions of rainfall and runoff at four sites in the Tianshan Mountains [16]. They aimed to distinguish between errors in timing and errors in magnitude of hydrograph peaks. They used a cross-wavelet technique to quantify timing errors and hence provided an empirical adjustment to model predictions of stream flow.

In this study, we have proposed a novel method for assessing short-term rainfall and stream flow models. The travel time between rainfall and stream flow gauges using cross-correlation functions [10,17]. They reported that the travel time was less than one day for the Onkaparinga catchment in South Australia. In this paper, we presume that there is a higher order relationship between rainfall gauge and stream flow data. It is, therefore, important in this study to construct the correlation structure. Linear regression models are commonly used for time series analysis [18], particularly for assessing evidence of trends, higher order changes and variability, including allowing for seasonality. We developed deseasonalized and detrended time series rainfall and stream flow models from deterministic regression models including linear, quadratic and cubic terms. These models take account of both lag rainfall and the influence of stream flow. The results of this study will be useful for water managers and policy makers involved in sustainable water resource management and climate change adaptation for the catchments used in this study. The approach is capable of modeling the non-linear relationships between inputs and outputs using ANNs [19]. The first advantage of ANN is that it only requires a small number of parameters and learns through a number of training iterations involving adjusting the parameters (weights) of the network [20]. A second advantage is that it is useful in situations where it is complex to build a physical or conceptual model, such as hydrological modeling of rainfall- stream flow processes [21–25]. ANN models were useful to find the relationships between rainfall and river flow data in a river basin in India [26]. We present a statistical approach that uses the deterministic features of a regression model to build many neural networks with a combination of different lagged input patterns. A wavelet based regression model for stream flow using the discrete wavelet transform (DWT) of the entire time series [27]. They also provided a comparison of their model performance with ANN. A chaotic stream flow model using an ensemble wavelet network [28]. Used wavelet analyses of rainfall and runoff and wavelet rainfall–runoff cross-analyses to investigate the temporal variability of the rainfall-runoff relationship [17,29]. They found that wavelet transforms provide a physical explanation of the temporal structure of the catchment response.

2. Data Collection and Preparation

The analysis is based on data from three rainfall and stream flow stations in South Australia, as presented in Figure 1. The Broughton River (BR) station is at Mooroola, which is located approximately 40 km north of Port Broughton and 20 km south west of Port Pirie. Torrance River (TR) station is located at Mount Pleasant, and its rivers and tributaries are highly variable in flow and together drain an area of 508 km^2. Wakefield River (WR) is an ephemeral river near Rhynie, with a catchment area of approximately 1913 km^2.

Figure 1. Location of Broughton River (BR), Torrance River (TR), and Wakefield River (WR).

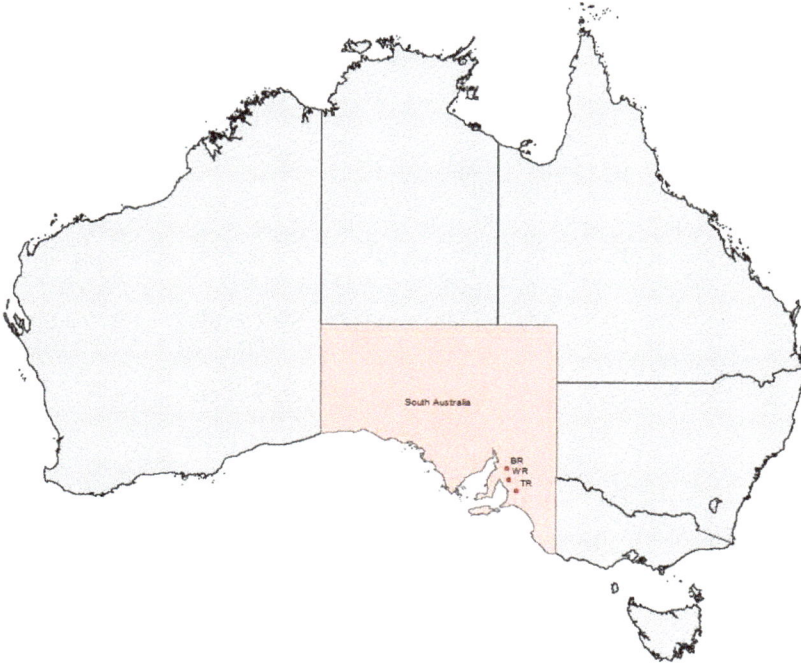

The elevation of the river may indicate the hydrological feature, presented in Table 1, Column 4. These stations were selected because they had long records of rainfall and stream flow and the highest quality control in terms of Australian Bureau of Meteorology, [30] and the Department for Environment, Water and Natural Resources [31] quality designations for rainfall and stream flow records. Information on these stations and data quality are presented in Table 1.

Table 1. Weather stations information, data quality and observations.

| Stations name | ID | Location | | Elevation | Variables | Data period | | % of |
		Latitude	Longitude			Start	End	Missing
Broughton River at Mooroola	A5070503	−33.53	138.51	196 m	Rainfall	Jun. 1989	Dec. 2011	0.1
					Stream flow	Jun. 1972	Dec. 2011	0.7
Torrance at Mount Pleasant	A5040512	−34.78	139.02	414.7 m	Rainfall	Jun. 1989	Dec. 2011	0.6
					Stream flow	May 1973	Dec. 2011	0.1
Wakefield river near Rhyine	A5060500	−34.13	138.63	202 m	Rainfall	Sep. 1985	Dec. 2011	0.9
					Stream flow	Jun. 1971	Dec. 2011	0.2

In this paper, there was less than 1% missing data and these were replaced by the mean of the series of rainfall and stream flow, to give an unbroken time series for analysis. Methods for replacing periods of missing values are discussed [18,32]. In this paper, we propose a dyadic signal time period (*i.e.*, 2^n where n is an integer and $n \geq 0$, for assessing the relationship between daily rainfall and stream flow during the period 1990–2012. We observe the discrete sequence of time series $\{y_t\}$ where $\{y_t\}$ is an integer ranging in length. We extract multi-level information of observed rainfall and stream flow series in three catchments in South Australia using the Haar wavelet decomposition. We split $\{y_t\}$ into 10 sub-time series of length power two $i.e.$, 2^n, where n is the level of the time series, starting from 0. We also investigate the correlation between rainfall and stream flow patterns for each sub-series from levels 0 to 8.

3. Statistical Analysis

3.1. Assessing the Relationship between Rainfall and Stream Flow

The open source software R [33] was used for the analyses in this paper. We calculate 10 subseries of rainfall and stream flow from 1990 to 2012 using the "wavethresh" R routine packages [34,35] for assessing the relationship between rainfall and stream flow. The length of time taken into account in 10 subseries for rainfall and stream flow is a period of 512 days.

The relationship between rainfall and stream flow within 10 subseries is presented in Figure 2. The maximum correlation coefficients are 0.08, 0.23 and 0.31 at Broughton River, Torrance River and Wakefield River, respectively. These values are between −1 and +1 in all cases, indicating the degree of linear dependence between rainfall and stream flow. For assessing short term spatial variability, a correlation coefficient of the sub-series of rainfall and stream flow less than 0.4 indicates a significant difference from 0 at each station. For example, in sub-series 2, the correlation coefficient was 0.04, 0.15 and 0.28 which indicates the independence of rainfall and stream flow at Broughton River, Torrance River and Wakefield River, respectively.

Figure 2. Correlation pattern subseries of rainfall and stream flow time series.

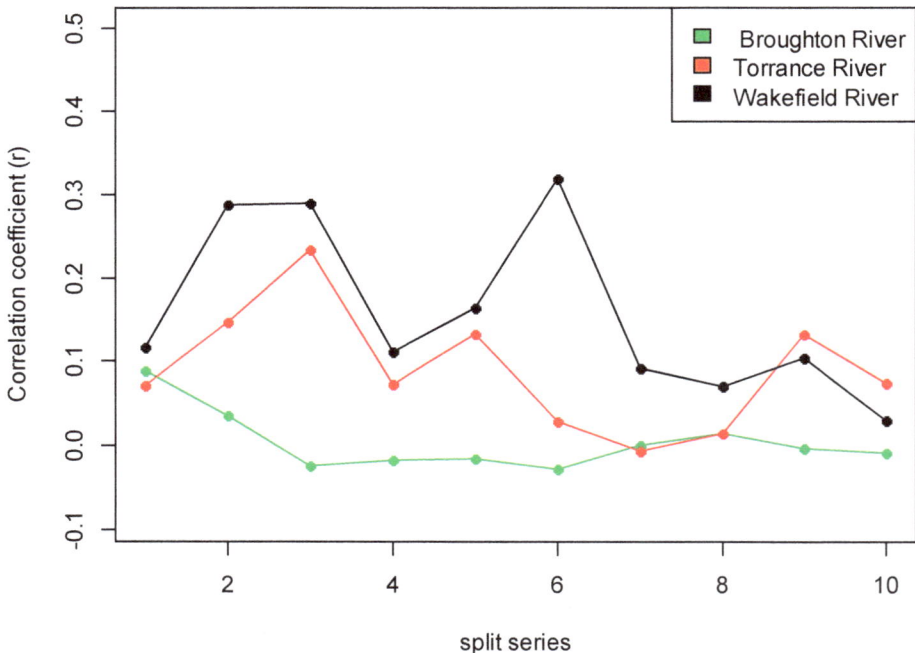

In order to understand stream flow availability under the climatic conditions in South Australia, we investigated the characteristics of rainfall and stream flow patterns, as categorized by climatic phenomena. A statistical measure of the dispersion of rainfall and stream flow patterns around the mean is defined as follows:

$$CV = \frac{S_x}{\mu_x}$$

(1)

where CV is defined as the coefficient of variation and is represented by the ratio of the standard deviation (S_x) to the mean (μ_x). Table 2 shows the degree of variation in rainfall and stream flow patterns.

Table 2. Rainfall and stream flow variability at Broughton River, Torrance River and Wakefield River in South Australia (SA) from 1990 to 2011.

Statistics	Broughton River		Torrance River		Wakefield River	
	Rainfall	Stream flow	Rainfall	Stream flow	Rainfall	Stream flow
Mean	1.653	9.817	1.530	5.396	1.282	25.333
Estimated standard deviation	0.385	4.075	0.296	4.201	0.223	21.300
Coefficient of variation (CV)	23.31%	41.51%	19.36%	77.85%	17.40%	84.07%

In Table 2, the CV for stream flow patterns indicates higher variability than for the rainfall series.

Figure 3 shows the variability of the wavelet coefficients from levels 0 to 8. The evidence of association between the rainfall and stream flow coefficient is strongly correlated at the 5% significance level in Table 1.

Figure 3. Standard deviations of wavelet coefficients of rainfall and stream flow from level 0 to 8. (**a**) Rainfall; (**b**) Stream flow.

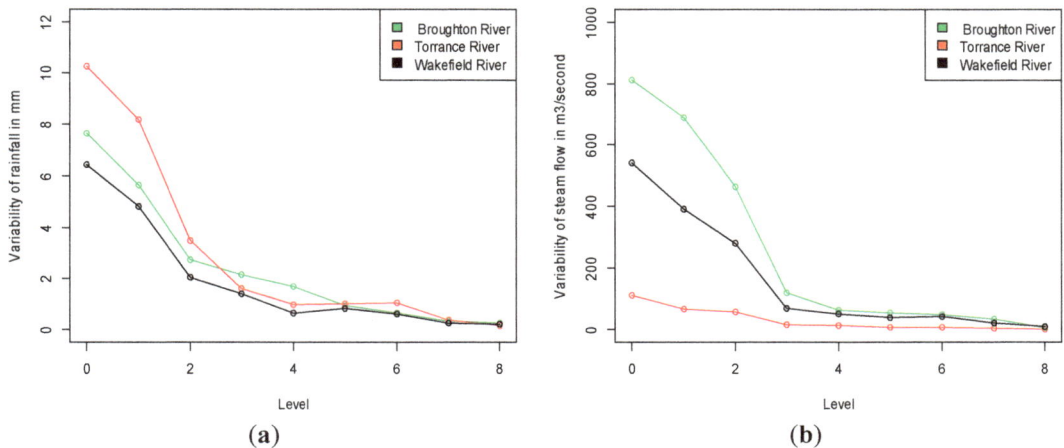

(**a**)   (**b**)

3.2. Correlation Structures between Rainfall and Stream Flow

In the previous sections, we calculated wavelet coefficients for each subset of the rainfall and stream flow series. In order to filter each of those series, we applied Haar wavelets.

The constructed correlation pattern for each rainfall and stream flow sub-series for levels 0 to 8 is given by:

$$r_k = \sum_{k=0}^{8} \frac{\sum_{j=1}^{10} (X_{wd_j,k} - \overline{X}_{wd,k})(Y_{wd_j,k} - \overline{Y}_{wd,k})}{S_{xwd} S_{ywd}} \tag{2}$$

and

$$\overline{X}_{wd} = \sum_{j=1}^{10} X_{wd_j} / 10, \ \overline{Y}_{wd} = \sum_{j=1}^{10} Y_{wd_j} / 10, \ (j = 1, 2, 3, \ldots 10, \text{ and } k = 0, 1, 2, \ldots 8) \tag{3}$$

where r_k is the constructed correlation with level n from 0 to 8 and X_{wd_j}, Y_{wd_j} is the *j*th sub-series of the rainfall and stream flow wavelet decomposed with the Haar procedure. The results are presented in Table 3.

Table 3. Constructed correlation pattern for different levels between (a) adjusted rainfall and adjusted stream flow; (b) squared adjusted rainfall and adjusted stream flow; (c) adjusted rainfall and squared adjusted stream flow; (d) squared adjusted rainfall and squared adjusted stream flow.

Day	Broughton River				Torrance River				Wakefield River			
s	a	b	c	d	a	b	c	d	a	b	c	d
1	0.71 **	0.89 ***	0.70 **	0.86 ***	0.76 **	−0.184	−0.513	0.447	0.76 **	−0.53 *	−0.59 *	0.86 ***
2	0.65 *	0.264	0.469	0.449	0.72 **	0.57	0.158	0.265	0.384	0.413	0.201	0.038
4	0.56 *	−0.189	0.257	0.146	0.63 *	0.061	−0.27	0.125	0.324	−0.28	−0.341	0.115
8	0.087	0.369	−0.296	0.103	0.032	0.55 *	0.173	−0.51 *	−0.369	0.483	0.191	−0.418
16	0.233	0.08	−0.009	−0.326	0.275	−0.15	0.009	−0.311	0.306	−0.652	0.081	−0.393
32	0.094	0.055	−0.059	−0.248	0.68 *	−0.84 **	−0.81 ***	0.97 ***	−0.116	0.002	−0.382	−0.121
64	0.411	0.036	0.292	0.238	0.488	0.67 *	0.456	0.71 **	−0.091	0.166	−0.005	−0.301
128	0.423	−0.409	0.604 *	0.68 *	0.299	−0.162	0.186	−0.223	0.279	−0.575	0.128	−0.51 *
512	0.456	0.354	−0.292	0.343	−0.218	−0.405	0.007	−0.405	0.094	0.117	0.098	−0.163

Notes: * Coefficients are statistically significant at 5%; ** Coefficients are statistically significant at 1%; *** Coefficients are statistically significant at 0.1%

The evidence of significant correlation ($r \geq 0.50$) between rainfall and stream flow wavelet coefficient series with at least a 5% significance level is shown in Table 3. Furthermore, to avoid co-linearity problems, the squared rainfall and stream flow wavelet coefficient series are also included. We found that a correlation structure ($r = 0.56$) such as stream flow is determined by rainfall on at least 4 days with 5% level at the Broughton River Basin and Torrance River Basin, as shown in Table 3. The adjusted squared stream flow and rainfall has a little evidence of correlations (*i.e.*, at 5% level) up to 64 days at Torrance River at, also a marginal correlation ($r = 0.51$) up to 128 days within squared adjusted rainfall and adjusted stream flow at Wakefield River. The rainfall and stream flow relationship was used to develop a response model for predicting stream flow.

3.3. Rainfall-Stream Flow Response Modeling

The constructed correlations described in the previous section may be partly due to common seasonal variations and trends, so a first step is to estimate these deterministic features with regression models for entire period from 1990 to 2010. The residuals from these regressions are reformed to the deseasonalized and detrended (dsdt) time series. For all three stations, a cubic trend gave a statistically improved fit over a linear or quadratic trend over the study period. The seasonal variation was reasonably modelled by a sinusoidal curve. Therefore, the regression models are of the form:

$$T_i = \beta_0 + \beta_1 \times time + \beta_2 \times time^2 + \beta_3 \times time^3 + \beta_4 \times C + \beta_5 \times S + \varepsilon_t \tag{4}$$

where, T_i represents either rainfall or stream flow; time is the mean adjusted time, that is $(t - \bar{t})$ where t is the number of days from the start of the record and $\bar{t}$ is the mean of t, $time^2$ and $time^3$, which allows for possible quadratic and cubic trends; C is $\cos(2\pi t/365.25)$ and S is $\sin(2\pi t/365.25)$ and together these allow for seasonal variation of period one cycle per year; β_j are the unknown coefficients to be estimated; and ε_t are random variations with mean 0 and constant standard derivation.

For the estimated coefficients, only a few values are significantly different from 0 even at the 5% significance level, as shown in Table 4. There is evidence of significantly different trends in rainfall at Wakefield River, which may have corresponded to increased stream flows if rainfall is increased. We have predicted the stream flow (Y_t) on day t from rainfall (X_t) with corresponding lags k. This is referred to as a Response Model (RM). The regression is defined as:

$$Y_t = \beta_0 + \beta_1 \times X_1 + \beta_2 \times X_2 + \ldots\ldots\ldots\ldots \beta_{128} \times X_{128} + \varepsilon_t \tag{5}$$

We assess stream flow in response to rainfall at lags 0 to 128. The best fitted model is selected based on the adjusted coefficient of determination; (R^2_{adj}); minimum sigma squared (σ^2) and the Akaike Criterion Information (AIC); The AIC is defined as:

$$\text{AIC} = 2 \times \text{number of parameters} - 2 \text{ Log(L)} \tag{6}$$

where L is the maximized value of the likelihood function for the estimated model. Comparisons of the AIC for different model is as shown in Table 5. The R^2_{adj} value significantly reduces and the estimated stream flow influence is close to zero after the exogenous rainfall at lag 7. Therefore, we reduced the exogenous rainfall at lags from 128 to 7 in the response model; referred to as RM0 in Table 5. This strategy is sub-optimal inasmuch as rejected terms might meet the retention criterion if added back individually. However; any small improvement in R^2_{adj} would be balanced by increased complexity in the model; which is undesirable if interaction and squared terms are added. The regression model is defined as RM:

$$Y_t = \beta_0 + \beta_1 \times X_1 + \beta_2 \times X_2 + \beta_3 \times X_3 + \beta_4 \times X_4 + \beta_5 \times X_5 + \beta_6 \times X_6 + \beta_7 \times X_7 \qquad (7)$$

In the second model, we add deterministic features to the regression model including linear, quadratic and cubic terms of t, allowing for seasonality effects. This model is defined as RM_D:

$$Y_t = \beta_0 + L + \sum_{l=1}^{7} \beta_{5+l} \times X_l \qquad (8)$$

where $L = \beta_1 \times time + \beta_2 \times time^2 + \beta_3 \times time^3 + \beta_4 \times C + \beta_5 \times S_t$.

The third model is defined as RMD_AR[1] and is an autoregressive model of order 1 (AR[1]) with RM_D. It can be written in the form:

$$Y_t = \beta_0 + L + \sum_{l=1}^{7} \beta_{5+l} \times X_l + \beta_{13} \times Y_{t-1} \qquad (9)$$

The fourth model is defined as RMD_AR[2], and is an autoregressive model of order 2 (AR[2]) with RMD_AR[1]. It can be written in the form:

$$Y_t = \beta_0 + L + \sum_{l=1}^{7} \beta_{5+l} \times X_l + \beta_{13} \times Y_{t-1} + \beta_{14} \times Y_{t-2} \qquad (10)$$

Table 4. Estimated coefficients of rainfall and stream flow variability from 1990 to 2012.

Station	Statistical Summary	Intercept (β_0)	Linear Term t	Quadratic Term t	Cubic Term t
Broughton River	Estimated rainfall	1.58	−0.000042	−0.000000001	−0.000000000003
	Variability of rainfall	0.106	0.00008	0.000000017	0.000000000008
	Estimated stream flow	52.08	−0.01244 *	0.0000031 *	−0.000000000258
	Variability of stream flow	6.18	0.004661	0.0000009	0.000000000485
Torrance River	Estimated rainfall	1.424	−0.00007	0.000000019	0.000000000004
	Variability of rainfall	0.077	0.00006	0.000000012	0.000000000006
	Estimated stream flow	3.47	−0.00174 *	0.0000003 *	0.000000000149 *
	Variability of stream flow	0.607	0.0004573	0.000000092	0.000000000048
Wakefield River	Estimated rainfall	1.226	−0.000123 *	0.0000000037	0.00000000001 *
	Variability of rainfall	0.067	0.000051	0.00000001	0.000000000005
	Estimated stream flow	15.95	−0.01144 *	0.0000007	0.0000000008 *
	Variability of stream flow	3.576	0.002694	0.0000005	0.000000000280

Note: * statistical significance at 5%.

Table 5. Fitted regression model for Broughton River, Torrance River and Wakefield River.

Model	Broughton River				Torrance River				Wakefield River			
	R^2_{*}	Std. Error	AIC	RMSE *	R^2_{*}	Std. Error	AIC	RMSE *	R^2_{*}	Std. Error	AIC	RMSE *
RMO	0.16	333.6	1104.9	3.5107	0.24	31.19	742.77	5.074	0.13	195.8	1023.53	1.1777
RM_D	0.18	331.3	1103.9	3.1507	0.26	31.02	741.9	5.012	0.14	195.3	1023.1	1.1777
RMD_AR[1]	0.35	292.9	1085.1	0.0353	0.42	27.35	722.7	0.052	0.21	187.4	1016.9	0.11777
RMD_AR[2]	0.36	291.7	1084.5	0.0313	0.43	27.35	722.5	0.0452	0.22	187.4	1016.8	0.10777
RMD_tau	0.39	285.8	1081.4	0.0035	0.42	27.32	722.1	0.0411	0.23	187.3	1016.1	0.10178

Note: Asterisk (*) units are in $m^3 s^{-1}$.

Finally, we develop a model for a benchmark comparison of stream flow on day t based on the entire previous period of stream flow and their influence (τ) adding with model RM_D. This model is defined as RMD_tau. Tau (τ) is 0 if there is no stream flow influence from the previous day's rainfall. We have demonstrated an example of count stream flow influence in Table 6.

Table 6. An example of count tau and stream flow influence rainfall over time.

Stream flow	y1	y2	y3	y4	y5	y6	y7	y8	y9	y10	y11	y12	y13
	8	9	0	0	0	2	9	22	3	5	8	8	6
Rainfall	x1	x2	x3	x4	x5	x6	x7	x8	x9	x10	x11	x12	x13
	3	2	5	3.2	3	2.8	2.6	2.4	2.2	2	1.8	1.6	1.4

In the Table 6, when the day t = 6, Y6 = 2, then we count tau = 3 (number of 0), and Y6-3-1 = 9, can be applied in the referred model RMD_tau.

The model RMD_tau can be written in the form:

$$Y_t = \beta_0 + L + \sum_{l=1}^{7} \beta_{5+l} \times X_l + \beta_{13} \times \tau + \beta_{14} \times Y_{t-\tau-1} \tag{11}$$

The fitted model for predicted stream flow in response to exogenous rainfall, deterministic features of the regression model, and previous stream flow influence, is presented in Table 5. The best fitting model selection was based on minimum AIC and minimum root mean square Error (RMSE). The RMSE is defined as:

$$RMSE = \sqrt{E(\hat{Y}_t - Y_t)^2} \tag{12}$$

where, $\hat{Y}_t$ is defined as the estimated stream flow and Y_t is the observed stream flow, respectively.

The response model RM0 has 128 predictor variables namely the rainfall lags at 0 to 128. Therefore, there are 129 parameters to estimate including the intercept. The estimated rainfall effects belong to 0 up to 7 days lag, therefore we reduced the rainfall lags from 128 to 7 days and the optimized R^2_{adj} values for this model are 0.16, 0.24 and 0.13 for Broughton River, Torrance River and Wakefield River, respectively, as presented in Table 5. We also offset the cross product term of lags to further reduce the complexity of this model. The second model included linear quadratic and cubic terms, and this model is denoted as RM_D. The number of parameters to be estimated is therefore 8 + 3 = 11 and the R^2_{adj} increased to 0.18, 0.26 and 0.14 for Broughton River, Torrance River and Wakefield River, respectively, which is a practical and statistically significant improvement. We then added a first order autoregressive term, referred to as a RMD_AR[1] model, and a second order autoregressive term referred to as a RMD_AR[2] model. We also made a benchmark comparison by using the entire stream flow record and this model is denoted RMD_tau, as presented in Table 5.

In Table 5, there is evidence of improvement of R^2_{adj} values, RMSE in m^3s^{-1} from RM to RM_D. Adding autoregressive order 1 (AR[1]) with RM_D results in substantially improved R^2_{adj} values (from 0.18, 0.26, and 0.14 to 0.35, 0.42 and 0.21 for Broughton River, Torrance River and Wakefield River, respectively. Furthermore, when adding autoregressive order 1 (AR[1]) with

10

RM_D, there is evidence of improvement but this may be offset by the increasing number of parameters that affect the complexity of the model. In addition, the RMD_tau model represents a small improvement for two of the three river basins. The best fitted models are RMD_tau for Broughton River, RMD_AR[2] for Torrance River and RMD_tau for Wakefield River, were selected based on the minimum Akaike Information Criterion (AIC) and minimum root mean square error (RMSE) in m^3s^{-1}. The residuals from the best fitted models were transformed to normalized form by factor multiplication. A factor was calculated, which allows for the fact that the mean of a non-linear function of a random variable is not equal to that function of the mean. The transform series follow an identically normalized form with mean (μ) of zero, standard deviation (σ^2) of 1 and a random disturbance term (ε_t) which is uncorrelated. The transformed series were used to predict the stream flow on day t based on the predicted stream flow influence over the short term, as shown in Figure 4.

Figure 4. Predicted stream flow based on dsdt rainfall for (a) Broughton River; (b) Torrance River; and (c) Wakefield River from 1990 to 2010.

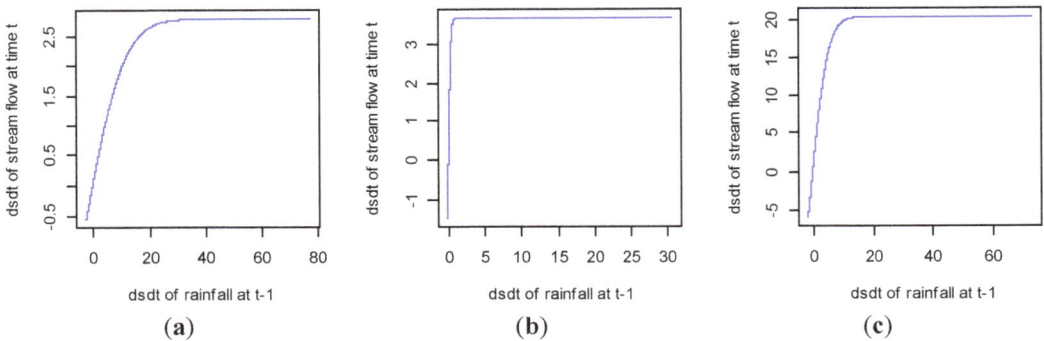

In Figure 4, we demonstrate the versatility of stream flow prediction. It can be seen that this is a non-linear relationship when expressed in terms of the physical interpretation of stream flow based on rainfall.

3.4. Modeling Stream Flow Using an Artificial Neural Network

Artificial neural network (ANN) techniques are motivated by the principles of biological nervous systems [36]. Although there are different types of ANN, the multilayer feed forward network is the most commonly used technique. For example, a common approaches of training using back-propagation in a multi-layer feed forward network [23]. The network consists of input, hidden and output layers. Each layer is fully connected with the proceeding layer with weights in each connection, as shown in Figure 5.

Figure 5. A schematic ANN including input, hidden and output layers.

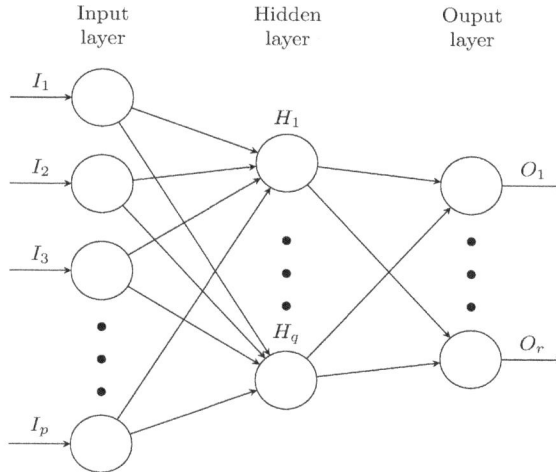

In Figure 5, the number of nodes in the input layer is p, the number of nodes in the hidden layer is q and the number of nodes in the output layer is r. The initial assigned random weights are updated during the training process by comparing the predicted output and the known output for errors. Errors are then back-propagated to adjust the weights. The dsdt of daily rainfall and stream flow data from the regression model developed in the previous section are considered for developing a prediction model for each of the three river basins for the years 1990 to 2010. A certain methods proposed such as input selection, model architecture selection, model calibration (training) and validation (testing) [37]. In addition, we emphasize the fact that ANN set-up has to be carefully achieved and described to get the reliable results. This study described the steps in building the prediction models for stream flow. We consider the prediction function as: $S_{t+1} = f(S_t, S_{t-1}, S_{t-2}, \ldots, S_{t-m}, R_t, R_{t-1}, R_{t-2}, \ldots, R_{t-n})$ where S represents stream flow, R represents rainfall, t is the current day, m = {3,...,8}, n = {3,...,8} and f represents the ANN as a regression function. We investigate necessary lagged inputs of rainfall and river flow for modeling the river flows at three locations in South Australia. We apply an artificial neural network (ANN) technique for modeling river flow. ANN models are developed with all combinations of rainfall and river flow input ranges. In addition, a standard range of nodes in the hidden layer are also considered. Among all models based on inputs and hidden nodes, the best model is selected based on mean absolute error criteria. This entire process is applied to all three locations. ANN models capture the non-linear relationships of rainfall and river flow patterns in modeling river flows from large time series data. For example, if we consider 3 days lag of stream flow and 5 days lag of rainfall, then the total number of input nodes in the ANN structure will be 8 and we consider the number of nodes in the hidden layers ranging from 1 to 10. To achieve the best model using ANN for each location, all inputs not only apply in combination, but we also consider setting a range of parameters, such as different number of nodes in the hidden layer, for each combination of inputs.

In predicting stream flow one day ahead as output, we consider stream flow and rainfall with combinations of consecutive lags where the minimum lag is 3 days and the maximum lag is 8 days. Thus, for each location, the total number of models to be trained becomes 36. As the data set is large, one year of data is considered initially for testing. For training ANN models at each location, we consider stream flow and rainfall data for the period 1990 to 2009. The remaining data for the year 2010 is used for testing the best model found in the training phase.

For the Multilayer Perceptron (MLP) function, the ANN stream flow prediction model was built using the RWeka package in R Language [38]. One of the important parameters to specify is the number of nodes in the hidden layer, which may vary for time series modeling in different locations. Using trial and error, the number of nodes in the hidden layer is considered from 1 to 10. This range is widely used in hydrological time series modeling [21]. We consider the learning rate (the amount the weights are updated) to be 0.3, momentum is 0.2 and the number of epochs to train is 500.

Application of back propagation in ANN with a sigmoidal function was used to set the normalized data in the MLP function. Furthermore, the mean absolute error (MAE) in m^3s^{-1} was minimized through an iteration process that varied the number of nodes in the hidden layer.

The best lag combination at each location is presented in Figure 6.

Figure 6. MAE for training data (1990–2009) using ANN with best lag combinations at each location, units in m^3s^{-1}.

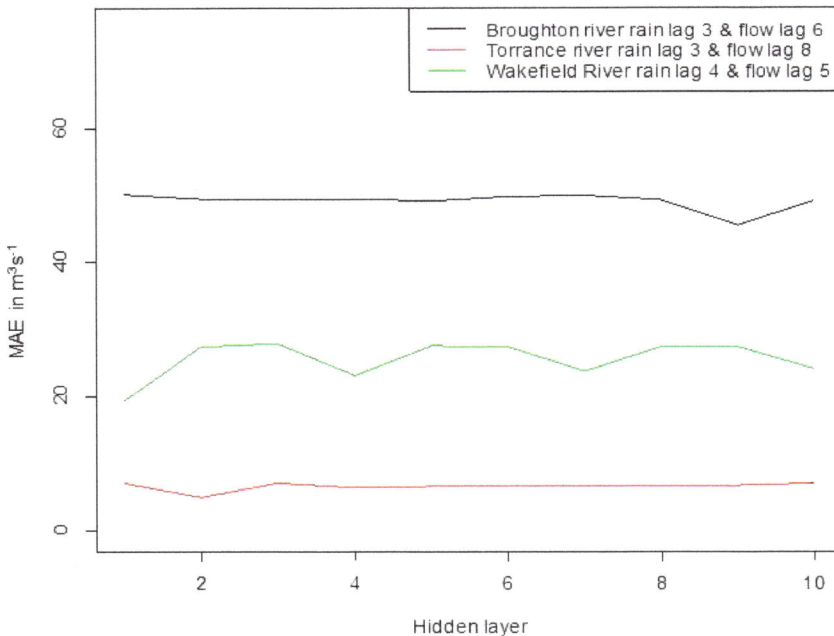

We find that both input lags and nodes in the hidden layer are different for each location. The best model based on correlation coefficient (R^2_{adj}) and the lowest root mean square error (RMSE)

and mean absolute error (MAE) for each location is presented in Table 7. For Broughton River, 3 days rainfall and 6 days stream flow as lagged inputs with 9 nodes in the hidden layers produces the lowest MSE. At Torrance River, 3 days rainfall and 8 days stream flow as lagged inputs with 2 nodes in the hidden layers produces the lowest MSE. For Wakefield River, 4 days rainfall and 5 days stream flow as lagged inputs with only one node in the hidden layer produces the lowest MSE. This indicates the variability in the ANN models for different locations.

When the best model is identified based on the training data for each location, we use this model on testing data prediction. This study show the prediction results for the testing data for each location. Figure 7 shows the predicted and observed stream flows using testing data for the locations Broughton River, Torrance River and Wakefield River, respectively.

Table 7. Best prediction model based on R_{adj}^2, lowest RMSE and MAE are in m^3s^{-1} on the training data.

Location	Input Lags	Nodes in Hidden Layer in ANN(H)	R_{adj}^2	RMSE *	MAE *
Broughton River	3 days rain, 6 days stream flow	9	0.68	270.33	45.53
Torrance River	3 days rain, 8 days stream flow	2	0.71	24.54	4.89
Wakefield River	4 days rain, 5 days stream flow	1	0.45	179.42	19.28

Note: Asterisk (*) units are in m^3s^{-1}.

Figure 7. Observed and predicted stream flow for **(a)** Broughton River; **(b)** Torrance River; and **(c)** Wakefield River for the year 2010.

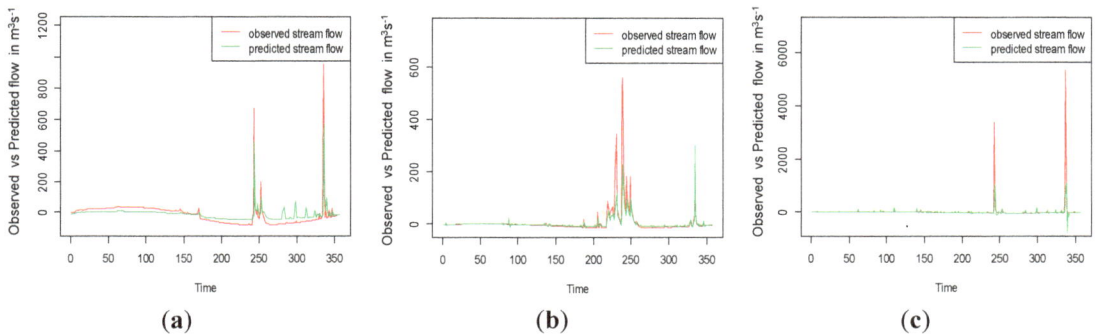

(a)   (b)   (c)

The MAE for training and testing data is shown in Figure 8 for all three locations. We observed that the MAE for the training and testing data at Broughton and Torrance Rivers do not vary significantly.

For Broughton, in training, the best ANN model structure includes 3 days lagged rainfall and 6 days lagged stream flow as inputs with 9 nodes in the hidden layer. This model has the lowest MAE, at 45.53 m^3s^{-1}. We further use this best model for testing and we find the MAE of 32.43 m^3s^{-1}. For Torrance, the ANN best model in training has 3 days lagged rainfall and 8 days lagged stream flow as inputs with 2 nodes in the hidden layer achieving the MAE of 4.89 m^3s^{-1}. For testing data, this model gives a MAE of 9.27 m^3s^{-1}. In case of Wakefield, the best ANN model has 4 days lagged rainfall and 5 days lagged stream flow as inputs with 1 node in the hidden layer

14

achieving the MAE of 19.28 m^3s^{-1}. For the testing data, this model achieves an MAE of 42.88 $m^3 s^{-1}$. The reason for the difference in MAE between the training and testing phases could be due to this river's ephemeral nature, and its substantial dependence on rainfall.

Figure 8. Comparison of MAE for training and testing data, units are in m^3s^{-1}.

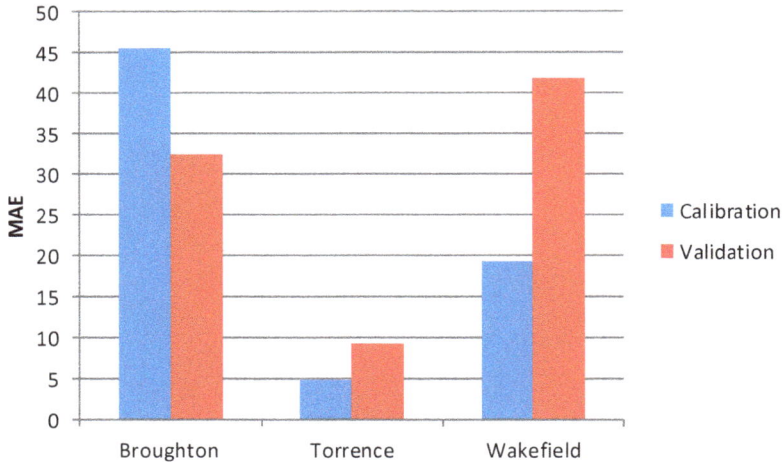

4. Conclusions

Initially, we split the whole series with a dyadic signal process for assessing the short term relationship between rainfall and stream flow including correlation using Haar wavelets. We have presented an innovative idea for the hydrological community for assessing stream flow for any catchment. In particular, the end user could assess the variability of changes and construct higher order correlations from 2 days up to as long as required. In addition, this study would be helpful for predicting stream flows using deterministic regression techniques, particularly where there is evidence of changes of statistical distribution characteristics, which is important for Water Sensitive Urban Design, as clearly demonstrated [39]. Using a deterministic regression based response model we found an increasing trend in stream flow when rainfall increased significantly. Predicted stream flow was more influenced by the previous few days' stream flows than when considering the entire previous period of stream flow. We also developed artificial neural network models for three locations. The results show that the influence of lagged rainfall and stream flow lies within a short temporal window. The results demonstrate that the ANN models perform better for Broughton and Torrance River in capturing the rainfall and stream flow relationships.

Acknowledgments

This study was funded by the Goyder Institute for Water Research under their Climate Change program. The researchers are grateful to the developers of the R project for software code and to the Australian Bureau of Meteorology for providing meteorological data.

Author Contributions

This manuscript, draft design by Mohammad Kamruzzaman and Md Sumon Shahriar, complete revised by Simon Beecham.

Conflicts of Interest

The authors declare no conflict of interest.

References

1. Moradkhani, H.; Sorooshian, S. General review of rainfall-runoff modeling: Model calibration, data assimilation, and uncertainty analysis. *Water Sci. Technol. Libr.* **2008**, *63*, 1–23.
2. Birkinshaw, S.J. Introduction. Available online: http://research.ncl.ac.uk/shetran/Introduction.htm (accessed on 26 July 2013).
3. Downer, C.W.; Ogden, F.L. GSSHA: A model for simulating diverse stream flow generating processes. *J. Hydrol. Eng.* **2004**, *9*, 161–174.
4. Perrin, C.; Michel, C.; Andreassian, V. Improvement of a parsimonious model for stream flow simulation. *J. Hydrol.* **2003**, *279*, 275–289.
5. Burnash, R.J.C.; Ferreal, R.L.; McGuire, R.A. *A Generalized Stream Flow Simulation System: Conceptual Modeling for Digital Computers*; U.S. Department of Commerce, National Weather Service and Department of Water Resources: Sacramento, CA, USA, 1973.
6. Chiew, F.H.S.; Peel, M.C.; Western, A.W. Application and testing of the simple rainfall-runoff model SIMHYD. In *Mathematical Models of Watershed Hydrology*; Singh, V.P., Frevert, D., Meyer, S., Eds.; Water Resources Publications: Littleton, CO, USA, 2002.
7. Beven, K.J. *Rainfall-Runoff Modelling: The Primer*; Wiley: Chichester, UK, 2011; pp. 1–360.
8. Castellano-Méndez, M.; Gonzàlez-Manteigo, W.; Febrero-Bande, M.; Prada-Sànchz, J.M.; Lozano-Calderon, R. Modelling of the monthly and daily behaviour of the runoff of the Xallas rivers using Box-Jenkins and neural networks methods. *J. Hydrol.* **2004**, *296*, 38–58.
9. Hipel, K.W. (Ed.) *Stochastic and Statistical Methods in Hydrology and Environmental Engineering: Time Series Analysis in Hydrology and Environmental Engineering, Water Science and Technology*; Springer: New York, NY, USA, 2010.
10. Kamruzzaman, M.; Metcalfe, A.; Beecham, S. Wavelet based rainfall-stream flow models for the South-East Murray Darling Basin. *ASCE J. Hydrol. Eng.* **2014**, *19*, 1283–1293.
11. Kamruzzaman, M.; Beecham, S.; Metcalfe, A. Climatic influence on rainfall and runoff variability in the South-East region of the Murray Darling Basin. *Int. J. Climatol.* **2013**, *33*, 291–311.
12. Kamruzzaman, M.; Beecham, S.; Metcalfe, A. Evidence for Changes in Daily Rainfall Extremes in South Australia. In Proceedings of 9th International Workshop on Precipitation in Urban Areas, IWA/IAHR, St. Moritz, Switzerland, 6–9 December 2009.

13. Solomatine, D.; See, L.; Abrahart, R. Data-driven modelling: Concepts, approaches and experiences. In *Practical Hydroinformatics*; Abrahart, R., See, L., Solomatine, D., Eds.; Springer Berlin: Heidelberg, Germany, 2008; Volume 68, pp. 17–30.

14. McIntyre, N.; Al-Qurashi, A.; Wheater, H. Regression analysis of rainfall-runoff events from an arid catchment in Oman. *Hydrol. Sci. J.* **2007**, *52*, 1103–1118.

15. Kim, Y.; Jeong, D.; Ko, I. Combining rainfall-runoff model outputs for improving ensemble stream flow prediction. *J. Hydrol. Eng.* **2006**, *11*, 578–588.

16. Liu, H.L.; Bao, A.; Chen, X.; Wang, L.; Pan, X.L. Response analysis of rainfall-runoff processes using wavelet transform: A case study of the alpine meadow belt. *Hydrol. Processes* **2011**, *25*, 2179–2187.

17. Kamruzzaman, M.; Beecham, S.; Metcalfe, A. Wavelet Based Assessment of Relationship between Hydrological Time Series in South East Australia. In Proceedings of 2nd Conference on Practical Responses to Climate Change, Engineers Australia, Canberra, Australia, 1–3 May 2012.

18. Kamruzzaman, M.; Beecham, S.; Metcalfe, A. Non-stationarity in rainfall and temperature in the Murray Darling Basin. *Hydrol. Process* **2011**, *25*, 1659–1675.

19. Hsu, K.-L.; Gupta, H.V.; Sorooshian, S. Artificial neural network modeling of the rainfall-runoff process. *Water Resour. Res.* **1995**, *31*, 2517–2530.

20. Sajikumar, N.; Thandaveswara, B.S. A non-linear rainfall-runoff model using an artificial neural network. *J. Hydrol.* **1999**, *216*, 32–55.

21. Kisi, O.; Shiri, J.; Tombul, M. Modelling rainfall-runoff process using soft computing techniques. *Comput. Geosci.* **2013**, *51*, 108–117.

22. Kisi, O. River flow forecasting and estimation using different artificial neural network techniques. *Hydrol. Res.* **2008**, *39*, 27–40.

23. Kisi, O. Stream flow forecasting using different artificial neural network algorithms. *J. Hydrol. Eng.* **2007**, *12*, 532–539.

24. Rezaeian Zadeh, M.; Amin, S.; Khalili, D.; Singh, V.P. Daily outflow prediction by Multi Layer perceptron with logistic sigmoid and tangent sigmoid activation functions. *Water Resour. Manag.* **2010**, *24*, 2673–2688.

25. Rezaeianzadeh, M.; Stein, A.; Tabari, H.; Abghari, H.; Jalalkamali, N.; Hosseinipour, E.Z.; Singh, V.P. Assessment of a conceptual hydrological model and Artificial Neural Networks for daily outflows forecasting. *Int. J. Environ. Sci. Technol.* **2013**, *10*, 1181–1192.

26. Sudheer, K.P.; Gosain, A.K.; Ramasastri, K.S. A data-driven algorithm for constructing artificial neural network rainfall-runoff models. *Hydrol. Processes* **2002**, *16*, 1325–1330.

27. Kisi, O. Wavelet regression model as an alternative to neural networks for river stage forecasting. *Water Resour. Manag.* **2011**, *25*, 579–600.

28. Dhanya, C.T.; Kumar, D.N. Predictive uncertainty of chaotic daily stream flow using ensemble wavelet networks approach. *Water Resour. Res.* **2011**, *47*, doi:10.1029/2010WR010173.

29. Labat, D.; Ababou, R.; Mangin, A. Rainfall-runoff relations for karstic springs. Part II: Continuous wavelet and discrete orthogonal multi resolution analyses. *J. Hydrol.* **2000**, *238*, 149–178.

30. Australian Bureau of Meteorology (BoM). Climate Data Online. Available online: http://www.bom.gov.au/climate/data/index.shtml (accessed on 12 January 2013).
31. Department for Environment, Water and Natural Resources (DEWNR) 2013. Available online: https://www.waterconnect.sa.gov.au/Systems (accessed on 12 March 2013).
32. Weerasinghe, S. A missing values imputation method for time series data: An efficient method to investigate the health effects of sulphur dioxide levels. *Environmetrics* **2010**, *21*, 162–172.
33. R Development Core Team. *Language and Environment for Statistical Computing*; R Foundation for Statistical Computing: Vienna, Austria, 2011.
34. Nason, G.P. *Wavelet Methods in Statistics with R*; Springer: New York, NY, USA, 2008.
35. R routine in open access. Wavethresh package, 2013. Available online: http://cran.ms.unimelb.edu.au/web/packages/wavethresh/index.html (accessed on 10 November 2013).
36. Haykin, S. *Neural Networks—A Comprehensive Foundation*, 2nd ed; Prentice-Hall: Upper Saddle River, NJ, USA, 1998; pp. 26–32.
37. Maier, R.H.; Jain, A.; Graeme, C.D.; Sudheer, K.P. Methods used for the development of neural networks for the prediction of water resource variables in river systems: Current status and future directions. *Environ. Model. Softw.* **2010**, *25*, 891–909.
38. R routine in open access. RWeka package, 2014. Available online: http://cran.r-project.org/web/packages/RWeka/index.html (accessed on 28 April 2014).
39. Beecham, S.; Chowdhury, R. Effects of changing rainfall patterns on WSUD in Australia. *Proc. ICE Water Manag.* **2012**, *165*, 285–298.

Understanding Irrigator Bidding Behavior in Australian Water Markets in Response to Uncertainty

Alec Zuo, Robert Brooks, Sarah Ann Wheeler, Edwyna Harris and Henning Bjornlund

Abstract: Water markets have been used by Australian irrigators as a way to reduce risk and uncertainty in times of low water allocations and rainfall. However, little is known about how irrigators' bidding trading behavior in water markets compares to other markets, nor is it known what role uncertainty and a lack of water in a variable and changing climate plays in influencing behavior. This paper studies irrigator behavior in Victorian water markets over a decade (a time period that included a severe drought). In particular, it studies the evidence for price clustering (when water bids/offers end mostly around particular numbers), a common phenomenon present in other established markets. We found that clustering in bid/offer prices in Victorian water allocation markets was influenced by uncertainty and strategic behavior. Water traders evaluate the costs and benefits of clustering and act according to their risk aversion levels. Water market buyer clustering behavior was mostly explained by increased market uncertainty (in particular, hotter and drier conditions), while seller-clustering behavior is mostly explained by strategic behavioral factors which evaluate the costs and benefits of clustering.

Reprinted from *Water*. Cite as: Zuo, A.; Brooks, R.; Wheeler, S.A.; Harris, E.; Bjornlund, H. Understanding Irrigator Bidding Behavior in Australian Water Markets in Response to Uncertainty. *Water* **2014**, *6*, 3457-3477.

1. Introduction

Water scarcity has emerged in many semi-arid regions of the world. This requires the development of mechanisms to efficiently reallocate available resources between competing extractive as well as in-stream uses. Water markets have been promoted as an efficient way of facilitating this process in a number of jurisdictions, such as Australia, USA and Chile [1–3] and more recently in Canada [4] and Spain [5]. As scarcity intensifies, demand for, and participation in, water markets is likely to increase. A continual review of market mechanisms will help to improve and facilitate greater market efficiency, through reducing transaction costs, improving product choice or reducing barriers to trade. Adoption of water market trading (where available) will represent one potential adaptation strategy for many irrigators in the face of climate change. Modeling by Adamson, Mallawaarachchi and Quiggin [6] demonstrates that adaptation will partially offset the adverse impact of climate change and suggests that improvements in the function of water markets could support adaptation.

In order to provide greater insights into how to best improve water market mechanisms and water management in general, a better understanding of irrigators' behavior in such markets and how this compares to behavior in other financial markets is necessary. In the Murray-Darling Basin (MDB) of Australia, irrigators' participation in the water market has been growing over the past two decades and this provides a unique opportunity to study irrigators' water market behavior. Two major forms of water markets exist in the MDB: the water allocation market (also known as temporary water markets, which involve the short-term right to use of water) and the water entitlement market (also known as permanent water markets involving the long-term right to access water—see

Wheeler *et al.* [7] for more detail). This paper focuses on the water allocation market.

Since the Council of Australian Governments water reform agenda in 1994, water markets have played a central role in allowing farmers to deal with increased volatility, risk and adjustment pressures by permitting them to alter their short and long-term access to water resources as well as allowing them to exit out of irrigation while realizing their water assets [8,9]. In 2011, the Murray-Darling Basin Authority (MDBA) released the MDB Plan, with a target of 2750 GL to be returned from consumptive to environmental use [10]. Water entitlements are to be sourced from willing sellers, and are bought by the Commonwealth of Australia. Increasingly, there are arguments that governments should also consider buying water from the allocation market (otherwise known as temporary water available in one season) to provide environmental flows [11]. The rationale for government utilizing the water allocation market is that benefits of carry-over, lower water allocation prices, and temporal demand can provide a more efficient and flexible supply of water to meet stochastic environmental flow requirements since the timing of entitlement releases does not correspond well with the volume and timing of water applications required to achieve environmental objectives [3,12].

In light of these policy arguments regarding government intervention in the allocation market, a more thorough understanding of irrigators' trading behavior in that market, particularly how they bid and offer for water, is needed. In particular, we need to understand how variability in climate conditions impacts on water market trading behavior. One way of analyzing water market trader behavior is to analyze the extent to which irrigator bids or offers exhibit price clustering (that is, the extent they cluster around particular numbers). The existence of clustering is important as it identifies a possible dead-weight loss that exists in water allocation markets. Utilizing bid and offer data also allows us to understand how differently buyers and sellers act in the water market, something that is difficult to do in other water market analysis. It also offers continuing insights into how irrigators behave in water markets, and how similar (or dissimilar) their behavior is to participation in financial markets. Understanding the similarity between irrigator bidding and offering behavior in a water market and a trader in a stock market may also offer insights into how well introducing other water market products (for example: option trading) will be received. For instance, Heaney *et al.* [13] discuss how missing options markets in storage and delivery might impact water trading. Addressing these issues is a function of market design. Hence, undertaking analysis on price clustering is informative for water management policies aiming to improve the efficiency and flexibility of resource allocation.

Price clustering in financial markets has been well documented in the literature (e.g., Chung and Chiang [14]). Clustering is found when indicative quotes for currencies end mostly around particular numbers, for example, those with either "zero" or "five". Round numbers are disproportionately represented in bid-ask spreads for major currencies. Typically, the economics and psychological literature identify different reasons for clustering. In economics, clustering is considered a rational response to trading impediments. In psychology, clustering is thought to occur due to a human bias for prominent numbers, such as zero and one. There have been very few studies that have analyzed clustering in other non-financial product markets. There are a number of similarities between financial markets and water markets, but there are also fundamental differences because water markets are dealing with common property resource issues. In addition there is an issue as to whether recent events in financial markets during the global financial crisis make such markets an appropriate benchmark for comparison of resource markets. The issue of the efficient markets hypothesis (EMH) and the global financial crisis (GFC) is discussed by Ball [15] and Brown [16] who suggests that the failure to predict the bursting of the real estate bubble-that lead to the GFC-is in fact consistent with the central idea in the EMH. This paper analyzes clustering in the water allocation market over the past decade. In doing so, we will be able to determine (a) the extent of price clustering in this market and (b) given the constraints that prevent traders having a precise valuation of water, whether clustering behavior is a response to uncertainty (either weather or policy changes) or a strategic behavior.

The only other paper that has examined price clustering in water markets [17] found robust evidence of clustering in the water market in northern Victoria from 2002 to 2007. Its' econometric modeling suggested uncertainty faced by irrigators is a major reason for clustering. This paper extends the work by Brooks, Harris and Joymungul [17] in four ways. First, a longer time span is used covering 10 trading seasons in the northern Victorian water market. Much of this data is not publicly available. Second, alternative price clustering definitions are employed to check the robustness of the findings. Third, a variety of other data are included to identify specific factors associated with irrigators' risk awareness that in turn influence the extent of price clustering. In particular, we are interested in assessing how government water policy changes, rainfall and evaporation influence bid and offer behavior in the water market. Finally, the extent to which price clustering is a result of traders' response to uncertainty and/or strategic behavior is examined.

2. Study Area

The Goulburn Murray Irrigation District (GMID) is Australia's largest irrigation district, located in northern Victoria, along the River Murray. It has one of Australia's longest running water markets with the bulk of trading taking place in three trading zones. The most active trading zone is Greater Goulburn, which provides the data source for this paper. Water allocations and entitlements have been traded since the early 1990s and irrigators are increasingly adopting water trading (in particular water allocation trading) over time [18]. Given that the majority of trades, especially bids and offers for water, are in the water allocations market, this is the market we chose to focus on for a study on price clustering in water markets. Dairy, fruit and, grape producers are the most significant buyers in the allocation market, whereas cereal, grazing and mixed farmers are the main sellers [19]. Over the past decade, MDB irrigators have faced considerable changes to their water allocation levels (which conversely influence the amount of land irrigated). An allocation level refers to the percentage of water entitlements that is available for the entitlement holder to use throughout a season. The resource manager manages seasonal allocation levels on behalf of all entitlement holders and regularly reviews the water budget calculations in the GMID. For example, Goulburn water allocation levels dropped from a consistently secure 200% in the early 1990s to around 30% in the mid 2000s. As a result, uncertainty for irrigators has increased considerably with opening allocations of 0% in eight consecutive years from 2002 and below 100% since 1998 and with closing allocations below 100% for five out of eight years from 2002 to 2010 [18]. In 2010 and 2011 higher than expected rainfall increased water allocations, this in turn increased the amount of land irrigated. Furthermore, water policy changes add to the climate uncertainties experienced by irrigators. There have been many government and institutional changes that impact on water markets in Australia over the time-period studied. This paper considers three of the most major ones that occurred, namely: (a) the lifting of the Cap (in 1994 the Victorian Government restricted the volume of water access entitlements that could be traded out of each irrigation district in Northern Victoria to no more than 2% annually of the volume of entitlement held in the district at the start of the irrigation season. On 1 July 2006 this was increased to 4%); (b) introduction of unbundling (this occurred on 1 July 2007 in northern Victoria and it is the legal separation of rights to land and rights to access water, have water delivered, use water on land or operate water infrastructure, all of which can be traded separately) and (c) the times when the Australian Government is conducting a tender in buying back water entitlements (the Federal government began a decade long policy of buying back water entitlements from willing sellers in February 2008 in order to return water from a consumptive to an environmental use—see [7,8] for more detail).

3. Price Clustering Literature and Applications to Water Markets

3.1. Price Clustering Theories

Empirical studies in the finance literature find that the degree of clustering in any market is a function of market structure, uncertainty, resolution costs and human preferences [14]. Several hypotheses have been developed to better understand why clustering occurs. These include: the negotiation hypothesis; the price resolution hypothesis (uncertainty); the attraction hypothesis and strategic behavior. We discuss briefly each of these hypotheses and their relevance in the context of the Australian water market.

A market's structure may bring about clustering and Harris [20] developed the negotiation hypothesis to explain these effects, arguing that regulatory restrictions can reduce negotiation costs for traders. These restrictions require quotes and transaction prices to be stated as some multiple of a minimum price variation, or trading tick. Negotiation costs fall because restrictions create a discrete price set around which traders bid and offer. In the absence of these restrictions, the number of possible offers and counter-offers widens so that negotiation time also increases, creating higher price risks for participants [21]. A discrete price set reduces the amount of information exchanged, leading price to converge more quickly than would otherwise occur. As a result, transactions costs are reduced. The bid prices in the Australian water market analyzed are not required to be some multiple of a trading tick greater than one cent. Therefore, the degree of clustering is expected to be small because irrigators bid on a continuous price set.

The method of trading can also influence the degree of clustering observed. For example, the use of electronic trade compared with floor trade (in person) alters the costs associated with precise valuation and, therefore, clustering. Chung and Chiang [14] found extreme clustering occurred on floor-traded futures compared with those traded electronically. Floor trade made precise valuation more costly because it takes more time to call out information to the accuracy of several digits and there is a wider margin for error in doing so [21]. The mechanism for water trading in the GMID creates a pool price that tends to decrease the costs associated with precise valuation, so a finer grid of numbers may be expected. However, a uniform pool price each trading week may also decrease the benefits of a precise valuation and the weekly trading frequency may be too long for traders to place more precise bids. Nevertheless, Brooks, Harris and Joymungul [17] found evidence of clustering on bid prices in the GMID water market.

The price resolution hypothesis contends that prices may be evenly clustered at particular points if valuation is indecisive [22]. Loomes [23] and Butler and Loomes [24] argued that economic decision makers do not measure utilities exactly but act in a sphere of haziness, which represents the degree of difficulty in precise valuation. In other words, the risk of taking certain actions increases with uncertainty. A greater sphere of haziness implies a higher clustering propensity due to people's risk aversion behavior. When uncertainty and volatility are high, precision valuation is costly, leading to greater clustering [25].

In the case of the water market, water availability uncertainty can be brought about by several factors, including rainfall variation, water allocations, demand fluctuations, government policy changes, and climate change [7,19,26]. Variable and unpredictable rainfall in the MDB system can be on a range of time scales and intra-season variations, making it difficult to forecast final closing allocations. Allocations are announced fortnightly during the water season, and as discussed often have started at 0%. Uncertainty in allocations can lead to miscalculations regarding seasonal allocations by irrigators at the time of planting decisions. If an irrigator overestimates what their expected allocations will be at the time of planting a crop, they may have to buy additional water later. Alternatively, if an irrigator underestimates the final allocations they will receive, they may have surplus temporary water available that can be sold in the market at a later point in the season or be carried forward into the next season (depending on storage availability). Climate information only becomes available as the season progresses, so depending on how accurate irrigators were in their water expectations and the watering requirements of their permanent or annual crops, changes (or lack of changes) in monthly seasonal allocations may cause relatively high price volatility in the market. Government intervention in water markets has increased considerably over the first decade of the 21st century [27]. Government intervention affects short- and medium-term price expectations, thereby increasing costs of precise valuations.

Ikenberry and Weston [28] demonstrate that clustering of U.S. stock price also stems from the psychological preferences of market participants. This is broadly referred to as the attraction hypothesis and it suggests that clustering is the result of behavioral idiosyncrasies (heuristics). Tversky and Kahneman [29] argued that individuals often rely on a number of heuristic principles that reduce complex tasks, such as valuation to simpler or even non-optimal judgment operations.

Brooks, Harris and Joymungul [17] use variables representing the price resolution hypothesis to explain price clustering in the GMID water market. Their results indicate a large proportion of the variation in price clustering cannot be explained by the price resolution hypothesis (the largest adjusted R^2 in their regression models is 0.61). Therefore, the attraction hypothesis is very likely to be able to explain some of the variation in price clustering that cannot be explained by the models of

Brooks, Harris and Joymungul [17]. Unfortunately, it is almost impossible to collect data on testable variables representing the attraction hypothesis.

An alternative explanation for clustering is that its existence is the result of strategic behavior—where people estimate the net benefits of their action [30]. Specifically, they weigh the benefits of increasing the precision of their bid/offer relative to the loss of value resulting from an imprecise estimate. In Victoria, the benefits of precise valuation are not obvious for individual traders on the water market because the water exchange Watermove used a pool price. Watermove was a trading organization in the GMID that conducted water exchanges within MDB trading zones, it operated by telephone and online. It closed down in August 2012, but still remains a valuable source of historical data, especially bid and offer data, and is used in the analysis here. Table 1 presents an example of how the Watermove exchange worked. For example, in the week of 8 September 2011, there were 35 sale offers with the offering price ranging from $14 to $100 and a total volume for sale of 4724.8 ML; and 21 buy bids ranging from $10 to $26.38 with a total volume for purchase of 5841 ML. A pool price of $21.15 (the average price of the last fulfilled sale offer, $20, and buy bid, $22.3—which is calculated after all bids and offers are received for the week) was found for the week in order to maximize the volume traded, namely 1441.5 ML. As a result, the last fulfilled buy bid had bought only 80.5 ML, instead of the full amount, 200 ML. This exchange mechanism results in the potential for price clustering to create a deadweight loss. The size of the deadweight loss depends on the pool price, the last fulfilled sale offer and buy bid prices and the amount of unsatisfied volume to sell or buy. It can be evident that the pool price could be quite different from their offer prices, which is likely to be caused by the weekly trading frequency. In this setting, the cost of rounding will be the lower likelihood of their orders being executed and the cost of not rounding will be the extra expenditure paid by buyers or the reduced revenue for sellers. A strategic bidder, therefore, would evaluate whether the cost of rounding outweighs the cost of not rounding in order to decide the bid price.

Traders who expect natural clustering can easily change their offer prices by a cent (penny) to avoid cluster points thereby increasing the probability of their offers being executed, described as the "pennying behavior" by Jennings [31] and also documented in Edwards and Harris [32]. This behavior is evident in the water market as demonstrated in Section 4.2. First, price clustering would decrease when traders seek a higher probability of their orders being executed. On the buyers' side, traders would require a higher probability of their orders being filled if they had overestimated seasonal allocations and therefore have experienced a deficit in available water. Assuming crop loss is a distinct possibility in this case; traders would avoid clustering to increase the likelihood that they will obtain water. On the sellers' side, greater precision could be used if surplus water could be sold at a premium price; for example, during times of protracted drought. The high returns available during these periods would encourage a greater determination for offers to be executed. Second, when buyers (sellers) consider the extra dollar expenditure (revenue) as more significant, that is, the costs of not rounding as considerable, price clustering is expected to increase. It is difficult to identify which of these effects will dominate strategic behavior in the water market, as this will depend on the market and biophysical conditions at specific times. The analysis here will investigate the effects of those conditions on the potential for strategic clustering.

Table 1. An example of Watermove weekly exchange bids and offers.

Seller Offer Price ($/ML)	Volume for Sale	Total Volume in Exchange	Buyer Bid Price ($/ML)	Volume for Purchase	Total Volume in Exchange
14.00	200	200	26.38	200	200
15.00	103	303	25.50	200	400
15.00	18.2	321.2	25.00	400	800
15.00	60	381.2	25.00	11	811
17.00	80	461.2	25.00	200	1011
18.00	55	516.2	25.00	150	1161
19.00	320	836.2	23.38	200	1361
19.90	100	936.2	22.30	200	1561
20.00	100	1036.2	22.00	20	1581
20.00	150	1186.2	20.38	200	1781
20.00	120	1306.2	20.00	100	1881
20.00	59	1365.2	20.00	100	1981
20.00	24.3	1389.5	19.85	1500	3481
20.00	52	1441.5	18.00	500	3981
28.00	50	1491.5	18.00	200	4181
28.00	210	1701.5	15.88	200	4381
29.00	92	1793.5	15.00	200	4581
30.00	379	2172.5	15.00	10	4591
30.00	150	2322.5	14.22	500	5091
30.00	150	2472.5	12.88	500	5591
30.00	60	2532.5	10.00	250	5841
30.99	300	2832.5			
30.99	140	2972.5			
35.00	500	3472.5			
42.38	490.7	3963.2			
42.38	192.6	4155.8	Date: 8 September 2011		
45.00	68	4223.8	Pool price: $21.15/ML		
45.00	20	4243.8	Total volume traded: 1441.5 ML (The shaded bids		
45.00	100	4343.8	and offer orders were executed, with the buy order		
50.00	20	4363.8	indicated by asterisk only fulfilled by 80.5 ML)		
50.00	70	4433.8			
58.00	100	4533.8			
60.00	50	4583.8			
60.25	46	4629.8			
100.00	95	4724.8			

3.2. Overall Water Market Clustering Hypothesis

In summary, we propose the reasons for price-clustering behavior in the water allocation market as: (1) attraction; (2) price resolution (or uncertainty); and (3) strategic behavior. Attraction suggests traders prefer certain price points to others for psychological reasons, which is discussed in Section 3.1. Price resolution proposes that traders are more likely to cluster when they perceive uncertainty in water markets is higher. We expect the following variables will be important influences on uncertainty: trading volume, water allocation price, water entitlement price, bid-ask spreads, water allocation level, climate conditions, seasonal factors, and government policy changes. Strategic behavior explanations for price clustering (where water traders will evaluate the cost and benefits of clustering, or the costs of rounding and not rounding) would also be influenced by many of these same factors. Farmers decide to trade one more unit of water allocations if the cost (revenue) from the trade is smaller (greater) than the value of the marginal product of their additional water using activities. Hence, the bid and offer prices are likely to reflect the farming enterprises and the associated risk levels for the farming enterprises if there is water scarcity. Since price clustering is measured for the whole Greater Goulburn region, we cannot consider farming enterprise variables. This question is left for future research that needs access to data across a variety of regions or access to bid and offer individual survey records (either entitlement or allocation records).

The following sections identify evidence that support the attraction hypothesis, as well as determining the extent to which price resolution and strategic behavior can explain price clustering in the water market.

4. Price Clustering Evidence in the Greater Goulburn Water Allocation Market

Before analyzing the drivers of clustering behavior in the water market, we first determined if there was evidence of clustering. We collected weekly data from Watermove on all individual buy and sell bids, including the volumes and prices of each bid for the period August 2001 to May 2011 for Greater Goulburn in Victoria—the most active trading zone. Most of these time series data are not publicly available. The data include quite a few weeks where the total number of orders is less than 20. In order to have a sufficiently large base of bids, we calculated price clustering at monthly intervals. Orders are fewer both at the start and toward the end of each season. The analysis covers ten years, and our monthly clustering series includes 100 observations, sufficient for the subsequent regression analyses. June and July are not included as there is usually no, or very scarce, trading in those two months. For the whole dollar amount clustering series of sell offers within the 10% range of the pool price, the number of sell offers is smaller than 30 for most of the months in the 2010/2011 season. This small number of observations makes the clustering calculation unreliable. Therefore, there are 90 months instead of 100 for this clustering series.

4.1. Evidence of Price Clustering

Table 2 provides an overview of the existence of price clustering in the Greater Goulburn water allocation market. We first examined the extent of clustering at whole dollar amounts, *versus* amounts at particular cents. Over the time period being considered, 80% of all water allocation buy bids and 96% of all sell bids were placed at whole dollar amounts. Moreover, Table 2 also illustrates that if percentages are weighted by the volume associated with each order, whole dollar clustering decreases to 73% and 92% for the buy and sell orders respectively.

Table 2. Water allocation price clustering (%) at whole dollars.

Water Trade Type	All		Within 10% of Pool Price		Within 5% of Pool Price	
	Number	ML	Number	ML	Number	ML
Buy bids	79.59	72.74	78.39	70.77	78.43	71.11
Sell offers	96.47	91.69	95.92	91.08	95.84	91.20

By including orders where prices are too distant from the pool price the extent of price clustering may be biased upward because it is less costly to be precise if a price offering is likely to be far away from the pool price. As a result, it is possible for an irrigator to be acting in a greater sphere of haziness. Therefore, we calculate the clustering at whole dollar amounts again but use only those orders whose prices are within 10% of the pool price range and then only within 5% of the pool price range. As expected, Table 2 indicates the extent of price clustering decreases when orders are constrained in a narrower range around the pool price. However, the decrease appears to be small and insignificant.

Table 3 explores the extent of clustering at specific whole dollar digits of the buy and sell offers. For buy offers that are whole dollar amounts, Table 3 shows more than half of them ended in zero, while about a fifth end in five. Results are similar if the percentages are weighted by the order volumes or if only those orders within 10% or 5% of the pool price range are used.

Table 3. Water allocation price clustering at whole dollar digits (%).

Whole Dollar Digits	All		Within 10% of Pool Price		Within 5% of Pool Price	
	Number	ML	Number	ML	Number	ML
			Buy offers			
0	54.17	44.82	52.14	44.53	51.61	42.99
1	9.87	11.21	10.63	11.86	11.24	11.66
2	4.58	8.64	5.32	5.39	5.83	6.60
3	1.89	2.42	2.16	2.57	2.09	2.59
4	0.76	0.88	0.94	1.20	0.83	1.19
5	20.01	20.64	18.94	20.65	17.97	20.61
6	3.99	5.08	4.27	5.61	4.22	4.88
7	2.13	2.78	2.06	2.88	2.11	3.24
8	1.74	2.62	2.39	3.97	2.92	4.94
9	0.87	0.91	1.14	1.34	1.18	1.29
			Sell offers			
0	71.38	59.16	67.28	56.02	67.02	55.00
1	0.48	0.84	0.53	0.86	0.49	0.74
2	0.87	1.81	0.96	1.78	1.01	1.80
3	0.66	1.32	0.87	1.36	1.05	1.78
4	1.36	2.63	1.75	2.66	1.60	2.48
5	14.52	19.76	16.54	21.00	15.86	20.31
6	0.57	1.07	0.76	1.29	0.84	1.30
7	1.04	1.67	1.16	1.68	1.27	1.77
8	3.12	4.77	3.54	4.91	4.36	6.36
9	6.01	6.96	6.62	8.44	6.50	8.46

The extent of clustering at whole dollar amounts and at specific whole dollar digits is similar to what is found for the Greater Goulburn trading zone in Brooks, Harris and Joymungul [17], where the authors use data from 2002 to 2007. Similar to Brooks, Harris and Joymungul [16], we used Chi-squared and HHI (Herfindahl-Hirschman Index) to test the significance of price clustering in our data. The results, which are available upon request, indicate the presence of significant price clustering. To further investigate price clustering over time, we present the buy and sell offer series for clustering at whole dollar amounts in Figure 1 and for clustering at the specific whole dollar digit zero in Figure 2. Figure 1 demonstrates that neither series exhibits a clear time trend but the buy offer series appears to have a greater variation over time. An augmented Dickey-Fuller unit root test indicates the absence of a unit root for both series. In Figure 2, both series appear to vary within a wider range, especially for the sell offers, compared to the results in Figure 1. The time series of clustering at the specific whole dollar digit zero also exhibits no clear time trend and does not have a unit root.

Figure 1. Water allocation price clustering at whole dollar amounts (all offers).

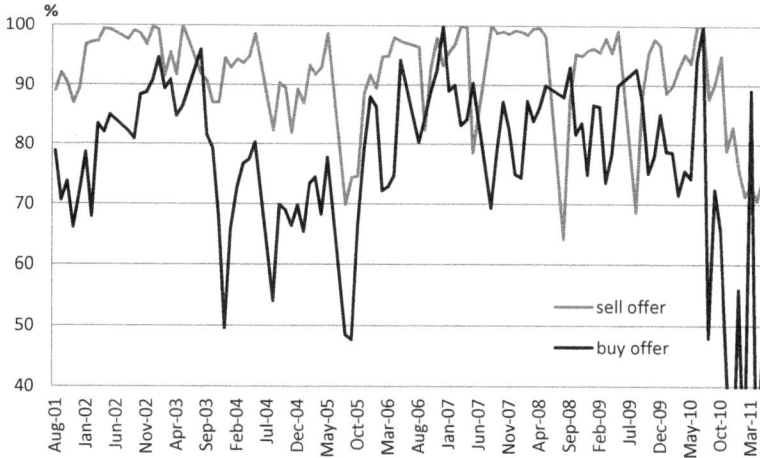

Figure 2. Water allocation price clustering at whole dollar digit ending in zero (all offers).

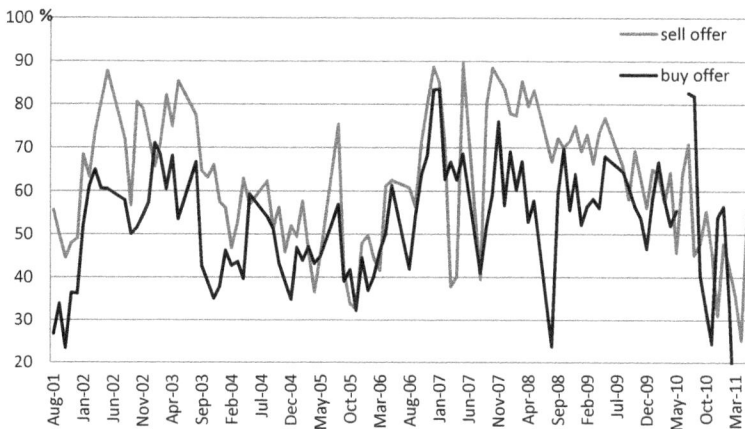

4.2. Evidence of Strategic Price Clustering Behavior

Niederhoffer [33] argues that asymmetry between ask and bid quotes around integer prices could exist because of strategic behavior where the intention is to exploit opportunities resulting from price clustering. Aşçıoğlu, Comerton-Forde and McInish [34] show that investors submit orders with one tick better than zero and five to avoid queuing orders at prices ending in these digits. Given prices cluster on round numbers, a water trader who places a bid and wants a higher probability of execution than a bid at the clustered price will tend to place the bid one cent away from the clustered price. Figures 3 and 4 investigate the evidence of strategic price-clustering behavior.

Figure 3. Distribution of buy offers not ending in whole or half dollar amounts.

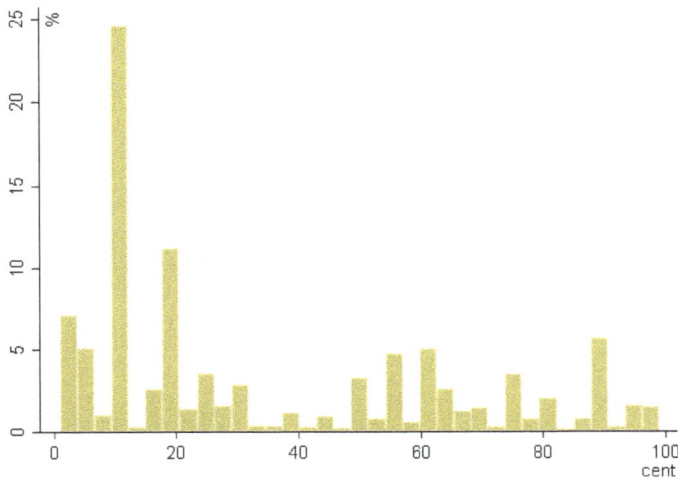

Figure 4. Distribution of sell offers not ending in whole or half dollar amounts.

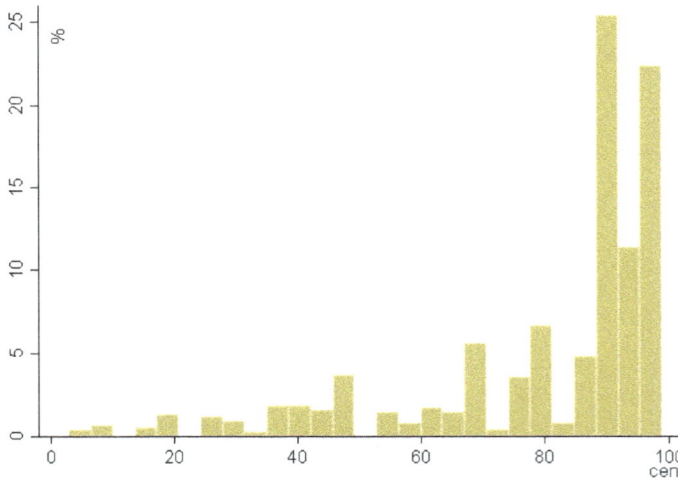

The figures respectively show the distribution of buy and sell offers that are not ended in whole or half dollar amounts. Clustering at half dollar amounts is also evident and much greater than its expected clustering. It is evident that the non-whole and half-dollar buy bids are most likely to be slightly greater than the price cluster, while the non-whole and half-dollar sell offers are mostly present slightly less than the price cluster. For those offers of whole-dollar, ending in other than zero or five, Figures 5 and 6 display the distribution across the remaining eight digits. If expecting clustering happens at zero, a buyer is most likely to place a bid with just one extra dollar. In fact, the probability of a buy bid ended in one is about 0.38-well above the probability of any other seven digits. On the sell side, a seller expecting clustering at zero is most likely to place a bid ended in nine. The probability of a sell bid ended in nine is about 0.43, well above the probability of any other seven digits.

Figure 5. Distribution of buy offers across eight digits.

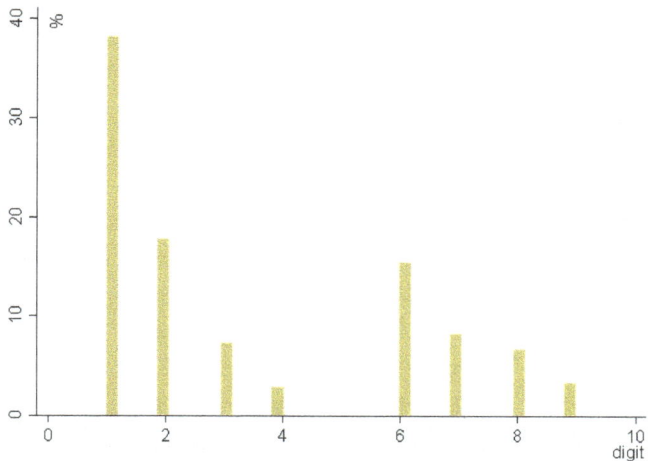

Figure 6. Distribution of sell offers across eight digits.

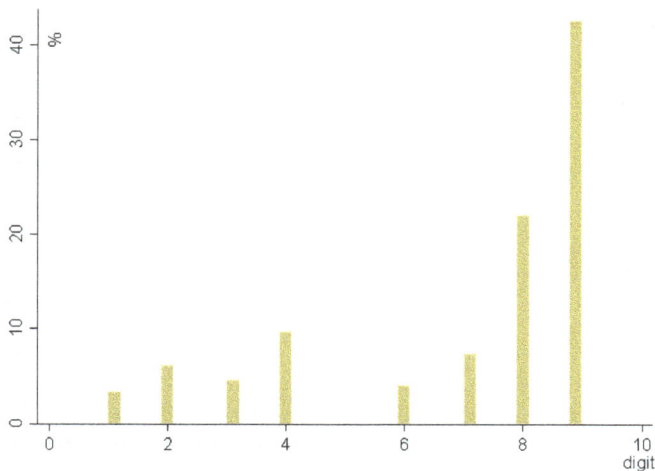

5. Methodology

Having observed substantial price clustering in the water allocation market, especially on the sell side, we now investigate the extent to which price clustering is driven by uncertainty and/or strategic behavior and if buyers and sellers' price-clustering behavior are influenced in the same way. The dependent variable, observed price clustering in a month, is defined as a proportion, which is bounded between 0 and 1. A linear probability model may not be appropriate as it can generate predictions outside the 0 and 1 interval. One way to take account of the bounded nature is the logit transformation and thus the fractional logit model, first used by Papke and Wooldridge [35]. The regression equation used was:

$$y_t^* = \log(\frac{y_t}{1-y_t}) = X_t \cdot \beta + \mu_t \tag{1}$$

where y_t is the observed price clustering in month t, X_t is a vector of regressors that potentially influence the dependent variable, and μ_t is the disturbance. The logit transformation of y_t results in a latent variable y_t^*, as a linear function of a set of regressors, X_t. The fractional logit model was executed by Stata 13's generalized linear model (GLM) command with the logit link function. We also used the type of standard error option that is heteroskedasticity- and autocorrelation-consistent to account for any heteroskedasticity and autocorrelation in the disturbance term μ_t.

y_t adopts two types of clustering weighted by order volume, namely: (1) clustering at whole dollar amounts *versus* fractions; and (2) clustering at whole dollar amounts ending in zero *versus* the remaining nine digits. For the first definition, the calculation is based on all offers and offers within the 10% range of the pool price. The clustering calculation based on offers within the 5% range of the pool price is not modeled as there is no significant difference in clustering between the 5% and 10% pool price range. For the second definition, the calculation is based on all offers since offers within the 10% range of the pool price in some months do not have enough observations to calculate a reliable clustering percentage.

Independent Variables

Table 4 lists the detailed definitions of the independent and dependent variables that were used in the price clustering models.

Table 4. Variable definitions.

Variable Name	Variable Definition
WholeBuy	Percentages of buy offers that are whole dollars in each month
WholeBuy_10	Percentage of buy offers that are whole dollars out of those within the plus and minus 10% range of pool price in each month
WholeSell	Percentages of sell offers that are whole dollars in each month
WholeSell_10	Percentage of sell offers that are whole dollars out of those within the plus and minus 10% range of pool price in each month
ZeroBuy	Percentage of buy offers that end in zero out of buy offers in whole dollars in each month
ZeroSell	Percentage of sell offers that end in zero out of sell offers in whole dollars in each month
Watervolume	Natural logarithm of volume traded for water allocations in Greater Goulburn in each month
Waterallocprice	Natural logarithm of average monthly price ($/ML) for water allocations in Greater Goulburn
Waterentprice	Natural logarithm of average monthly price ($/ML) for water entitlements in Greater Goulburn
Ln_spread	Natural logarithm of the spread between the last outstanding buyer and seller offering water allocation prices
Allocationlevel	Allocation level for Goulburn at the beginning of each month (%)
Evapminusrainfall	Monthly evaporation minus rainfall at Kerang station (mm)
Feedbarley	Natural logarithm of export price for feed barley ($/ton)
Wholemilkprice	Natural logarithm of export price for whole milk powder ($/kg)
Cattleprice	Natural logarithm of export price for cattle ($cent/kg)
Carryover %	Percentage of water entitlement allowed for carryover (note for 2010/11 season all the allocation in linked Allocation Bank Account on 30 June 2011 is eligible for carryover—there is no maximum)
Govpolicy	1 for the months when major water market policies were introduced/ongoing in the GMID (namely the lifting of the Cap, introduction of unbundling and the times when the Government is conducting a tender in buying back water). For Cap and unbundling introduction, the dummy is coded for the first three months after policy introduction
Govpolicy10/11	Interaction variable between Govpolicy variable and season 2010/11
Monthindex	Monthly index from 1 to 10 for August to May, respectively
Monthindexsqrd	Monthly index squared

Note: The first six variables are the respective dependent variables for the six regression models presented in Table 5.

Our final choice of independent variables was influenced by other studies that have studied influences on water market trade (e.g., Wheeler *et al.* [19] and Brooks *et al.* [17]). It was also determined by statistical issues, such as serious multicollinearity (discussed in Section 6). There are a number of potential relationships our independent variables could have with price clustering, and these impacts will vary depending on whether we are looking at buyer or seller behavior. For example, weather, measured by net evaporation in millimeters, may be positively related to price clustering according to the price resolution (uncertainty) hypothesis. Net evaporation is calculated as total evaporation minus total rainfall for the month in question. *Ceteris paribus*, drier weather increases water prices and, in turn, increases the uncertainty perceived by irrigators who are trying to buy water, resulting in a higher level of price-clustering behavior in the water market. However, drier weather presents a greater need for water in general. In turn, at the margin some buyers will have a greater need to have their orders executed, and therefore act more strategically in the market.

This will reduce price clustering overall. Alternatively, as water prices increase sellers' risk decreases so there is less need for strategic behavior to sell their water. As a result, the overall effect of weather on price clustering depends on whether water buyers or sellers are behaving more risk aversely or strategically. Other independent variables that may be influenced by the price resolution (uncertainty) hypothesis for both buyers and sellers include water allocation and water entitlement prices, trading volume, the spread between the offer prices, feed barley prices, carryover level and government policy. Our government policy variable represents either (a) a time of uncertainty, namely three months after major policy changes, such as unbundling of land and water and the changing water trade restriction policies; and/or (b) a time when the government is purchasing water entitlements in the market. Victoria has had annual restrictions on the amount of entitlement trade allowed out of a district for years. In January 2006, the cap on entitlement trade was eased from 2% to 4%. The unbundling of land and water occurred in the GMID on 1 July 2007. Unbundling reduced the transaction costs associated with trading water, and allowed irrigators to own shares in different rivers (reducing risks). The unbundling aimed to facilitate trading in water entitlement and allocation and make trading more efficient.

Variables that may be influenced primarily by the strategic behavior hypothesis include whole-milk powder prices, cattle prices and water allocations received by irrigators, but risk averse behavior may also play a part in influencing price clustering. Whole-milk powder represents a production output of dairy farmers, feed barley represents an input substitute for watering pasture for dairy production, and cattle represents an alternative output production substitute. The overall influence of each variable will be determined by the strength of each hypothesis in determining behavior. Wherever model statistics allow, we have included all the same independent variables in every model to examine whether there are any differences between the influences on buying and selling clustering behavior.

6. Results and Discussion

Results for our buy and sell price clustering models in the Greater Goulburn water allocation market are presented in Table 5. Since the coefficient results produced by the fractional logit model are not practically meaningful, we report the marginal effect estimates. Multicollinearity was an issue in some of the models, with the variance inflation factors (VIFs) of water allocation price, water entitlement price, spread, allocation level and government policy variables being greater than five. The potential consequence is to make the variables involved insignificant where they should be significant. In order to verify whether collinearity caused this problem, we dropped the variables with insignificant coefficients one by one and checked whether the coefficients of the remaining variables became significant. If this was the case, the involved insignificant variables were dropped. However, if it was not the case they were kept in order to minimize omitted variable bias.

Table 5. Buy and sell offer monthly water allocation price clustering.

Variable	WholeBuy	WholeBuy_10	ZeroBuy	WholeSell	WholeSell_10	ZeroSell
Watervolume	−0.003	−0.034	−0.035 ***	0.016 ***	0.026	−0.032 *
Waterallocprice	0.016	0.033	-	−0.005	-	0.045
Waterentprice	0.067	0.079	-	−0.165 ***	-	-
Ln_spread	−0.015	−0.096 ***	0.050 ***	0.046 ***	-	0.063 ***
Allocationlevel	0.001	0.002	-	−0.0002	-	-
Evapminusrainfall	0.001 ***	0.001 ***	0.0003	0.00004	−0.0001	0.00002
Feedbarley	0.240 ***	0.377 ***	0.090	−0.010	0.052	−0.020
Wholemilkprice	−0.122	−0.225 *	−0.130 **	0.094 ***	0.026	0.179 **
Cattleprice	−0.059	0.011	−0.039	−0.363 ***	−0.478 ***	−0.389 *
Monthindex	−0.055 **	−0.128 ***	0.016 ***	0.009 ***	-	0.015 ***
Monthindexsqrd	0.005 **	0.011 ***	-	-	-	-
Carryover	−0.001	0.000	0.0002	−0.0003	−0.001	−0.001 **
Govpolicy	0.076 *	0.135 **	0.008	−0.028	−0.070 **	−0.090
Govpolicy10/11	−0.255 ***	−0.248 ***	−0.277 ***	−0.029	-	0.075
Observations	100	94	100	100	90	100
Log likelihood	−35.80	−32.66	−44.52	−19.34	−18.89	−42.65
BIC	−386.85	−356.94	−405.79	−393.45	−360.68	−401.31

Note: Marginal effects are reported. * $p < 0.1$; ** $p < 0.05$; *** $p < 0.01$ indicate significance at the 10%, 5% and 1% levels, respectively.

6.1. Buy Offer Price Clustering

Positive coefficients for net evaporation, feed barley price and the government policy dummy suggest that uncertainty (from the price resolution hypothesis) is able to explain clustering by buyers in the water market. Higher net evaporation loss increases water uncertainty and increases clustering. Higher feed barley prices augment water demand because it is an input substitute for on-farm feed production. In turn, as feed barley prices rise dairy farmers will find it more costly to replace water to grow their own pasture with purchased feed. This increases water market demand, and the costs of precise bids thereby causing greater clustering in buy offers.

The government policy dummy represented periods of uncertainty and significant government intervention in the market (e.g., the first three months following significant government changes) and is associated with greater uncertainty in water prices; especially in the short-term after the policy introduction. For two of three buy models; periods of policy uncertainty were positively and significantly associated with price clustering. This implies that water allocation buyers are using price clustering as a response to policies that add to market uncertainty. In times of change; irrigators will be operating in a greater sphere of haziness; with higher levels of uncertainty and volatility being experienced; so buyers exhibit a higher clustering propensity.

A surprising finding regarding the government policy variable is the result of its interaction with season 2010/11, when water was plentiful due to the record rainfall during the season, when prices dropped accordingly. Contrary to the positive impact of government policy on price clustering observed for previous seasons, government policy had a significantly negative impact in 2010/2011. Two influences (government intervention and rainfall) may explain this result. The Commonwealth was in the market buying entitlement water from November 2010 to May 2011, which was a time of flooding and falling water prices. The flooding reduced irrigator buyers' risk and their water demand, thereby reducing their clustering.

The price resolution hypothesis also predicts that the trade volume is negatively associated with price clustering, while price is positively related to price clustering. Our results, however, only offer a very weak support for this. The volume of trade has a significantly negative impact on clustering in the zero buy model, while a negative but insignificant impact on clustering at whole dollar. Neither water allocation nor entitlement prices have significant impact on clustering although their impacts are estimated as positive.

The coefficients of our time variable—months in the year (and its squared term)—suggests buyer price clustering generally decreases from the start of the season (August) until the month of January and then increases until the end of the season. Brennan [27] argued irrigators are generally risk averse and will hold more water than required at the start of a season when climate and allocation information is yet to be revealed, creating price premiums. As a result, some buyers may be more concerned with having their orders executed, increasing the costs of rounding. If, in the aggregate, all buyers behave this way, clustering will fall over the season. This result could also be explained by buyers' aversion to the sequential resolution of uncertainty suggesting a preference for uncertainty to be resolved all at one time rather than sequentially [36]. Hence, facing limited and uncertain climate information at the beginning of the season, buyers intend to secure the water they need at one time rather than through multiple orders as the season progresses. Later in the season (e.g., January onwards in our results) when climate and allocation information are revealed, uncertainty will diminish, the costs of rounding will decrease and therefore, clustering will increase again. The results presented in Table 5 demonstrate this outcome for most of the buy models, whereas in the sell models the opposite is true: clustering tends to increase throughout the water season.

The results that suggest strategic behavior as a reason for clustering by water buyers include the negative coefficient for whole milk powder price. When the milk powder price increases, irrigators have greater incentive to produce milk to take advantage of the higher returns. In turn, they are more determined to have their buy offers executed, so the costs of rounding increases thereby decreasing price clustering. We would expect to see the opposite effect on clustering if the price resolution hypothesis applied in this case.

But overall, it appears that buyer bid behavior in water markets is most influenced by price resolution (uncertainty) rather than strategic behavior. In light of the fact that our data-set includes years during which irrigators were learning how to use the new water market, it is not surprising that, on balance, uncertainty would create costs associated with precision thereby leading to greater clustering. The continuing tendency for clustering on the buyers' side of the market may well reflect the ongoing uncertainty caused by the combined effects of Australia's highly variable climate and changes in government policy.

6.2. Sell Offers Price Clustering

Both the price resolution hypothesis and strategic behavior can also be identified from significant variables in our seller price clustering models. The results for volume could support either hypothesis with positive coefficients in the WholeSell models and negative coefficients for the ZeroSell model. An increase in clustering in the whole sell models reflects strategic behavior where the costs of rounding are low because sellers may be less determined to have their trades executed. The price resolution hypothesis better explains the decrease in clustering in the ZeroSell model because greater trade intensity creates higher liquidity levels and produces more information with regard to value, allowing for greater precision. In combination, these factors reduce volatility and clustering. Alternatively, these mixed signs could suggest that the attraction hypothesis better explains the effects of volume on clustering for water sellers and that these traders are simply drawn to particular numbers.

The positive significant coefficients for spread lend support to the price resolution hypothesis because a wider bid-ask spread indicates precise valuations are more difficult. This adds to market volatility, so clustering will increase. The negative and significant coefficient on cattle prices is also consistent with the price resolution hypothesis. An increase in cattle prices (which is a dryland output substitute for irrigated production) would lead to a reduction in water demand and price. Falling water prices increase the costs of rounding thereby causing the clustering levels to fall also.

Water entitlement price is significantly negative in the WholeSell model, which suggests that strategic behavior, rather than the price resolution hypothesis, explains price clustering at whole dollars. When some buyers replace water entitlements with water allocations due to increasing water entitlement prices, the demand for water allocations increases and this pushes up water allocation prices. Water allocation sellers may consider the loss in revenue from pennying behavior is compensated by the higher allocation price and a greater chance of offer execution. Hence price clustering decreases and pennying behavior increases.

A positive impact of whole milk powder price, or a negative impact of carry-over level on price clustering, would suggest that strategic behavior may be playing a role in seller behavior. Our results support these hypotheses. Whole milk powder price has a positive estimate in all sell models and significant in the WholeSell and ZeroSell models, while carry-over level has a negative estimate in all sell models but is only significant in the ZeroSell model. As whole milk powder price rises, demand for water also increases so that higher returns from selling water accrue and sellers may expect to trade a higher volume. This magnifies the extra dollar per megaliter from clustering at whole dollars ending in zero, indicating strategic behavior may be utilized by sellers in these situations.

A higher carry-over percentage potentially increases the demand for water allocations in the market, especially later in the season, as risk-averse farmers can carry-over water that they have not used, and buy extra supplies to cover potential shortfalls the following season. This is a more dynamic explanation of the impact of carry-over in the water market, where irrigators are adjusting their practices over seasons. Water allocation prices are therefore higher than otherwise and price clustering decreases.

The government policy variable had a significant negative impact on price clustering in the WholeSell_10 model. In general, periods of policy uncertainty decrease price-clustering behavior by sellers, indicating that perhaps price increases are expected, there is a lower risk of entering the market for sellers and hence price-clustering behavior falls.

The relationship between most of the variables and clustering outcomes on the sellers' side of the market runs in the opposite direction to that which would be expected under the price resolution hypothesis. Therefore, it appears that strategic behavior influences seller bid behavior more than buyer bid behavior.

7. Conclusions

This paper has provided evidence to show there are a range of influences impacting buyer and sellers' water allocation market behavior in the Greater Goulburn trading zone in Victoria. While there are similarities between irrigators' behavior in the water market and general investors' behavior in the financial product markets, such as strong evidence of price clustering present in both markets, differences between two markets exist in terms of the explanations for price clustering, which we have investigated in the current study. Understanding irrigators' water market clustering behavior allows us to gain a range of possible insights about how buyers and sellers may respond to uncertainty and policy changes in the market. These insights are useful for achieving more efficient resource allocation. Our analyses indicate that buyer-clustering behavior is for the most part explained by the price resolution hypothesis—where uncertainty tends to increase risks and decrease the costs of rounding. The cost of precision valuation increases when water allocation prices are difficult to predict and are volatile. For buyers, times of severe climate conditions (e.g., hotter and drier conditions), commodity price volatility, and government policy introduction increases the risk associated with trading and, thereby, their price-clustering behavior.

Conversely, the models' results seem to reflect that sellers' clustering behavior is more reflective of strategic behavior than uncertainty. Strategic behavior in water markets prevails when the benefit of clustering does not outweigh its cost. These costs may include a reduction in the chance of order execution; an increase in the purchase cost for buyers; or an associated loss of sale revenue for water sellers. Correspondingly, the cost of unsuccessful sale offers is high if buyers are in greater need of water or if sellers keenly anticipate the revenue from water sales. Under such circumstances of high costs, traders are likely to consider carefully the cost of clustering and bid/offer strategically, which our results suggested happened the most in the seller clustering models. Hence, our results suggest sellers are acting in a more sophisticated manner in water markets than water buyers, and most of the costs of clustering are therefore borne by buyers.

In terms of policy implications from this research, it is clear that there is a need, wherever possible, for governments to attempt to reduce irrigator uncertainty. This will be of most importance for buyers. More effective farmer adaptation to external impacts, such as water variability is driven by timely and useful information. Water price, climate, commodity forecasts, allocation information and certainty in government policy are all important influences of water market strategies. Incomplete and fragmented information, as well as uncertain policy, decreases farmers' ability to manage their water needs.

Acknowledgments

This research was supported by an Australian Research Council Discovery Project DP140103946.

Author Contributions

Alec Zuo conducted the majority of the analysis, and wrote the paper with Sarah Wheeler. Robert Brooks and Edwyna Harris provided the original idea for the paper, and Henning Bjornlund provided historical water market data.

Conflicts of Interest

The authors declare no conflict of interest.

References

1. Bjornlund, H.; McKay, J. Aspects of water markets for developing countries—Experiences from Australia, Chile and the US. *Environ. Dev. Econ.* **2002**, *7*, 767–793.
2. Grafton, R.Q.; Libecap, G.; McGlennon, S.; Landry, C.; O'Brien, B. An integrated assessment of water markets: A cross-country comparison. *Rev. Environ. Econ. Policy* **2011**, *5*, 219–239.
3. Wheeler, S.; Garrick, D.; Loch, A.; Bjornlund, H. Evaluating water market products to acquire water for the environment in Australia. *Land Use Policy* **2013**, *30*, 427–436.
4. Nicol, L.; Klein, K.; Bjornlund, H. Permanent transfers of water rights: A study of the southern Alberta market. *Prairie Forum* **2008**, *33*, 341–356.
5. Giannoccaro, G.; Pedraza, V.; Berbel, J. Analysis of stakeholders' attitudes towards water markets in Southern Spain. *Water* **2013**, *5*, 1517–1532.
6. Adamson, D.; Mallawaarachchi, T.; Quiggin, J. Declining inflows and more frequent droughts in the Murray-Darling Basin: Climate change, impacts and adaptation. *Aust. J. Agric. Resour. Econ.* **2009**, *53*, 345–366.
7. Wheeler, S.; Loch, A.; Zuo, A.; Bjornlund, H. Reviewing the adoption and impact of water markets in the Murray-Darling Basin, Australia. *J. Hydrol.* **2014**, *518*, 28–41.
8. Crase, L.; Pagan, P.; Dollery, B. Water markets as a vehicle for reforming water resource allocation in the Murray-Darling Basin. *Water Resour. Res.* **2004**, *40*, 1–10.
9. National Water Commission. *Impacts of Water Trading in the Southern Murray—Darling Basin Between 2006–07 and 2010–11*; Commonwealth of Australia: Canberra, Australia, 2012.
10. Murray-Darling Basin Authority (MDBA). *Proposed Basin Plan*; MDBA: Canberra, Australia, 2011.
11. *Market Mechanisms for Recovering Water in the Murray-Darling Basin*; Final Report for Productivity Commission: Canberra, Australia, 2010.
12. Loch, A.; Bjornlund, H.; Wheeler, S.; Connor, J. Trading in allocation water in Australia: A qualitative understanding of irrigator motives and behavior. *Aust. J. Agric. Resour. Econ.* **2012**, *56*, 42–60.
13. Heaney, A.; Dwyer, G.; Beare, S.; Peterson, D.; Pechey, L. Third-party effects of water trading and potential policy responses. *Aust. J. Agric. Resour. Econ.* **2006**, *50*, 277–293.
14. Chung, H.; Chiang, S. Price clustering in E-mini and floor-traded index futures. *J. Futur. Mark.* **2006**, *26*, 269–295.

15. Ball, R. The global financial crisis and the efficient markets hypothesis: What have we learned? *J. Appl. Corp. Financ.* **2009**, *21*, 8–16.

16. Brown, S. The efficient markets hypothesis: The demise of the demon of chance? *Account. Financ.* **2011**, *51*, 79–95.

17. Brooks, R.; Harris, E.; Joymungul, Y. Price clustering in Australian water markets. *Appl. Econ.* **2013**, *45*, 677–685.

18. Bjornlund, H.; Wheeler, S.; Rossini, P. Water Markets and Their Environmental, Social and Economic Impact in Australia. In *Water Trading and Global Water Scarcity: International Perspectives*; Maestu, J., Ed.; Francis Taylor: Gloucester, UK, 2013; pp. 68–93.

19. Wheeler, S.; Bjornlund, H.; Shanahan, M.; Zuo, A. Price elasticity of allocations water demand in the Goulburn-Murray irrigation district of Victoria, Australia. *Aust. J. Agric. Resour. Econ.* **2008**, *52*, 37–55.

20. Harris, L. Stock price clustering and discreteness. *Rev. Financ. Stud.* **1991**, *4*, 389–415.

21. Grossman, S.; Miller, M.; Cone, K.; Fischel, D.; Ross, D. Clustering and competition in asset markets. *J. Law Econ.* **1997**, *40*, 23–60.

22. Ball, C.; Torous, W.; Tschoegl, A. The degree of price resolution: The case of the gold market. *J. Futur. Mark.* **1985**, *5*, 29–43.

23. Loomes, G. Different experimental procedures for obtaining valuations of risky actions: Implications for utility theory. *Theory Decis.* **1988**, *25*, 1–23.

24. Butler, D.; Loomes, G. Decision difficulty and imprecise preferences. *Acta Psycholog.* **1988**, *68*, 183–196.

25. Capelle-Blanchard, G.; Chaudhury, M. Price clustering in the CAC 40 index options market. *Appl. Financ. Econ.* **2007**, *17*, 1201–1210.

26. Mallawaarachchi, T.; McClintock, A.; Adamson, D.; Quiggin, J. Investment as an Adaptation Response to Water Scarcity. In *Water Policy Reform: Lessons in Sustainability from the Murray-Darling Basin*; Quiggin, J., Mallawaarachchi, T., Chambers, S., Eds.; Edward Elgar: Cheltenham, UK, 2012; pp. 101–126.

27. Brennan, D. Water policy reform in Australia: Lessons from the Victorian seasonal water market. *Aust. J. Agric. Resour. Econ.* **2006**, *50*, 403–423.

28. Ikenberry, D.L.; Weston, J.P. Clustering in US stock prices after decimalisation. *Eur. Financ. Manag.* **2008**, *14*, 30–54.

29. Tversky, A.; Kahneman, D. Judgement under uncertainty: Heuristics and biases. *Science* **1974**, *185*, 1124–1131.

30. Mitchell, J. Clustering and psychological barriers: The importance of numbers. *J. Futur. Mark.* **2001**, *21*, 395–428.

31. Jennings, R. Getting "pennied": The effect of decimalization on traders' willingness to lean on the limit order book at the New York Stock Exchange. *NNYSE Doc.* **2001**, *1*, 1–24.

32. Edwards, A.; Harris, J. *Stepping Ahead of the Book*; Securities and Exchange Commission: Washington, DC, USA, 2002.

33. Niederhoffer, V. Clustering of stock prices. *Oper. Res.* **1965**, *13*, 258–265.

34. Aşçıoğlu, A.; Comerton-Forde, C.; McInish, T.H. Price clustering on the Tokyo Stock Exchange. *Financ. Rev.* **2007**, *42*, 289–301.

35. Papke, L.E.; Wooldridge, J.M. Econometric methods for fractional response variables with an application to 401(K) plan participation rates. *J. Appl. Econ.* **1996**, *11*, 619–632.

36. Palacios-Huerta, I. The aversion to sequential resolution of uncertainty. *J. Risk Uncertain.* **1999**, *18*, 249–269.

Impact of Climate Change on the Irrigation Water Requirement in Northern Taiwan

Jyun-Long Lee and Wen-Cheng Huang

Abstract: The requirement for irrigation water would be affected by the variation of meteorological effects under the conditions of climate change, and irrigation water will always be the major portion of the water consumption in Taiwan. This study tries to assess the impact on irrigation water by climate change in Taoyuan in northern Taiwan. Projected rainfall and temperature during 2046–2065 are adopted from five downscaled general circulation models. The future evapotranspiration is derived from the Hamon method and corrected with the quadrant transformation method. Based on the projections and a water balance model in paddy fields, the future crop water requirement, effective rainfall and the demand for water for irrigation can be calculated. A comparison between the present (2004–2011) and the future (2046–2065) clearly shows that climate change would lead both rainfall and the temperature to rise; this would cause effective rainfall and crop water requirement to increase during cropping seasons in the future. Overall, growing effective rainfall neutralizes increasing crop water requirement, the difference of average irrigation water requirement between the present and future is insignificant (<2.5%). However, based on a five year return period, the future irrigation requirement is 7.1% more than the present in the first cropping season, but it is insignificantly less (2.1%) than the present in the second cropping season.

Reprinted from *Water*. Cite as: Lee, J.-L.; Huang, W.-C. Impact of Climate Change on the Irrigation Water Requirement in Northern Taiwan. *Water* **2014**, *6*, 3339-3361.

1. Introduction

The fourth assessment report of the Intergovernmental Panel on Climate Change indicates that the observations of global average temperature during 1995–2006 have increased, and heavy rainfall events have become much frequent. This report also predicts the global average surface temperature during 2080–2099 may rise between 1.1 °C and 6.4 °C more than the period during 1980–1999, and cause crop productivity to increase [1]. It clearly shows the affection of climate change.

General circulation models (GCMs) are the most advanced tools available to simulate the response of the global climate system to increasing greenhouse gas concentrations. With the models, an assessment of the future climate would be possible [2,3], and the problem of uncertainty may be mitigated by considering multiple models [4,5]. According to the results of the cited researches, the rainfall distribution would be different and temperature would rise under climate change. It would make challenges for water resources management.

Taiwan is a small island in the north-west Pacific. Analyzing the historical meteorological data in Taiwan over the past hundred years, the annual rainfall increased in the northern regions, decreased in central and southern regions, and exhibited no clear tendency in the eastern regions [6]; moreover, the surface temperature rose 0.8–1.6 °C in each region [7]. There is 22.7% of the area that has been cultivated in Taiwan in recent years. The annual water consumption in Taiwan is 17,064 million m^3, of which is 11,088 million m^3 consumed by irrigation. The proportion of irrigation water is about 65% of the total consumption; in other words, the irrigation requirement is the main demand factor.

Under climate change, the variation of rainfall and temperature would also impact the irrigation water demand. There are many methods to determine the irrigation water requirement, for example: the Erosion Productivity Impact Calculator [8,9], the Global Irrigation Model [10,11], the CROPWAT model [12,13], and the Stochastic Crop Water Production Functions [14]. The basis of these models is to capture the characteristics of crop water consumption in different periods. Therefore, in the given growth characteristic of crops, rainfall and temperature distribution, and geology of a region, according to the water balance model, the irrigation water requirement would be determined by simulation.

Although the average annual rainfall is about 2500 mm in Taiwan, high rainfall intensity along with a steep slope of river makes water resource storage difficult. Since the supply of irrigation water is one of the most significant tasks for water management, an impact evaluation on irrigation water under climate change in Taiwan is essential. That is the purpose of this study.

2. Materials and Method

First of all, daily meteorological data such as rainfall and temperature either from observation or projection are needed for estimating the evapotranspiration of crops. Second, the effective rainfall and irrigation water on paddy fields could be estimated by simulation method based on the water balance. In addition, data concerning the crop coefficient, percolation rate, conveyance loss rate, and farming area are collected. In this study, the present and future are represented by the periods 2004–2011 and 2046–2065, respectively. The flowchart is shown in Figure 1.

2.1. Study Area

In this study, the irrigation district in northern Taiwan, governed by the Taoyuan Irrigation Association (TIA), is chosen as the study area (see Figure 2). In addition, Shihmen reservoir in the upper reach supplies TIA's irrigation water. The meteorological station adopted here is located in the middle of the irrigation district and operated by the Agricultural Engineering Research Center. The meteorological data include air temperature, dewpoint temperature, solar radiation, sunshine duration, wind speed and rainfall. On average, the air temperature is 22.49 °C, solar radiation equals 10.07 MJ/m^2·day, sunshine duration lasts for 7.12 h, wind speed is 2.25 m/s and the annual rainfall is 1876 mm.

According to the irrigation plan, the area available for farming in TIA is 24,233 ha. There are four types of soil: clayey loam, sandy loam, sand clay loam and light clay. The percentages of each soil are about 41%, 22%, 18% and 19% respectively. In the area, average percolation rate on paddy fields is 8.14 mm/day, and average water conveyance loss is 12.6% [15].

Paddy is the main crop in Taiwan, the proportion of paddy fields to total farming area of TIA is about 95%. The subtropical climate makes two harvests of paddy rice in a year in Taiwan possible.

This study assumes the first and second cropping season start between 1 March to 28 June (from the 7th to 18th day) and 1 August to 28 November (from the 22th to 33th day), respectively. That is, a 120 day period for paddy growth is required in each cropping season.

Figure 1. Flow chart of this study.

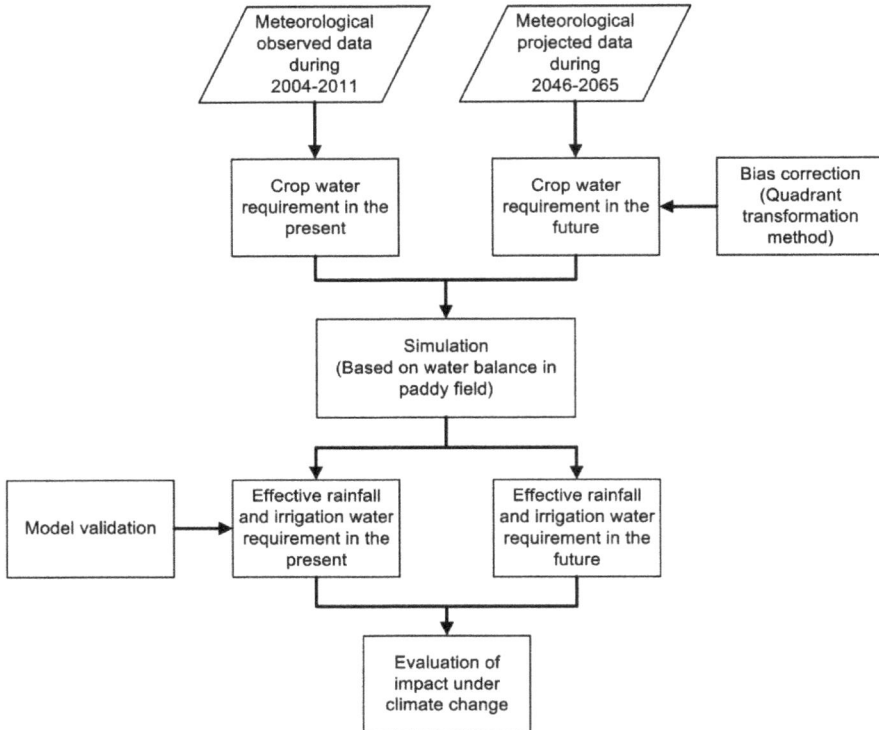

2.2. Projected Rainfall and Temperature

The projected rainfall and temperature under climate change in the period of 2046–2065 came from five GCMs: CGCm3 from the Canadian Center for Climate Modeling and Analysis (CCCma), Cm3 from the Center National de Recherches Meteorologiques (CNRM), Mk3.0 from Australia's Commonwealth Scientific and Industrial Research Organization (CSIRO), Cm2.0 from the Geophysical Fluid Dynamics Laboratory (GFDL) and FGOALS-g1.0 from the State Key Laboratory of Numerical Modeling for Atmospheric Sciences and Geophysical Fluid Dynamics (LASG), which are based on SRES A1B scenarios. The A1B scenario describes a future world of very rapid economic growth, global population that peaks in mid-century and declines thereafter, and the rapid introduction of new and more efficient technologies. Moreover, the A1B scenario is distinguished by its technological emphasis: a balance between fossil and other energy sources [16].

Figure 2. Location of study area and meteorological station.

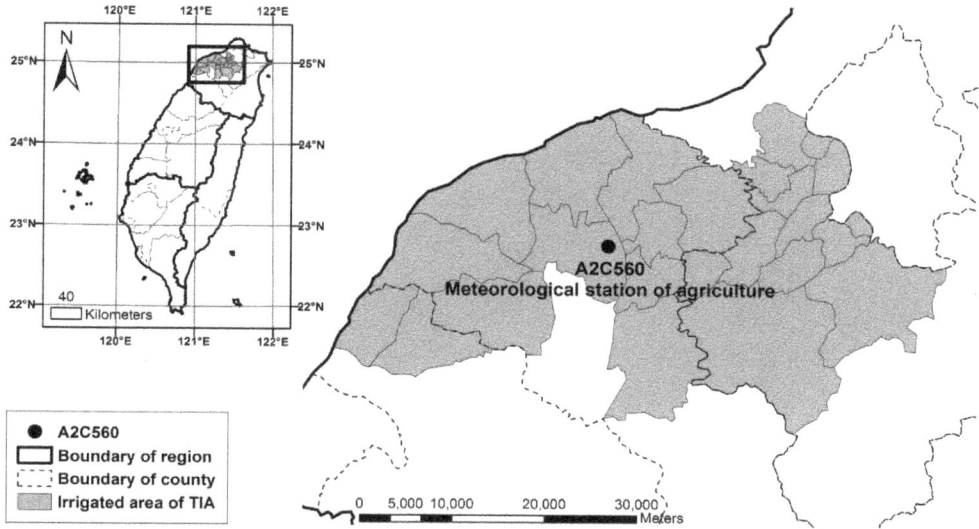

Because of coarse resolution from GCM projection, statistical downscaling of GCM scenario-run outputs to local climate stations were needed and applied. All of the data have been downscaled by the Global Change Research Center of National Taiwan University. Briefly, the process of the downscaling technique would be done in three stages [17]: first, the GCM outputs near Taiwan were adjusted with respect to the NCEP reanalysis data [18] during the training period by linking the normalized probability distribution functions of the mean climate parameters; second, a transfer function (*i.e.*, a multiple-variant linear regression) was established to link NCEP reanalysis variants with local climatic observations during the training period; third, the projected temperature and precipitation data at each station during the verification period were adjusted with respect to the local observation data by the procedure in the first stage. The linkage established was then extended to adjust outputs for the years of projections. If more details about the downscaling technique are needed, please refer to [17,19].

2.3. Paddy Water Requirement

The paddy evapotranspiration in this study is assumed under standard conditions, which means the paddy is grown in large fields with disease-free and well-fertilized conditions. The crop water requirement equals crop evapotranspiration under standard conditions, and it is expressed as [20]:

$$ET_c = K_c \times ET_o \tag{1}$$

where ET_c is crop evapotranspiration under standard conditions (mm/day); K_c is crop coefficient (dimensionless); ET_o is reference evapotranspiration (mm/day).

Notice that K_c varies during the cropping season and depends also on the type of crops. The K_c value that is commonly used in Taiwan at different growth stages for the first and second cropping seasons is shown in Figure 3 [21].

Figure 3. Crop coefficient (K_c) of paddy at each day during cropping season.

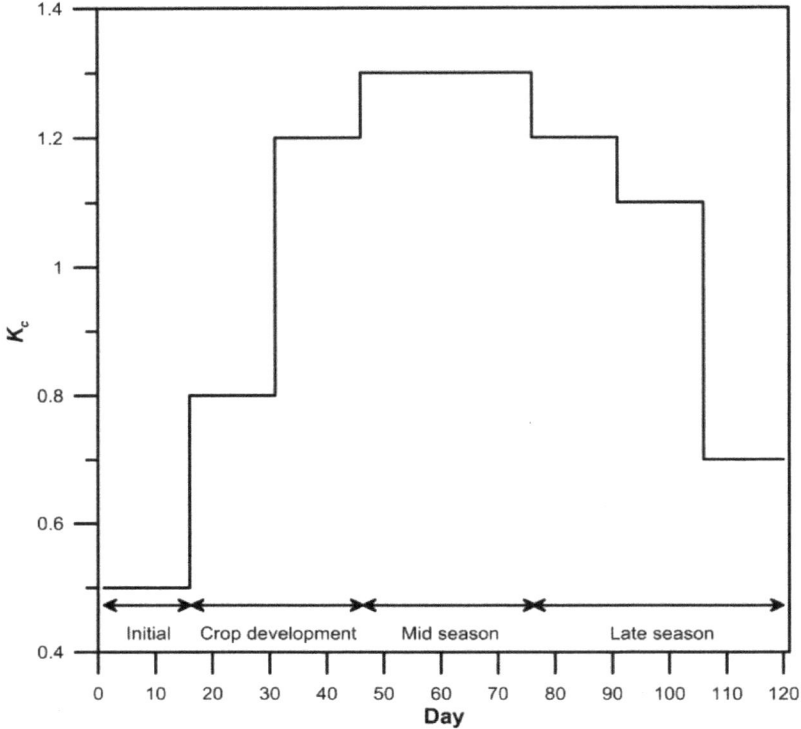

2.3.1. FAO Penman-Monteith Equation

The FAO Penman-Monteith (PM) equation is adopted here for estimating the reference evapotranspiration. The use of the PM equation is recommended as a standard for reference evapotranspiration to provide more consistent values with actual crop water use data worldwide [20]. The equation is expressed as:

$$ET_o = \frac{0.408\Delta(R_n - G) + \gamma \dfrac{900}{T + 273} u_2(e_s - e_a)}{\Delta + \gamma(1 + 0.34u_2)} \tag{2}$$

where Δ is slope vapor pressure curve (kPa/°C); R_n is net radiation at the crop surface (MJ/m^2·day); G is soil heat flux density (MJ/m^2·day); γ is psychrometric constant (kPa/°C); T is mean daily air temperature at 2 m height (°C); e_s is saturation vapor pressure (kPa); e_a is actual vapor pressure (kPa); u_2 is wind speed at 2 m height (m/s). A detailed explanation of this equation can be found in the literature [20].

2.3.2. Hamon Method

Since only projected temperature and rainfall are available from the output of GCMs in this study, the PM equation could not be used for estimating evapotranspiration in the future. In order to solve this problem, the Hamon method, a temperature-based equation, is adopted. Hamon considered temperature and vapor pressure are the important factors that affect evapotranspiration [22]. The modified Hamon equation is expressed as [23]:

$$PE = 29.8N \frac{e_s}{T + 273.2} \tag{3}$$

where PE is potential evapotranspiration by Hamon (mm/day); N is sunshine duration (h).

Daily sunshine duration N can be calculated through the sunset hour angle (ω_s) in theoretically [20]:

$$N = \frac{24}{\pi} \omega_s \tag{4}$$

2.3.3. Bias Correction

As mentioned above, the PM equation and Hamon method are adopted for estimating the evapotranspiration of the present and projected respectively. The basic assumption of these method are different; hence, there would be a mismatch between ET_o and PE. In addition, the evapotranspiration of PE may need correction. A conversion method, the quadrant transformation method [5,24], is applied here for bias correction. The concept of quadrant transformation method is shown in Figure 4. Here the difference between the 1st and 4th quadrants is estimated either by ET_o or by PE, while the difference between the 3rd and 4th quadrants is due to climate change. The duration curve for the corrected PE of the future (2046–2065) in the 2nd quadrant is built by the other three quadrants' conversion.

The procedure for correction would be expressed by following steps: (1) constructing the daily duration curves, a cumulative frequency curve that show the percent of time specified rainfall were equaled or exceeded during a given period [25], for the 1st, 3rd and 4th quadrant by the evapotranspiration of PM in the present, Hamon in the future and Hamon in the present, respectively; (2) confirming the corresponding percentile of evapotranspiration at the specific day for the duration curve of the 3rd quadrant; (3) finding the corresponding percentile for the 4th quadrant by the evapotranspiration for the 3rd quadrant; (4) using the percentile by above to find a new corresponding evapotranspiration for the 1st quadrant, and which would be the corrected daily value.

By repeating steps (2) to (4) we would obtain the duration curve for the corrected data in 2nd quadrant. Considering the seasonal variation, this correction method is based on monthly duration curves of evapotranspiration.

46

Figure 4. Bias correction of evapotranspiration through quadrant transformation.

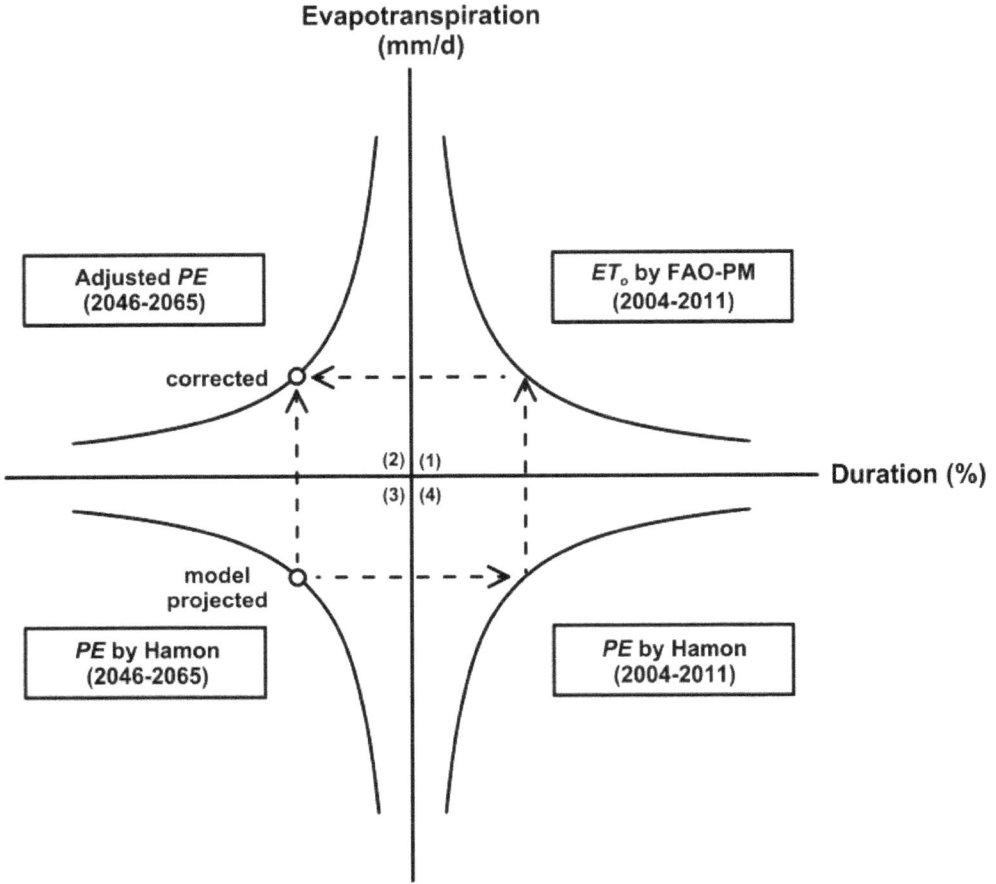

2.4. Calculation of Irrigation Water Requirement

Since the rainfall, paddy water requirement and geology are understood, the irrigation water requirement can be calculated with a water balance model of the paddy fields. Figure 5 shows the factors affecting water balance in the paddy fields. Irrigation should supply the deficiency of water that paddy growth needs. Considering different soil types on paddy fields and water conveyance loss of irrigation canals, the continuity equation in paddy fields is expressed as:

$$S_{t+1} = S_t + P_t - ET_{ct} - \sum_j f_j A_j^\circ + \frac{IR_t}{\sum_i A_i^* (1+CL_i)} \tag{5}$$

where S_t, P_t, ET_{ct}, IR_t, respectively, indicate the daily water storage, rainfall, crop evapotranspiration, and irrigation water requirement in the paddy fields at time t (mm/day); f_j is the percolation rate for jth soil type (mm/day), A_j° is the percentage of jth soil type area (%); A_i^* gives the percentage of total farming area controlled by the ith canal (%), and CL_i shows the average water conveyance loss for the area controlled by ith canal (%).

The irrigation water requirement must be externally supplied to fill the deficit for paddy growth when rainfall and storage do not satisfy the water consumption in paddy fields; in contrast, the irrigation water requirement will be 0 while the consumption has been satisfied. Therefore, it could be rewritten as:

$$IR_t = \sum_i A_i^* (1 + CL^i) \cdot (ET_{ct} + \sum_j f^j A_j^\circ + S_t^{\min} - S_t - P_t) \qquad if \quad S_t + P_t - ET_{ct} - \sum_j f^j A_j^\circ < S_t^{\min} \tag{6}$$

The rainfall stored in paddy fields for growth is called effective rainfall (P_t^*), and it used in the paddy fields at time t equals:

$$P_t^* = \min\left\{ S_t^{\max} - S_t + ET_{ct} + \sum_j f_j A_j^\circ, P_t \right\} \tag{7}$$

where, $S_t^{\max}$ and $S_t^{\min}$, respectively, are the maximum and minimum ponding storage in the paddy fields during different growth stages (mm). Here, we assume the water is abundant for a continued irrigation to keep the fields in an appropriate state of water depth. The maximum and minimum storage during different growth stages for paddy are shown in Figure 6 [26,27]. Please note that emptying out the storage would be recommended for root growth at the specific time. Therefore, the minimum ponding storage in some days would be 0.

Figure 5. Concept of water balance in paddy field.

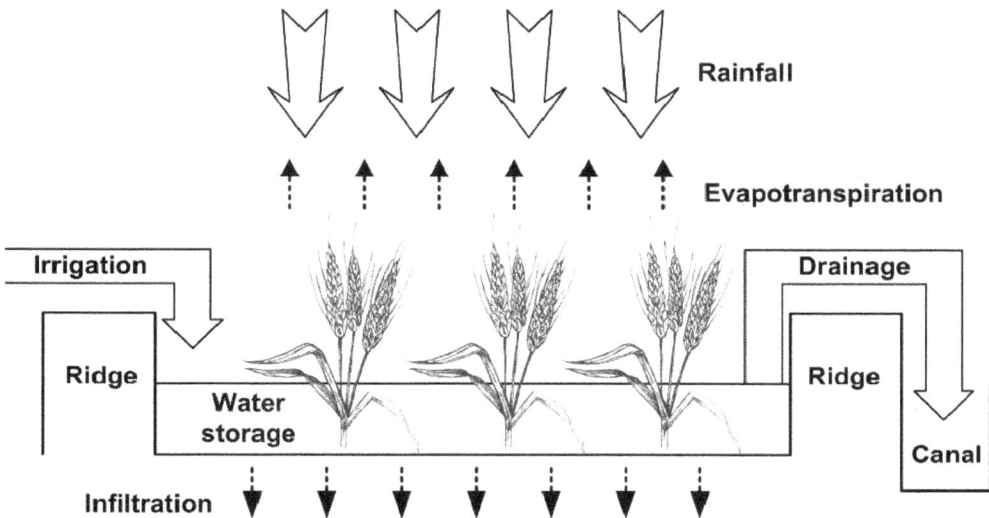

Figure 6. Suggested water depth in paddy field during cropping season.

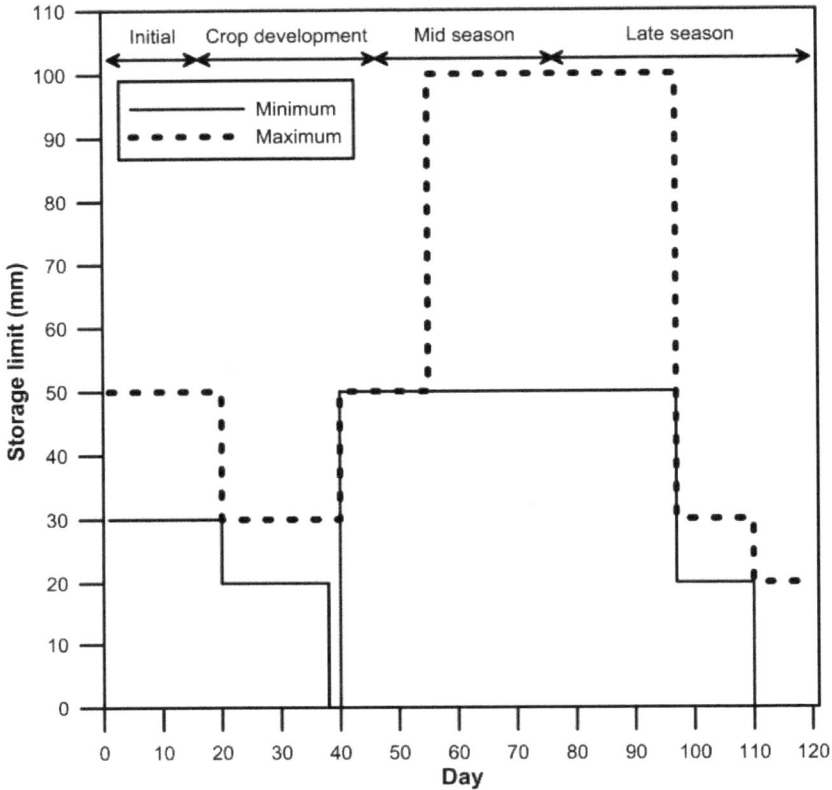

3. Result and Discussion

For the A1B emissions scenario in the period of 2046–2065, five GCMs produce different patterns of change on temperature and rainfall. In this study, the mean value of models is used to show the result for the following. It represents the average of the results by five GCMs, instead of the result of average temperature and rainfall by these GCMs.

3.1. Result of Bias Correction

As seen in Figure 7, ET_o values in the 1st quadrant are inconsistent with PE values in the 4th quadrant. PE assessed by the Hamon method seems underestimated as evapotranspiration exceeds 2.9 mm/day, and vice versa. Apparently, the biases between the 1st/4th quadrants in the present (2004–2011) and the 2nd/3rd quadrants in the future (2046–2065) are similar. It shows the LASG-based PE projections in 3rd in March quadrant can be appropriately adjusted in the 2nd quadrant by the quadrant transformation method.

Figure 7. Comparison of duration curve for evapotranspiration at each quadrant in March (LASG GCM).

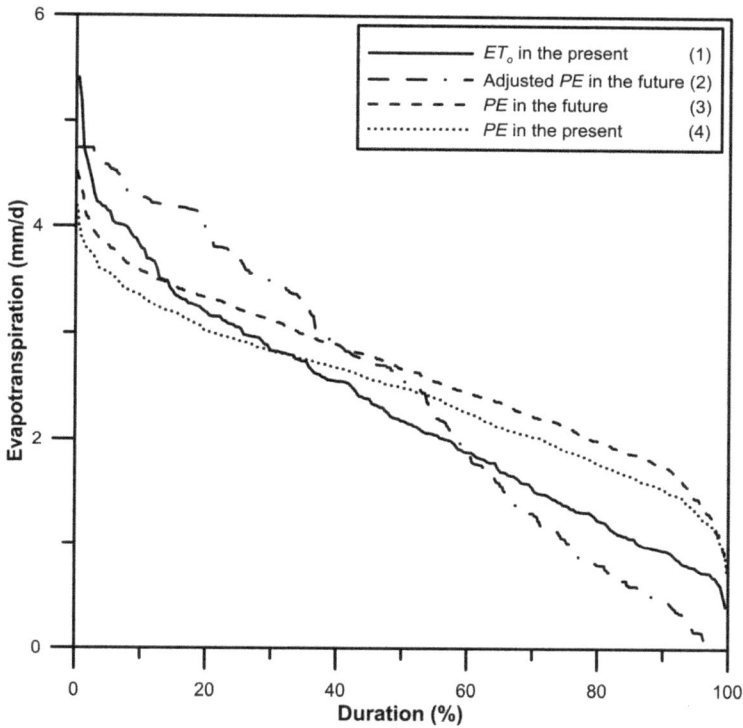

Here, Figure 8 shows the comparison with before/after bias correction for these GCMs. While focusing on the different between ET_o and PE in the present, we would find the PE calculated by Hamon method is always overestimated during January to September, especially June to August. After bias correction, most of the month will be adjusted to be lower, except some months for GCMs. The bias correction does not seem to work during June/July/August/September for some GCMs. The main reason is that the extreme high ET_o always happened in summer. If the extreme value of ET_o in the present is more than the value of PE in the future, and the future PE is always more than the present PE. Then, the adjusted evapotranspiration would be more than the non-adjusted probably.

3.2. Comparison of Model Estimation and Actual Investigation

Meteorological factors like temperature, wind speed and net radiation during 2004–2011 have been adopted for estimating irrigation demand by the model. Before starting the process to estimate future irrigation requirements, the methods mentioned in Section 2 have to be compared with the actual irrigation consumption, as shown in Table 1. For each year, the difference of model estimation and actual investigation is always over 20%, except for the first cropping season in 2005. The result seems to show that the model has failed. Why has this happened? It would be summarized by the following two reasons:

Figure 8. Comparison with before/after bias correction: (**a**) CCCma GCM; (**b**) CNRM GCM; (**c**) CSIRO GCM; (**d**) GFDL GCM; (**e**) LASG GCM.

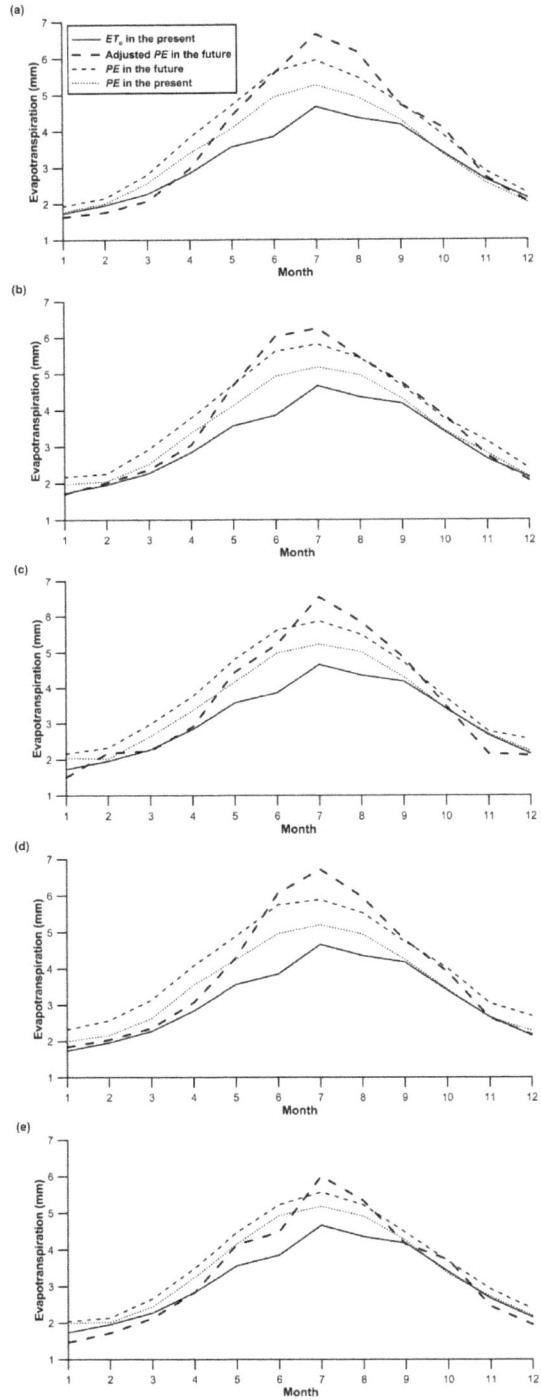

First, the actual investigation of irrigation consumption is according to the water received by canals, and it there would be interference from the discharge of the river, storage of reservoir, operation of canals, *etc.* However, the model estimation depends on the meteorology and the growth stages of paddy. There is a difference in the foundation between the actual investigation and model estimation.

Second, the strategies of the government would also influence the supply of irrigation water significantly. In January 2002, Taiwan became a member of the World Trade Organization. According to the membership commitment, Taiwan has to import 144,720 metric tons of rice per year. To achieve this goal, the cultivation area in TIA was reduced significantly (probably reduced from 20,000 to 6000 ha). After 2002, the amount of agricultural water consumption is not only for irrigation, it involves multi-purpose uses like water resources scheduling, groundwater recharge and environmental conservation.

Since comparing the difference between model estimation and actual investigation in the same years is not proper for evaluating the model, we try to use the data before 2002 for model validation. In this study, the two-sample t test is adopted because the data of validation is not the same as the period.

At the 0.01 level of significance, the absolute value of threshold value t for 16 sample size is 2.98 (a double-tailed test). The result of two-sample t test is shown in Table 2. In the comparison of the actual investigation between 1994–2001 and 2004–2011, the absolute values of t of the first and second seasons are 2.05 and 3.71, respectively. The t of the second season is greater than 2.98, and it implies that the difference in two means of 2nd season between 1994–2001 and 2004–2011 is significant, it proves the inference that irrigation consumption during 2004–2011 is not only for farming paddy and not proper for evaluating the model. In the comparison of actual investigation during 1994–2001 and model estimation during 2004–2011, the absolute value of t of 1st and 2nd seasons are 1.57 and 2.43 respectively, they are both smaller than 2.98. It implies that the difference in two means is insignificant between actual investigation and model estimation. That is, the proposed process could be accepted and applied to evaluating the impact on irrigation in the future.

Table 1. Comparison of irrigation water requirement between model estimation and actual investigation. (Unit: mm).

	Actual Investigation						Model Estimation	
Year	First Season	Second Season	Year	First Season	Second Season	Year	First Season	Second Season
1994	907	1145	2004	1512	1512	2004	1169	1204
1995	1191	1136	2005	995	1967	2005	983	1358
1996	839	1046	2006	1260	2672	2006	985	1457
1997	1242	1128	2007	840	828	2007	1029	1229
1998	1294	1108	2008	1762	2201	2008	1099	1362
1999	1343	1361	2009	1977	2219	2009	1130	1172
2000	1245	1218	2010	1741	1867	2010	867	1158
2001	1253	1152	2011	1949	1859	2011	1114	1337
Mean	1164	1162		1505	1891		1047	1285
Standard deviation	186	94		432	547		100	108

Table 2. The absolute value of tow sample t test for different periods.

Title	Actual Investigation (2004–2011)	Model Estimation (2004–2011)
Actual investigation	2.05	3.71 *
(1994–2001)	1.57	2.43

Note: *: Exceed the threshold value $t = 2.98$.

3.3. Projected Evapotranspiration Analysis

According to the projections of temperature from the five GCMs, the average increments of temperature during the first and second cropping seasons over TIA are 2.2 °C and 1.1 °C, respectively. This will cause evapotranspiration, ET_o, to increase in the future. As Table 3 shows, no matter whether the first or second season, evapotranspiration grows by all selected GCMs. The GFDL model yields higher projections of evapotranspiration, as LASG model gives lower ones. The average increments during the first and second are 133 mm and 95 mm, respectively, about 35.8% and 16.8% more than the present period (2004–2011).

Table 3. The assessment of water consumption and supply for paddy field in the present and future.

Assessment Factors	Cropping Seasons	Present	Future					
			CCCma	CNRM	CSIRO	GFDL	LASG	Average
Evapotranspiration	First season	372	504	506	511	532	472	505
(mm)	Second season	564	665	662	654	677	635	659
Crop water	First season	383	521	523	525	549	487	521
requirement (mm)	Second season	547	638	632	625	648	607	630
Rainfall (mm)	First season	793	1262	762	904	1105	1344	1075
	Second season	637	629	480	670	658	527	593
Effective rainfall	First season	465	769	503	579	659	788	660
(mm)	Second season	308	533	433	580	567	507	524
Effectiveness (%)	First season	62.6	64.2	70.7	67.7	61.0	62.0	65.1
	Second season	50.8	85.9	91.2	87.0	88.0	96.5	89.5
Irrigation water	First season	923	803	1087	1000	945	742	915
requirement (mm)	Second season	1,208	1179	1285	1110	1152	1172	1180

Furthermore, Figures 9 and 10 show the exceedance probability distribution of evapotranspiration distribution over time for the present and future (ensemble of GCMs). Since evapotranspiration is a cost factor in paddy fields, both figures give the representative values from an optimistic 90% to a pessimistic 10% for the first and second cropping seasons. Obviously, evapotranspiration increases in May and June of the plum period (first cropping season) and gradually decreases within the typhoon period from August through October (second cropping season). The comparison between 2004–2011 and 2046–2065 indicates the temporal distribution of the future is similar to the present, but the variance of the future is much lower than the present.

Figure 9. Exceedance probability distribution of evapotranspiration at each ten-day during the first cropping season. (**a**) present (2004–2011); (**b**) future (2046–2065).

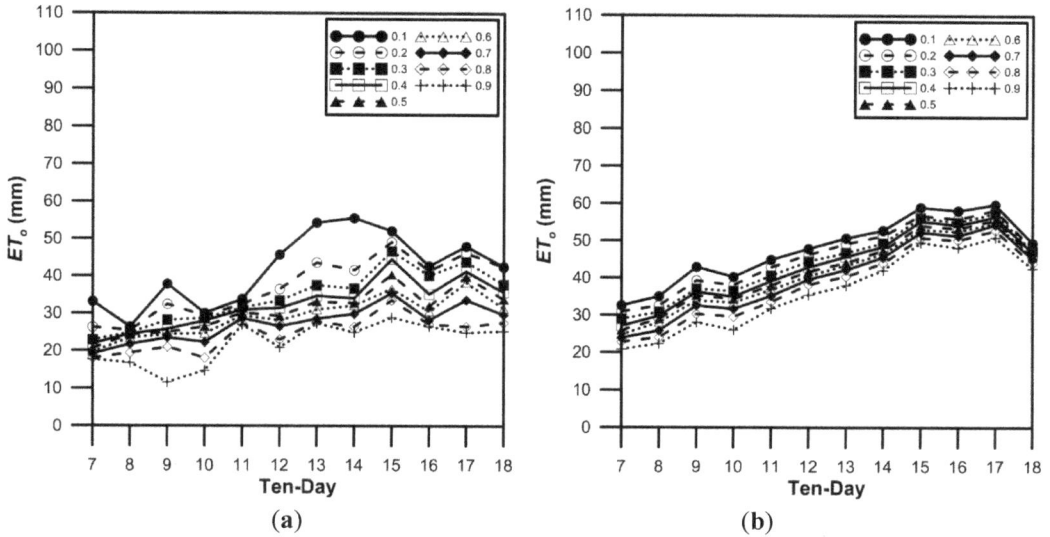

Figure 10. Exceedance probability distribution of evapotranspiration at each ten-day during the second cropping season. (**a**) present (2004–2011); (**b**) future (2046–2065).

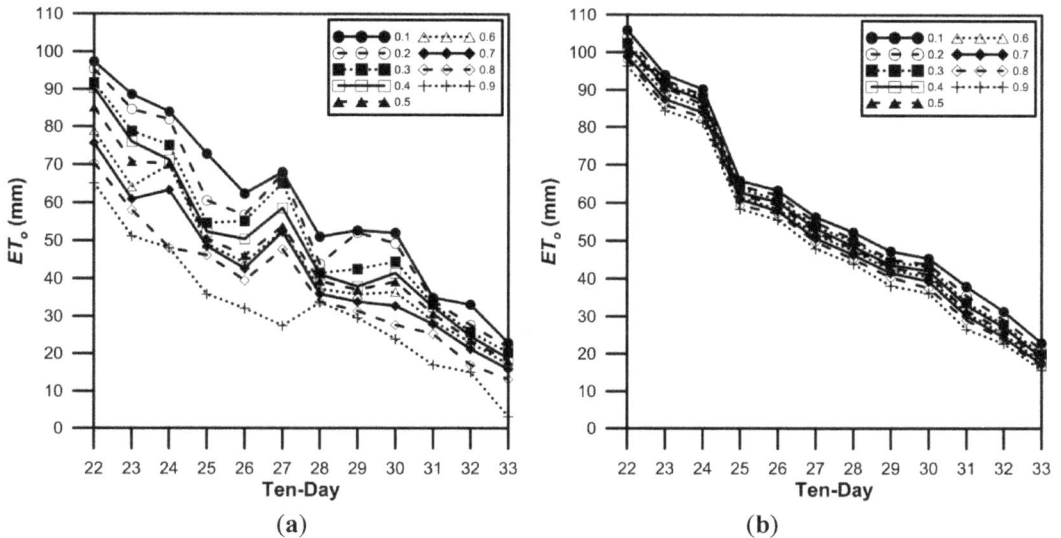

3.4. Crop Water Requirement Analysis

With evapotranspiration multiplied by the time-varying crop coefficient (K_c), as shown in Figure 3, the crop water requirement for rice cultivation could be obtained. Because of rising evapotranspiration in the future, it will result in an increase in the crop water requirement during the first and second cropping seasons. Table 3 shows the comparison of paddy water requirement between the present and future. Clearly, the second cropping season needs more water than the first season. Future also requests more water than the present, based on the GCMs' outputs. On average, 138 mm and 83 mm more water would be needed in the future during the first and second seasons, an increment of nearly 36.0% and 15.2%, respectively. This is due to the increase of temperature in the future, and would increase the possibility of water deficit if rainfall could not supply essential crop water requirement in the future. In addition, as illustrated in Figures 11 and 12, the patterns of exceedance probability distribution of crop water requirement seems to be a great difference between the present and the future in the second cropping season (ensemble of GCMs). The periods with maximum water requirements occur in May during the first cropping season and in September during the second cropping season. Notice that the representative values of exceedance probability distribution are from an optimistic 90% to a pessimistic 10%.

3.5. Projected Rainfall Analysis

Table 3 presents the comparison of the rainfall between observation (2004–2011) and projected (2046–2065). It shows, except for the CNRM model, that the GCMs project much more rainfall than the present in the first cropping season. The average rainfall increases 282 mm (35.6%) in the future. However, in the second cropping season, the average decreases 44 mm (6.9%). In particular, CNRM and LASG models produce much less rainfall than the present. Plus, Figures 13 and 14 give the representative values of rainfall exceedance probability distribution from a pessimistic 90% to an optimistic 10% for the first and second cropping seasons, because rainfall is a benefit factor in paddy fields. The figures show that the period in May and June in the first cropping season has more rainfall in the future (see Figure 13). However, a lower quantity of rainfall occurs in July to October in the second cropping season (see Figure 14).

Figure 11. Exceedance probability distribution of crop water requirement at each ten-day during the first cropping season. (**a**) present (2004–2011); (**b**) future (2046–2065).

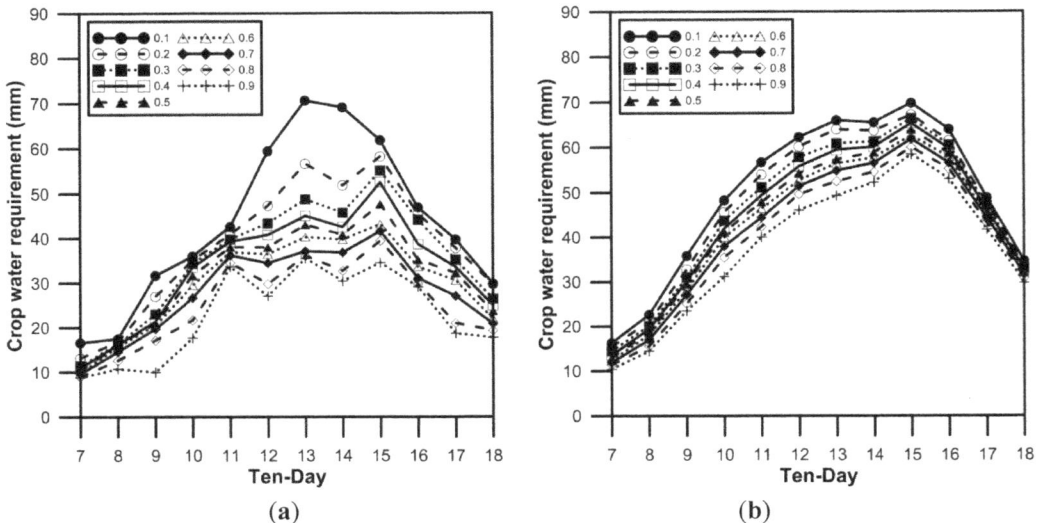

(a)   (b)

Figure 12. Exceedance probability distribution of crop water requirement at each ten-day during the second cropping season. (**a**) present (2004–2011); (**b**) future (2046–2065).

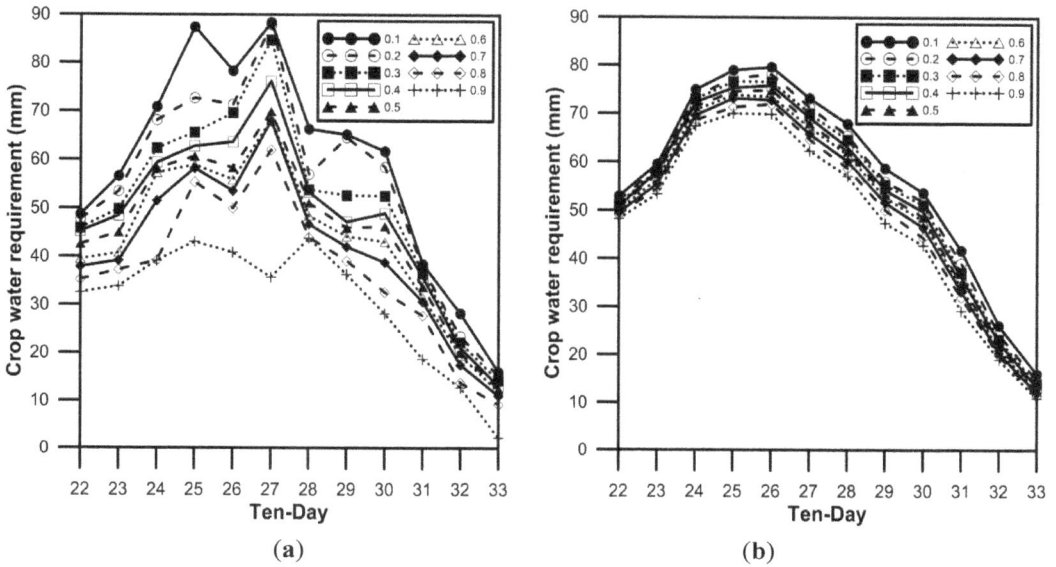

(a) (b)

Figure 13. Exceedance probability distribution of rainfall at each ten-day during the first cropping season. (**a**) present (2004–2011); (**b**) future (2046–2065).

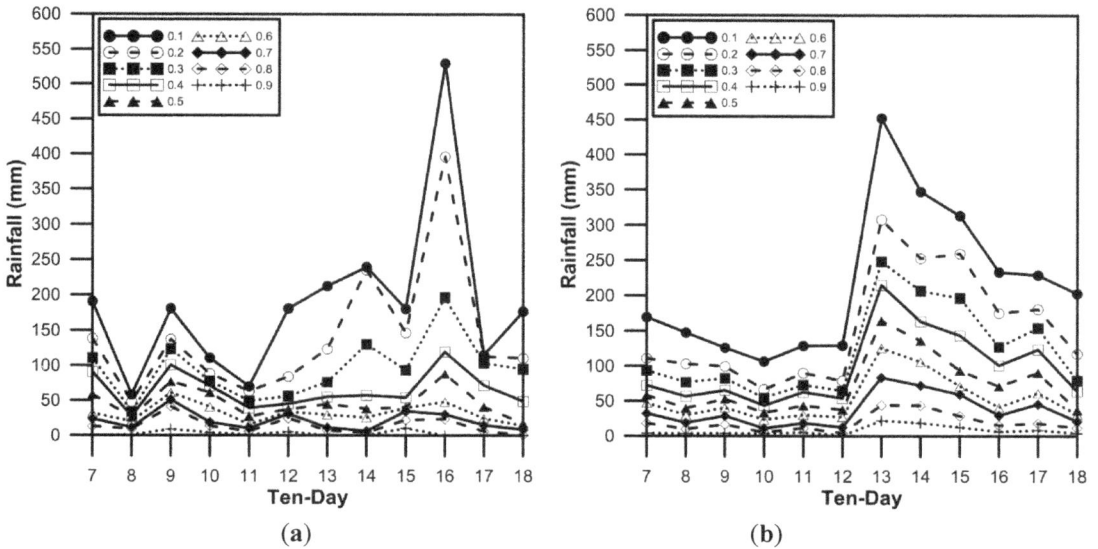

(a) (b)

Figure 14. Exceedance probability distribution of rainfall at each ten-day during the second cropping season. (**a**) present (2004–2011); (**b**) future (2046–2065).

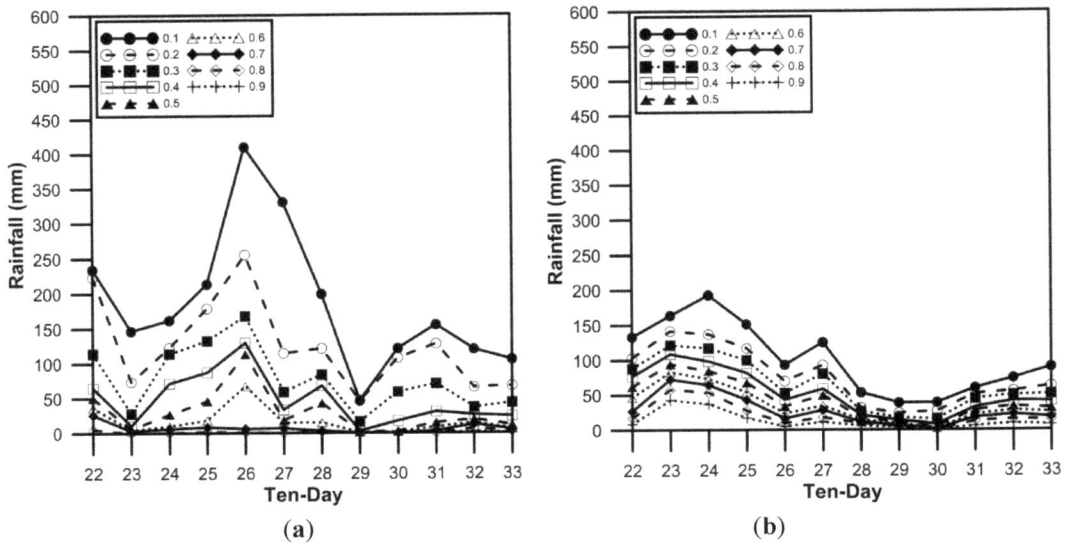

(a)   (b)

3.6. Effective Rainfall Analysis

As shown in Figures 13 and 14, rainfall distribution of the future at each ten-day period seems more even than the present. This may increase the occurrence of effective rainfall in the future. As a benefit factor in paddy fields, Figures 15 and 16 show representative values of exceedance probability distribution of effective rainfall from a pessimistic 90% to an optimistic 10% for the first and second cropping seasons. In the future, more effective rainfall is obtained in May–June during the first cropping season, and in August-September within the second cropping season. Moreover, from Table 3, we can find that all the five GCMs project more effective rainfall than the present during the cropping seasons. The average increments are 195 mm (41.9%) and 216 mm (70.1%), respectively, in the first and second cropping season. Certainly, this is helpful to paddy cultivation and reduces the irrigation requirement.

Effectiveness is defined by this study as a ratio of effective rainfall to total rainfall during cropping season. The difference of effectiveness between the observation and the projection is insignificant in the first cropping season. In contrast, although the projected rainfall is smaller in the second cropping season in the future, the effectiveness appears much better than the present because rainfall distribution become more even. Rainfall effectiveness in the future increases 2.6% and 38.7%, respectively, during the first and second cropping seasons. Apparently, rainfall is more effectively utilized during the second cropping season.

Figure 15. Exceedance probability distribution of effective rainfall at each ten-day during the first cropping season. (**a**) present (2004–2011); (**b**) future (2046–2065).

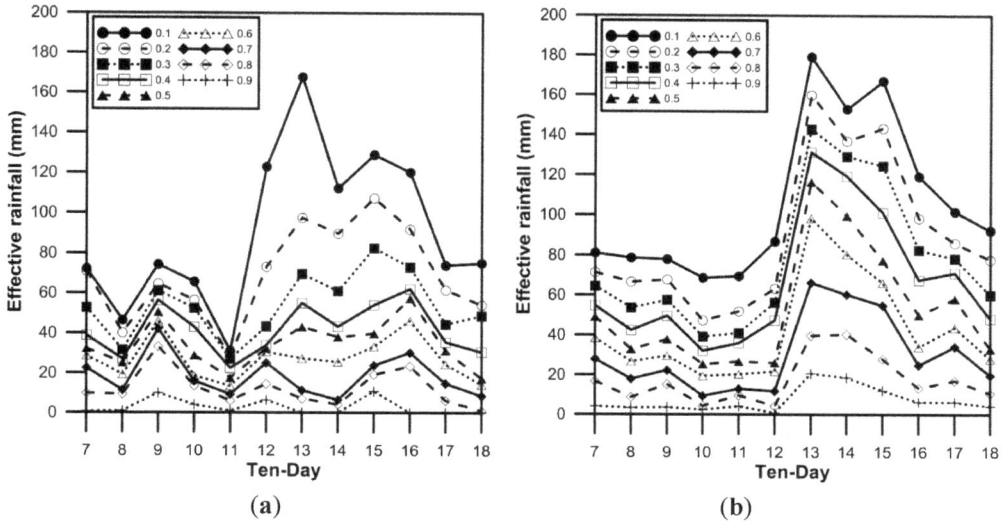

(a)   (b)

Figure 16. Exceedance probability distribution of effective rainfall at each ten-day during the second cropping season. (**a**) present (2004–2011); (**b**) future (2046–2065).

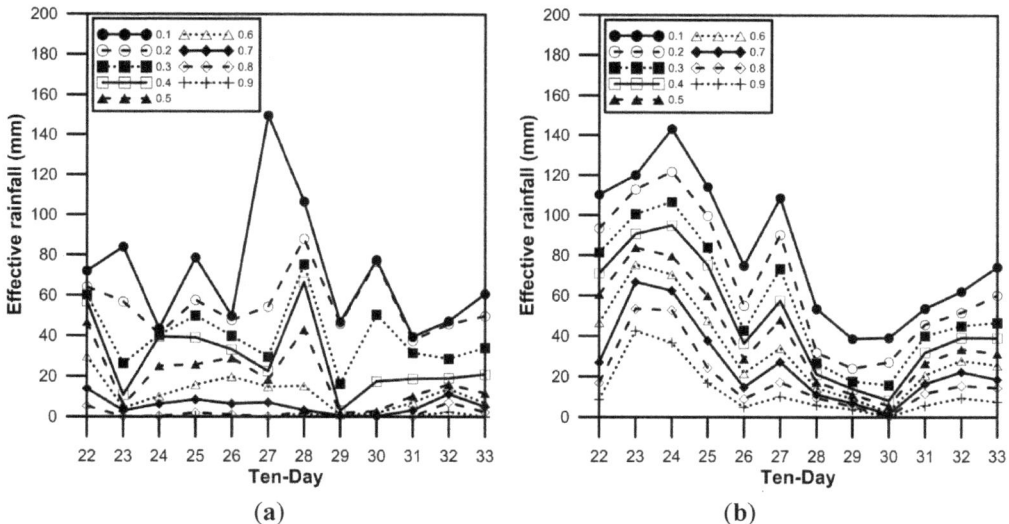

(a)   (b)

3.7. Irrigation Water Requirement Analysis

Linking a higher crop water requirement with more effective rainfall in the future, the impact on irrigation water requirement would probably be neutralized. The process for estimating irrigation requirement can be done by Equations (6). As seen in Figures 17 and 18, the pattern of exceedance probability distribution on irrigation water requirement, by comparing the present with the future, is similar. Notice that the exceedance probability distribution from 90% to 10% represents optimistic to pessimistic, because irrigation requirement is a cost factor. Overall, in the present, the agricultural sector needs to supply 923 mm and 1208 mm water, respectively, for irrigation. That is, 9230 m³/ha and 12,080 m³/ha during the first and second cropping seasons. In fact, the estimation of the future irrigation requirement depends on a chosen GCM. For example, in the first cropping season, CCCma and LASG produce less requirement, but CNRM, CSIRO and GFDL models request more irrigation water (see Table 3). On average, future requirements, respectively, reach 915 mm and 1180 mm in the first and second cropping seasons. In contrast to the present, future needs less irrigation water, though the difference is not significant.

Figure 17. Exceedance probability distribution of irrigation water requirement at each ten-day during the first cropping season. (**a**) present (2004–2011); (**b**) future (2046–2065).

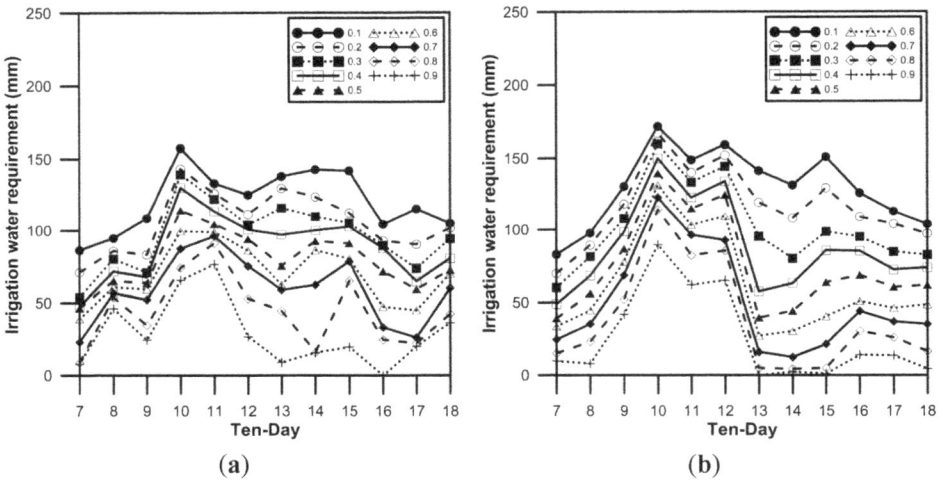

(a)

(b)

Figure 18. Exceedance probability distribution of irrigation water requirement at each ten-day during the second cropping season. (**a**) present (2004–2011); (**b**) future (2046–2065).

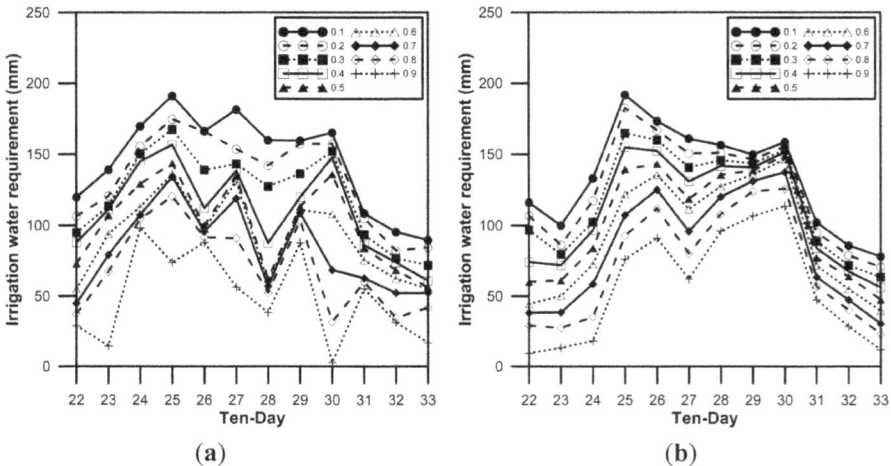

(a)

(b)

Generally, the threshold value for the water requirement of an irrigation plan can be determined on the basis of a 5-year deficit, a deficit event that has a 20% probability of occurring in any given year. That is, the exceedance probability of greater than the threshold value is 20%. By using Log-Pearson Type III frequency analysis [28,29], Table 4 shows the required irrigation water in accordance with a different return period. For the first cropping season, with the exception of a 2-year event, the future would need more irrigation water. In contrast, the required irrigation water at the second cropping season is less in the future. By choosing a 5-year return period event as the standard for the irrigation plan, future planning irrigation water requirement in the first cropping season would increase 73 mm (730 m^3/ha), and decrease 28 mm (280 m^3/ha) in the second cropping season. Multiplying the planned farming area 24,236.93 ha, extra 17.7 million m^3 of water would be needed in the first cropping season. In contrast, water would be 6.8 million m^3 less than the present in the second cropping season.

Table 4. Frequency analysis of irrigation water requirement in the present and future.

Irrigation Water Requirement (mm)	Return Period (year)	Present	Future					
			CCCma	CNRM	CSIRO	GFDL	LASG	Average
First season	2	933	824	1072	1000	931	721	910
	5	1021	1004	1220	1153	1146	946	1094
	10	1060	1078	1308	1231	1266	1076	1192
	20	1089	1127	1386	1295	1369	1189	1273
Second season	2	1199	1215	1299	1109	1134	1161	1183
	5	1307	1296	1373	1192	1277	1260	1279
	10	1371	1319	1404	1236	1365	1318	1328
	20	1426	1330	1426	1272	1445	1371	1369

4. Conclusions

This study investigates the impact on the irrigation water requirement under climate change between the present (2004–2011) and future (2046–2065). Impacts in terms of five selected GCMs under the SRES A1B scenario were assessed for the paddy fields of the Taoyuan Irrigation Association (TIA) in northern Taiwan. Projected meteorology in the future would be different because of GCM features and downscaling methods; therefore, considering several GCMs to reduce the uncertainty of models is necessary.

The FAO-PM equation is mostly used for evapotranspiration assessment, but it would not be suitable for evaluating the projection of evapotranspiration. This paper tries to combine the Hamon method and the Quadrant transformation method for assessing evapotranspiration in the future, and it is a possible and effective way to solve the problem.

Due to the rising temperature, the estimated evapotranspiration will increase in both cropping seasons in the future. Meanwhile, estimated crop water requirement would increase 36.0% and 15.2% in the 1st and 2nd seasons respectively. On the other hand, projected rainfall increases 35.6% in the 1st cropping season, but decreases 6.9% in the 2nd cropping season.

The impact of irrigation water requirement under climate change would not be easily assessed by crop water requirement and rainfall, although they play an important role. In the paddy field, storage, percolation and conveyance loss also influence the magnitude of irrigation water requirement. For evaluating the impact, this study simulates the water balance in paddy fields day by day.

As mentioned above, the projected rainfall decreases in the second cropping season, but estimated effective rainfall augments by 41.9% and 70.1% during the first and second seasons, respectively. This is because of the rainfall distribution, which becomes more even in the future.

Increased effective rainfall neutralizes the augmented crop water requirement, and causes the difference of irrigation requirement between the future and present to be insignificant. Estimated irrigation water requirements decrease 0.9% and 2.3% in the 1st and 2nd seasons, respectively. The decrements are equal to 1.9 million m^3 and 6.8 million m^3 by multiplying the planned farming area 24,236.93 ha.

In addition, this study uses frequency analysis to analyze the change of irrigation requirement. Based on the 5-yr threshold value, the estimated irrigation water requirement would increase by 7.1% in the first cropping season, and decrease by 2.1% in the second cropping season. By multiplying the planned farming area, the difference of irrigation requirement based on 5-yr return period between the future and present would increase by 17.7 million m^3 in the first cropping season but decrease by 6.8 million m^3 in the second cropping season.

The variation of the irrigation water requirement of TIA in the future would be insignificant. Nevertheless, since the projected meteorology of the basin of Shihmen reservoir would probably change [2,5,30], which is the main facility to supply the irrigation water for the TIA. This could be crucial for water resource planning of the Taoyuan area. The risk of water shortage for future irrigation demand needs further study. In addition, some possible adaptations to changing conditions either on the supply side or demand side is worthy of concern.

Acknowledgments

The authors appreciate the National Taiwan University Global Change Center and the Agricultural Engineering Research Center for providing relevant GCMs and meteorological data. This study is sponsored by the project of the Council of Agriculture under Executive Yuan of Taiwan (101AS-8.2.5-IE-b1).

Author Contributions

Jyun-Long Lee analyzed the data and drafted the manuscript; Wen-Cheng Huang revised the manuscript. The study concept and design was to come up with both authors.

Conflicts of Interest

The authors declare no conflict of interest.

References

1. IPCC Fourth Assessment Report: Climate Change 2007. Available online: http://www.ipcc.ch/publications_and_data/ar4/syr/en/contents.html (accessed on 4 November 2014).
2. Yang, T.C.; Yu, P.S.; Wei, C.M.; Chen, S.T. Projection of climate change for daily precipitation: A case study in Shih-Men reservoir catchment in Taiwan. *Hydrol. Process.* **2011**, *25*, 1324–1354.
3. Hsu, H.H.; Chen, C.T.; Lu, M.M.; Chen, Y.M.; Chou, C.; Wu, Y.C. *Climate Change in Taiwan: Scientific Report*; National Science Council, Executive Yuan: Taipei, Taiwan, 2011.
4. Liu, T.M.; Tung, C.P.; Ke, K.Y.; Chuang, L.G.; Lin, C.Y. Application and development of a decision-support system for assessing water shortage and allocation with climate change. *Paddy Water Environ.* **2009**, *7*, 301–311.
5. Huang, W.C.; Chiang, Y.; Wu, R.Y.; Lee, J.L.; Lin, S.H. The impact of climate change on rainfall frequency in Taiwan. *Terr. Atmos. Ocean. Sci.* **2012**, *23*, 553–564.
6. Yu, P.S.; Yang, T.C.; Kuo, C.C. Evaluating long-term trends in annual and seasonal precipitation in Taiwan. *Water Resour. Manag.* **2006**, *20*, 1007–1023.
7. Chang, L.N.; Lin, P.L.; Chang, C.M.; Shen, H.S. *Assessment of Climate Change in Taiwan (II)*; Environmental Protection Administration, Executive Yuan: Taipei, Taiwan, 1998. (In Chinese).
8. Williams, J.R.; Jones, C.A.; Dyke, P.T. A modeling approach to determining the relationship between erosion and soil productivity. *Trans. Am. Soc. Agric. Eng.* **1984**, *27*, 129–144.
9. Eheart, J.W.; Tornil, D.W. Low-flow frequency exacerbation by irrigation withdrawals in the agricultural midwest under various climate change scenario. *Water Resour. Res.* **1999**, *35*, 2237–2246.
10. Döll, P. Impact of climate change and variability on irrigation requirements: A global perspective. *Clim. Chang.* **2002**, *54*, 269–293.
11. Döll, P.; Siebert, S. Global modeling of irrigation water requirements. *Water Resour. Res.* **2002**, *38*, 8-1–8-10.
12. Smith, M. *CROPWAT: A Computer Program for Irrigation Planning and Management*; Food and Agriculture Organization of the United Nations: Rome, Italy, 1992.
13. Chung, S.O.; Nkomozepi, T. Uncertainty of paddy irrigation requirement estimated from climate change projections in the Geumho river basin, Korea. *Paddy Water Environ.* **2012**, *10*, 175–185.
14. Kloss, S.; Pushpalatha, R.; Kamoyo, K.J.; Schütze, N. Evaluation of crop models for simulating and optimizing deficit irrigation systems in arid and semi-arid countries under climate variability. *Water Resour. Manag.* **2012**, *26*, 997–1014.
15. Taoyuan Irrigation Association. *Irrigation Plan of Taoyuan Irrigation Association in 2005*; Council of Agriculture, Executive Yuan: Taipei, Taiwan, 2005. (In Chinese)
16. Contribution of Working Group I to the Fourth Assessment Report of the Intergovernmental Panel on Climate Change, 2007 (WG1). Available online: http://www.ipcc.ch/ipccreports/tar/wg1/029.htm (accessed on 4 November 2014).

17. Lin, S.H.; Liu, C.M.; Huang, W.C.; Lin, S.S.; Yen, T.H.; Wang, H.R.; Kou, J.T.; Lee, Y.C. Developing a yearly warning index to assess the climatic impact on the water resources of Taiwan, a complex-terrain island. *J. Hydrol.* **2010**, *309*, 13–22.

18. Kalnay, E.; Kanamitsu, M.; Kistler, R.; Collins, W.; Deaven, D.; Gandin, L.; Iredell, M.; Saha, S.; White, G.; Woollen, J.; *et al.* The NCEP/NCAR 40-year reanalysis project. *Bull. Amer. Meteor. Soc.* **1996**, *77*, 437–470.

19. Lin, S.H.; Chen, Y.C.; Lin, W.S.; Liu, C.M. Climate Change Projection for Taiwan based on Statistical Downscaling on Daily Temperature and Precipitation. In Proceedings of the 15th International Joint Seminar on the Regional Deposition Processes in the Atmosphere and Climate Change, Taipei, Taiwan, 12–14 November 2009.

20. Allen, R.G.; Pereira, L.S.; Raes, D.; Smith, M. *Crop Evapotranspiration*; FAO irrigation and drainage paper No.56; Food and Agriculture Organization of the United Nations: Rome, Italy, 2006.

21. Kan, C.E. *Studies on Water Distribution for Irrigation System*; Water Resources Agency, Ministry of Economic Affairs: Taipei, Taiwan, 1979.

22. Hamon, W.R. Estimating potential evapotranspiration. *J. Hydraul. Div.* **1961**, *87*, 107–120.

23. Shaw, S.B.; Riha, S.J. Assessing temperature-based PET equations under a changing climate in temperate, deciduous forests. *Hydrol. Process.* **2011**, *25*, 1466–1478.

24. Tsai, A.Y.; Huang, W.C. Estimation of regional renewable water resources under the impact of climate change. *Paddy Water Environ.* **2012**, *10*, 129–138.

25. Searcy, J.K. Flow-Duration Curves. *Manual of Hydrology: Part 2. Low-Flow Techniques*; United States Department of the Interior: Washington, DC, USA, 1959.

26. Cheng, M. *Irrigation and Drainage Management for Paddy*; Council of Agriculture, Executive Yuan: Taipei, Taiwan, 1977. (In Chinese)

27. Kan, C.E. *Water Conveyance Loss and Irrigation Efficiency in Agriculture*; Tsao-Jin Memorial Foundation for R&D for Agriculture and Irrigation: Kaohsiung, Taiwan, 2000. (In Chinese)

28. Haan, C.T. *Statistical Methods in Hydrology*; Iowa State University Press: Hertfordshire, UK, 1982.

29. Linsley, R.K.; Kohler, M.A.; Paulhus, J.H.L. *Hydrology for Engineers*; McGraw-Hill book Co.: New York, NY, USA, 1988.

30. Yu, P.S.; Wang, Y.C. Impact of climate change on hydrological processes over a basin scale in northern Taiwan. *Hydrol. Process.* **2009**, *23*, 3556–3568.

Assessing Climate Change Impacts on Water Resources and Colorado Agriculture Using an Equilibrium Displacement Mathematical Programming Model

Eihab Fathelrahman, Amalia Davies, Stephen Davies and James Pritchett

Abstract: This research models selected impacts of climate change on Colorado agriculture several decades in the future, using an Economic Displacement Mathematical Programming model. The agricultural economy in Colorado is dominated by livestock, which accounts for 67% of total receipts. Crops, including feed grains and forages, account for the remainder. Most agriculture is based on irrigated production, which depends on both groundwater, especially from the Ogallala aquifer, and surface water that comes from runoff derived from snowpack in the Rocky Mountains. The analysis is composed of a Base simulation, designed to represent selected features of the agricultural economy several decades in the future, and then three alternative climatic scenarios are run. The Base starts with a reduction in agricultural water by 10.3% from increased municipal and industrial water demand, and assumes a 75% increase in corn extracted-ethanol production. From this, the first simulation (S1) reduces agricultural water availability by a further 14.0%, for a combined decrease of 24.3%, due to climatic factors and related groundwater depletion. The second simulation (S2-WET) describes wet year conditions, which negatively affect yields of irrigated corn and milking cows, but improves yields for important crops such as non-irrigated wheat and forages. In contrast, the third simulation (S3-DRY) describes a drought year, which leads to reduced dairy output and reduced corn and wheat. Consumer and producer surplus losses are approximately $10 million in this simulation. The simulation results also demonstrate the importance of the modeling trade when studying climate change in a small open economy, and of linking crop and livestock activities to quantify overall sector effects. This model has not taken into account farmers' adaptation strategies, which would reduce the climate impact on yields, nor has it reflected climate-induced shifts in planting decisions and production practices that have environmental impacts or higher costs. It also focuses on a comparative statics approach to the analysis in order to identify several key effects of changes in water availability and yields, without having a large number of perhaps confounding assumptions.

Reprinted from *Water*. Cite as: Fathelrahman, E.; Davies, A.; Davies, S.; Pritchett, J. Assessing Climate Change Impacts on Water Resources and Colorado Agriculture Using an Equilibrium Displacement Mathematical Programming Model. *Water* **2014**, *6*, 1745-1770.

1. Introduction

The agricultural economy in Colorado is dominated by livestock production and sales, which account for 67% of total receipts. Crops, including feed grains and forages, account for the remainder. Most cropping receipts are based on irrigated production, which is sourced from groundwater, especially from the Ogallala aquifer, and surface water, which comes from runoff of snowpack in the Rocky Mountains. Currently, about 86% of water resources are in agriculture, but this is projected to decline due to demographic factors that lead to increased Municipal and Industrial (M&I) water demand, economic factors related to higher costs of irrigation, increased water demand for oil shale mining, and geographic factors such as climatic changes and groundwater depletion. Moreover, hydrologic studies point to an expected decline in runoff from 6% to 20% by 2050, and also a shift in the timing of that runoff to earlier in the spring. These studies also showed that late-summer flows may be reduced [1–4].

Colorado agriculture has blossomed with the development of water resources used for growing crops, which, in turn, spurs value-added production in the meat and dairy subsectors. Yet, increasing urban development is expected to create a reallocation of 740 million m³ (hereafter million = M) of agricultural water to new municipal and industrial demands by 2030 [5]. Another challenge to the agricultural sector is a possible expansion of ethanol production in Colorado. Shifting corn to ethanol use rather than animal feed could place livestock production, Colorado's dominant agriculture industry, at a disadvantage as the key input becomes more expensive, even though dry distillers' grains mitigate some of the constraint. These pressures on agriculture may be exacerbated by the presence of climate change, particularly its effect on water availability and yields. Stakeholders thus seek ways to better understand the implications of climate change on statewide water availability and requirements for crops and livestock, in the presence of a larger population and other new demands such as ethanol production. This research evaluates these issues with illustrations on how resources might be reallocated and how prices respond in the future.

The research uses a positive mathematical programming model specified to represent the Colorado agricultural sector, which is simulated to examine impacts of selected future constraints on water and yields resulting from climate change. First, this model was calibrated to 2007 quantities and prices. Then, a "Base" scenario was constructed, which reflects several future drivers of change affecting the state's agriculture: (1) increasing competition for water due to population growth, especially shifts in the resource from agricultural to municipal uses in the South Platte and Arkansas River basins; and (2) we also add two ethanol plants into the South Platte River Basin, which leads to a 75% increase in corn extracted-ethanol production there, and provides competition to the cattle feeding industry's use of a key input, corn for grain.

The changes incorporated into the Base scenario are related to the anticipated growth in the local economy, but to do not include effects of climate change. With this Base established, we run three simulations to explore the implications of climate change. The first one further reduces water availability based on forecasts of reduced runoff, while the second and third simulations introduce yield changes that might arise due to higher temperatures and increased variability of rainfall. Results for these scenarios are reported in terms of acreage changes, total value of production, exports and imports from the state, and prices. The overall changes in consumer and producer surpluses across the simulations are also reported. This modeling effort does not attempt to capture the full set of dynamic effects that will in fact occur, because for a small region, the range of possible outcomes over the next several decades is high, and is dependent on an equally extensive set of possibilities. Our approach is thus to focus on important outcomes with regard to climate change using a comparative statics method.

The document is organized into a series of sections. The current status of Colorado agriculture and its dependence on irrigation water supplies is reviewed in Section 2. This section also includes a review of expected climate change impacts on the availability of water and effects on commodities. Section 3 provides a literature review with regard mathematical programming methodology, while Section 4 lays out our particular model. Section 5 provides a discussion of the simulations and results, and Section 6 gives conclusions and thoughts for further research.

2. Colorado Agriculture and Water Use: Current and Projected Changes

This section contains two parts: the first covers the current size and structure of Colorado agriculture and describes key changes that might occur over the next decades; the second looks at the current pattern of water use and reviews forecasts of water reallocation.

2.1. Agriculture in Colorado

The agricultural economy in Colorado is dominated by livestock (almost $5.8 billion in sales during 2007, the year used to calibrate our model), which accounts for 67% of total receipts from the sector. The 2007 commodity balances are contained in Table 1. Colorado agriculture is heavily traded outside the state and abroad, as we learned when building commodity balance sheets used in the model. Fed beef, the largest economic sector, produced $3.4 billion in 2007 and traded 82% of its production out of state. The cattle feeding industry creates a substantial derived demand for corn production ($463 Million hereafter M) and corn imports, which reached $703 M in the same year. In 2007, 75% of the total value of Colorado's crops came from irrigated acreage, as most of hay, corn, and pasture for livestock were produced on irrigated land [6].

Table 1. Production, in-state sales, exports, and imports of key Colorado agricultural commodities.

Crop or Commodity	Production (M $)—Column 1	In State Sales (M $)—Column 2	Exports (M $)—Column 3	Imports (M $)—Column 4	% Exports/ production % of column 3 /column 1	% Imports/ production % of column 4 /column 1
Corn*	462.8	1051.7	113.7	702.6	24.6	151.8
Wheat	483.5	61.0	474.8	52.3	98.2	10.8
Barley	185.3	67.2	163.9	45.9	88.5	24.7
Sorghum	383.7	400.3	0.0	16.6	0.0	4.3
Dry beans	24.7	7.2	17.6	0.0	71.0	0.0
Beef	3382.5	905.6	2748.4	0.0	81.3	0.0
Cow calf	135.8	278.2	0.0	142.4	0.0	1.0
Hogs	170.9	204.0	0.0	33.1	0.0	19.3
Dairy	522.3	566.4	0.0	44.1	0.0	8.5
Sheep	488.5	27.1	461.4	0.0	94.5	0.0
Broilers	145.7	205.4	0.0	59.6	0.0	40.9
Eggs	74.1	85.6	0.0	11.5	0.0	15.5

Note: * Corn sales includes ethanol production.

It is not possible to say how much imported corn went into ethanol production, but ethanol used the equivalent of 23% of the state's production, while 67% of corn was imported. The value of wheat production equaled that of corn output, but 98% was exported across state boundaries. The sheep and lamb industry is also heavily export-oriented, with slightly less than $500 M in revenues during 2007, and 94% exported. Sorghum was the largest feed grain produced after corn, with revenues in excess of $380 M and imports totaling about $17 M. Instate sales of corn, excluding the ethanol industry, exceeded $800 M, while sorghum was $400 M. Colorado's dairy and hog sectors sold output within the state and required imports to meet demand, totaling 8.4% and 19% of production respectively. Imports of cows and calves were 50% of instate calf sales ($278 M) with buyers almost exclusively being feedlots. Barley and dry beans were relatively small agricultural subsectors and produced mostly for exports (88% and 71% of their production respectively). At the other end, 30% of broilers' sales in Colorado (about $60 M) and 13% of egg sales were imports.

Ethanol production may play a key role in Colorado's energy future and plans therefore exist to expand production capacity. Yet, Colorado is a small producer of ethanol, with just three plants located in the South Platte River Basin. The average plant capacity in Colorado is 215 M liters per year, or about 1.3% of the nation's ethanol capacity. The "corn footprint," or demand by these plants, is approximately 1.6 M tons each year, which requires about 130,000 hectares of irrigated corn production.

Expected Climate Change Effects on Colorado Agriculture. A consensus of climate change models suggest temperature in Colorado is expected to increase by up to 9–11 degrees Fahrenheit in the worst case scenario. The timing of seasons is likely to shift as well, with an earlier spring and longer fall. Midwinter precipitation should occur later in the calendar year, while less rain is expected to fall in late-spring and summer. As temperatures rise, runoff will peak earlier in the spring and be reduced significantly in late summer. Earlier run off could result in an 8.5% reduction of in-stream flows by midcentury in the Colorado River basin and a 5%–10% possible reduction in the Arkansas and Rio Grande basins. Little work has been done for the South Platte in terms of the impact of climate change on winter snow runoff [7]. The variability year to year is also likely to grow.

Climate change will have effects on crop yield and water requirements. The main climate factors affecting agriculture are temperature, availability of water, and the concentration of atmospheric CO_2. Soil water availability depends on the above three factors as they interact with soil properties, while field humidity, clouds and solar radiation also influence plant water requirements. The major commodities in Colorado agriculture are affected variously by these climate factors. For corn, the yield loss associated with increased temperature exceeds the positive effects of increasing carbon dioxide levels, so yields are expected to decline [8]. Also, high temperatures earlier in the season lead to less pollen germination and lower yields [9]. The changing precipitation patterns suggest increased yields for non-irrigated wheat in Colorado given the increase rainfall in winter and early spring.

High temperatures also extend the number of growing degree days in the crop season, which has a positive effect on yields and overall production for hay. However, few studies exist on the effects of climate change for this crop. In a review of three studies, depending on the assumed increase of

CO_2 concentration, alfalfa yields were estimated to change from a 16.7% increase to a decrease of 19.4%. However, this added growth and length of season may lead to lower nutritional content, depending on soil quality constraints [10,11]. On the other hand, productivity may be higher than previously expected in semi-arid grasslands, and thus additional forage may become available [12].

Warmer temperatures increase plant evapotranspiration, while CO_2 concentration partially offsets this process by increasing plant water-use efficiency. Wheat and hay are more sensitive to CO_2 than corn [8,13]). Although there is great uncertainty about the future CO_2 concentration, it is unlikely to neutralize the effect of anticipated, protracted droughts on crop production.

Increasing heat also affects livestock growth and performance. Higher temperatures reduce livestock production in the summer but increase it in winter. Under heat stress, animals reduce grazing to stay in the shade, thus reducing their feed intake and suffering from weight loss. Reduced quality of forage and digestibility leads to reduced dairy productivity. The greater the stress, the easier is the spread of parasites and disease pathogens. For dairy cows, heat stress reduces the milk fat and protein content in milk, and the quantity of milk produced is reduced up to 10%; moreover, other factors may also lead to lower yields as high-producing dairy cows are the most susceptible to heat stress due to breeding selection for high productivity, and reproduction rates are also adversely impacted [3,14–16].

2.2. Colorado's Outlook for Water Resources

Competition for water is increasing in the West. Colorado is a headwater state, supplying water through river systems to eighteen downstream states. Interstate compacts mean that Colorado is not entitled to all surface water flows, and may only retain six billion m^3 in an average year. This water is allocated among users according to the Prior Appropriation Doctrine, and, as nearly all of Colorado's rights have been appropriated, new users must obtain rights from others through voluntary transactions. Agriculture is the largest diverter and consumptive user of these surface flows. Agriculture also makes use of groundwater resources so that, on average, 1.0 M hectares of cropland are irrigated via groundwater or surface water. As noted earlier, irrigated crops comprise three-quarters of cropping receipts in Colorado, with two-thirds of these receipts bound for Colorado's livestock feeding industry [17].

Irrigation water depends on both groundwater, especially from the Ogallala aquifer, and surface water, which comes from runoff due to snowpack in the Rocky Mountains. Currently, about 86% of the state's water resource is used in agriculture, but this amount is projected to decrease. Causes for decline include demographic factors, such as increased Municipal and Industrial (M&I) water demand, economic factors related to higher costs of irrigation, increased water demand for oil shale mining, and geographic factors such as climatic changes and groundwater depletion.

While agriculture holds the majority of water rights, new demands for water resources come from a growing population and environmental uses. Population forecasts are for an increase of more than 50% in the next twenty years, so a gap between existing municipal water supplies and demand from the larger population is anticipated. The Colorado Water Conservation Board's Statewide Water Supply Initiative (SWSI) predicts that Colorado's South Platte Basin will experience a 61.9% increase in water demand, or about 505 M m^3, by 2030, which will continue to rise thereafter. With water already appropriated in the South Platte, an estimated 73,000 irrigated hectares will need to be permanently fallowed to supply these increasing demands. The plans for nearly all South Platte water providers include significant water rights transfers [1,18].

Great variation exists among findings of hydrologic studies regarding expected decline in runoff, from 6% to 20% by 2050, although there is consensus on the persistence of the shift of runoff to earlier in the spring, and a change in precipitation to a greater intensity during winter and lesser in spring and summer [1–3]. The topography of the state and other factors make projections particularly complex [19].

3. Literature Review

The model used in this research is an optimization model using mathematical programming in a manner that has a long history in economics and engineering. The approach chooses activity levels that maximize an objective function in the face of physical constraints on resources. Positive Mathematical Programming (PMP) improves on earlier techniques by allowing perfect calibration to a base and additions of more realistic behavior into such models [20,21]. As an activity based approach, PMP simplifies communication across disciplines and is particularly suited to study bio-physical and environmental features of agricultural systems.

Over the last 10 years, the PMP approach has been object of extensive review, critique and extensions [22–25], as policy makers increased their reliance on quantitative economic models to understand effects of agricultural policies. As such, the method has been widely used in sectoral and regional analysis. In the European Union (EU), several models analyzed policy instruments within the EU's Common Agricultural Policy (CAP), especially the effects of the CAP reform starting in 2003–2004, where a switch to decoupled payments to farmers was made. Some examples of these models include the FAL, Parma and Madrid models, which use PMP to calibrate to observed values, and also apply the maximum entropy approach to estimate total variable costs [26–36].

The PMP method is thus versatile enough to model policy scenarios in a straightforward fashion, and has been adopted as especially well-suited to examine animal feed requirements and land constraints [25], and to study jointly agricultural outputs and environmental externalities [31]. Howitt *et al.* [32] applied the methodology to estimate effects of climate change on irrigated agriculture in California using the State Water and Agricultural Production model (SWAP). SWAP improves on traditional PMP models by allowing for large policy shocks and enhanced flexibility in handling input substitutions. These models are often linked to hydrological network models and other biophysical system models.

The equilibrium displacement modeling approach [33,34] represents an economic system of demand and supply relationships, and can show the effects of exogenously determined shifts of supply and demand from an initial equilibrium (a displacement). Changes in market prices and quantities resulting from the displacement determine changes in consumer and producer surpluses. This follows originally from Samuelson [35], who shows that maximizing profits is equivalent to maximizing the total surplus when markets are competitive.

The Equilibrium Displacement Mathematical Programming (EDMP) model originally developed by the USDA Economic Research Service Harrington and Dubman [36] is a sector-wide, comparative statics model of the U.S. agricultural sector, applying a mathematical programming approach to the equilibrium displacement methodology, with specific farm sector relationships and policies reflected. They used values estimated by econometric studies and applied the asset-fixity theory of Johnson and Quance [37] to estimate slopes of supply functions. The Harrington and Dubman model is similar to the general PMP approach, but the supply and demand curves are explicit, and the base calibration is achieved by shifting intercepts until they match initial values with as much precision as is needed. Thus this approach is termed an "equilibrium displacement mathematical programming" model.

Regional and Climate Change Studies. Connor *et al.* [38] noted that an increasing number of analyses assess the impacts of climate change on irrigated agriculture in arid and semi-arid regions of the world, especially those that face a projection of drier weather. The objective function of their irrigation sector model maximizes profits across three sub-regions in the Murray-Darling River basin, Australia, subject to land and water constraints. The scenarios included a base case, a water scarcity model, a water variability model, and full effects model. The latter model includes both water variability and implications for changes in salinity. They concluded that ignoring the combined water-climate effects, along with salinity, leads to results that understate costs and impacts on output. Moreover, using the analysis of salinity, they identify various thresholds of climate change that create structural change in productivity and costs related to levels of salinity.

Henseler *et al.* [39] studied global change in the Upper Danube basin using an agro-economic production model, with two climate change scenarios. The first scenario assumed a significant increase in temperature, while the second one showed effects of a moderate increase. This study's results showed large differences in agricultural income and land use between the two scenarios and shifts that lead to increases in cereal production and extensive grassland farming due to the increased temperature in the first scenario. Qureshi *et al.* [40], Whitney and van Kooten [41], and Wolfram *et al.* [42], studied climate change impacts on agriculture at the regional levels in Canberra Australia, Western Canada, and California respectively. These studies reached conclusions that are similar to the studies discussed above. Whitney and van Kooten [41] expanded the model to include impacts on pasture and wet-land.

Finally, with regard to previous Colorado analyses, Bauman *et al.* [43] estimated the economic impacts of the drought in 2011 using an Input–Output (I/O) model and a variant of the current Colorado Equilibrium Displacement Model. The authors found that the 2011 Colorado accounted for $83 to $100 M in economic impact, when all economic sectors of the state economy were included. Schaible *et al.* [44] argued that the gradual warming in the Western United States is expected to shift the precipitation pattern and alter the quantity and timing of associated stream flows. In addition, the effects of climate change will move bio-energy growth to the Ogallala aquifer in the Western States, which demand that careful optimization of water use is needed to choose irrigation technologies. They underline the importance of further research to understand economic implications of climate change at the regional level.

Thus, previous studies agreed that there are likely to be significant shifts in land use and crop mix due to climate changes at the regional level. These studies also agreed on the importance of understanding possible structural changes, and noted that there will be significant income and price effects due to climate change. Furthermore, the above review suggests that a lack of studies investigating the impact of climate change at the regional level exist, in particular those that trace out impacts in a small, open economy via trade with the Rest of the World (ROW) and include livestock and crop interactions. Previous studies also agree that positive mathematical modeling fits the research problem and unveils opportunity to simulate possible production and cost changes due to climate change, which should enable a better understanding of welfare implications at the regional level.

4. Structure of the Colorado Equilibrium Displacement Positive Mathematical Programming (Colorado EDMP) Model

The Colorado Equilibrium Displacement Positive Mathematical Programming model (Colorado EDMP) is a variant of the EDMP model by Harrington and Dubman, which the authors adapted for Colorado's agricultural sector [45]. This model maximizes the sum of producer and consumer surpluses across most major products in Colorado's agricultural sector, subject to a number of spatial market and resource constraints. The Colorado EDMP is calibrated to Colorado's agricultural economy, and adds other natural resource dimensions (*i.e.*, Colorado agricultural sector demand for water). Spatial constraints consist of three regions with separate water availability for irrigation in each basin (South Platte River basin, Arkansas Basin, and other Colorado basins) along with differing crop water requirements in each basin. These requirements were developed using irrigation water requirement (IWR) coefficients per crop per region from the Colorado Decision Support System (CDSS) weather and soil characteristics databases [46]. The optimization model selects food and feed crops, water supplies, and other inputs to maximize the sum of producer and consumer surpluses, subject to constraints on water and land, and subject to economic conditions regarding prices, yields, and variable costs. In the following paragraphs, we describe the Colorado EDMP and its basic dimensions.

The particular function given below is a second order Taylor series expansion as first introduced by Takayama and Judge [47], which permits an approximation of an unknown functional form for the cost function:

$$\text{Max: } Z = F'x - 1/2\ x'\,H\,x \qquad (1)$$

with $x > 0$, where x is a vector of endogenous variables that relate to sector demand and production processes. In the following expanded form of the Equation (1), the vectors x are divided into five groups. In the notation below the vectors of variables are written in lower case, while the vectors of parameters are in upper case, and indices under the summation operators are simplified as:

$$Z = \sum_j (F'' - .5H\ q_j)q_j - \sum_i \sum_b (F'' + .5Hcl_i)cl_i - \sum_n u + \sum_g (F'' - .5Heg)e_g - \sum_s (F'' - .5HM_s)M_s \qquad (2)$$

where, q_j = domestic sales of j agricultural commodities (in M tons) and livestock products (M head, tons, or dozens of eggs); cl_i = feed and food crop activities i identified by river basin (for selected crop activities, in M hectares) and livestock activities (head counts, live weight, milk tons and dozens); u = dollar value of n inputs (in M dollars); e_g = exports of t agricultural commodities (in M tons) and livestock products (M tons, dozens of eggs); M_s = imports of s agricultural commodities (in M tons) and livestock products (M tons, dozens of eggs); F'' = a vector of intercepts indexed under each set above, which are determined in the calibration phase; H = the diagonal elements of the Hessian matrix flowing from the First Order Conditions. H is assumed to be negative semi-definite.

In Equation (2), the first term is the function of total revenue, where $(F'' - .5H\ q_j)$ = p is the vector of price dependent domestic demand functions, and p is the vector of output prices. The H_j elements are derived from predetermined elasticities of demand for j commodities and livestock products. The second element is a non-linear total variable cost function, where H_{ib} are elements of the Hessian of supply functions; they are calculated as the ratios of capital replacement costs over excess capacity for i activities in b river basins. The term $(F'' + .5Hcl_i)$ = Marginal Cost provides the supply side equivalent to a price dependent demand function in the first term. The third element is the sector's sum of inputs used in the sector, entered in value terms. The last two elements represent the export and import functions (these include out-of-state trade as well as international trade), which are included in the sector's the objective function (see also Helming [48]). H_t and H_s are also exogenously calculated. Examples of the constraints included in the mathematical program are presented in Appendix.

The agricultural activities in the model cover 91% of total agricultural production in Colorado, including thirteen crop and nine livestock commodities, which are sold to local consumers or out-of-state exports. Imports for nine products are present and compete with local production. The nine livestock sectors are cow calf, fed beef, hogs, dairy, sheep, broilers and layers, turkeys, and horses. Some of these livestock activities produce multiple products, including meat, milk, and/or eggs. Demand for feed crops and forages are derived from livestock activities through demand for rations. Food crops are wheat, potatoes, sunflower, and dry beans. Calf imports go directly into the cattle feeding industry. The commodities included, their acreage and production values, and a comparison of how our calibrated model compares to historical 2007 values is given in Appendix Table A1.

The model also includes accounting costs for all activities. Inputs are categorized in the following categories: genetic inputs, such as seed or calves; specialized technology; mineral fertilizers (without manure applications); other chemicals; fuel and lube; electricity; irrigation energy and other irrigation costs; other variable purchased inputs; fixed cash costs; and capital replacement costs. Farm production costs reflect various yields and cost structures in different basins. Irrigated and non-irrigated crop costs are derived from enterprise budgets created by extension professionals in Colorado and the High Plains. Currently, the relationship between inputs and outputs is fixed, with no substitution, so that corn production, for example, has a fixed yield of 8.3 tons per hectare and each hectare uses a certain quantity of fertilizer, other chemicals, and irrigation energy (when irrigated).

Demand elasticities from the literature provide the values for the Hessian's elements related to demand, which help the model, provide reasonable responses when used in scenario analyses. The F values, or intercept terms, are estimated by repeated adjustments until the prices and quantities are calibrated to a desired level of accuracy.

It is possible, with enough time, to exactly calibrate prices and quantities by shifting demand and supply intercepts. While this can be a tedious process, it provides an examination of the relationships and tendencies in the model, which cannot be achieved as intuitively when using a large set of cross price elasticities that, in any case, cannot be reliably identified for a small region like Colorado. We show the results of our efforts at calibration in Appendix Table A1, where the table shows the calibrated quantities *versus* actual values for selected products. It also provides estimates of the intercepts and slopes (Hessian elements) of the associated supply curves.

5. Base Scenario and Climate Change Simulations

This research includes three climate change simulations that are compared to the Base simulation, where the "Base" is designed to represent selected features of the sector several decades in the future. The two main features included are reduced water availability in the South Platte and Arkansas River basins, and added demand for ethanol, which represents a competing demand for corn. Three simulations then are created to show incremental effects of climate change on the Base model. The first simulation (hereafter S1) reduces agricultural water availability by a further 14.0% across all basins, for a combined decrease of 24.3%. This reduction comes from climatic factors and related groundwater depletion, as detailed in the Colorado Water Conservation Board's Statewide Water Supply Initiative study (CWCB) [2]. There are no changes in yields or other factors.

In addition to the direct water reduction, the effects of increased heat and an extreme dry year are reflected in the second and third simulations. First, climate change models suggest up to a 9–11 degree Fahrenheit increase in temperature, as a high end case [23]. The average rise in temperature also affects the variability and likelihood of years with more extreme weather, as illustrated in Figure 1. This figure illustrates how the increase in average temperature leads to a greater likelihood of extreme weather events, such as droughts, but also to years with higher precipitation. Simulation two represents a warm and wet year (hereafter S2-WET), with shifts in the pattern of precipitation, but with an increase in average temperature included as well. The third simulation reflects a drought year (S3-DRY) with dry conditions, in addition to the temperature increase and shifts in precipitation found in S2-WET.

Table 2 summarizes the percentage changes in crop yields and dairy productivity from those used in S1. The irrigated corn yield in S3-DRY decreases due to higher July temperatures, and from lack of rain and cloud cover, which hampers pollination [8]. Yields for non-irrigated wheat increase as sufficient winter rainfall is present s during the critical growing period in S2-WET, and decline by an equal amount in S3-DRY to reflect the effect of less rainfall and higher temperatures [13].

Both irrigated hay and corn silage yields surge with higher temperatures, which result in a longer growing season and more cuttings, in the case of hay, and help biomass growth in silage. Because both are grown on irrigated land, decreased rainfall does not have an effect, and yields are kept high in both scenarios. Yields in rangeland and pasture increase in S2-WET year, as sufficient rainfall supports germination and growth, but like other non-irrigated crops, these sources of feed see reduced yields in S3-DRY. Dairy sector productivity plummets in both the second and third simulations, reflecting animal stress from high temperatures in absence of mitigating strategies. (These impacts of climate change are presented in more detail in Section 2).

Table 2. Percent yield and productivity changes in S2-WET and S3-DRY, relative to S1. Sources: [13] (pp. 34–48, 56–61, 77–82).

Simulation	Irrigated corn	Dryland wheat	Irrigated hay	Silage	Pasture	Rangelands	Dairy
S2-WET	−0%	13%	18%	13%	8%	8%	−18%
S3-DRY	−15%	−13%	18%	13%	−13%	−13%	−18%

In summary, the following conditions are analyzed in the next sections:

Base: This scenario examines the economic impacts of shifting water resources from agricultural to municipal uses in the South Platte and Arkansas River basins by 22% and 18% respectively;

S1: This simulation alters the Base scenario by reducing agricultural water availability by a further 14.0% across all basins for a combined decrease of 24.3% based on expected climate change effects;

S2-WET: This simulation represents a warm and wet year, with shifts in the pattern of precipitation and an increase in average temperature;

S3-DRY: A drought year is simulated in the third example, using dry conditions along with the temperature increase and shifts in precipitation found in S2-WET.

Figure 1. Climate change scenarios in Colorado Economic Displacement Mathematical Programming (EDMP).

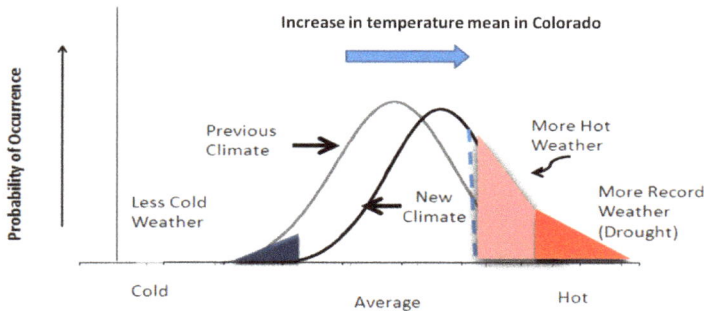

Note: Relatively small shift in the average climate can substantially increase risk of extreme events such as drought [4].

Base Scenario. The Base scenario reflects selected supply and demand factors for agricultural inputs and outputs in the future. First, it includes expected implications of competition between the agricultural sector and other sectors (e.g., M&I) for water at the basin level. In particular, this scenario shifts water resources from agricultural to municipal uses in the South Platte and Arkansas River basins by reducing water availability to agriculture by 22% and 18% respectively, with respect to calibrated values for 2007. This reduction follows estimates by the Colorado River Water Availability Study—CRWAS-report [49], and results in the fallowing of a proportional amount of irrigated land in each basin, although individual crops can vary without constraint aside from the overall reduction in irrigated acreage. Overall a net decrease of 10.3% in total water availability occurs because nearly 50% of annual volume is in river systems outside of these two basins.

Additionally, this simulation adds two ethanol plants in the South Platte River Basin, thereby increasing Colorado ethanol production from 175 M gallons annually to 308 M gallons. The Base scenario values are found in Tables 3 and 4.

Base Scenario Results. Due to an anticipated reallocation of water from agriculture to municipal uses in the South Platte and Arkansas basins in the Base scenario, crop acreage shifts relative to the calibrated values of 2007, particularly for those commodities that are produced on irrigated land. Also, adding two ethanol plants raises annual production from 662 M liters annually to 1165 M liters. This increase raises demand for corn by about 1168 k tons (hereafter thousand = k), which must be supplied from various sources. On the one hand, other uses of corn can decrease, which in this model are feed, final consumption and exports. Also, supplies can come from added production or greater imports.

Table 3. Area harvested in Base scenario and Climate Change Simulations.

Crop/Livestock product	Base	Simulation			Percentage change from Base		
		S1	S2-WET	S3-DRY	S1	S2-WET	S3-DRY
South platte dry corn	0.07	0.07	0.06	0.1	0.0%	−14.3%	42.9%
South platte irrigated corn	0.27	0.26	0.25	0.17	−3.7%	−7.4%	−37.0%
Arkansas irrigated corn	0.07	0.07	0.07	0.04	0.0%	0.0%	−42.9%
Arkansas dry corn	0.16	0.17	0.15	0.04	6.3%	−6.3%	−75.0%
All corn	0.58	0.57	0.53	0.36	−1.7%	−8.6%	−37.9%
South platte dry wheat	0.62	0.65	0.65	0.33	4.8%	4.8%	−46.8%
South platte irrigated wheat	0.03	0.03	0.04	0.16	0.0%	33.3%	433.3%
Arkansas dry wheat	0.29	0.29	0.3	0	0.0%	3.4%	−100.0%
All wheat	0.94	0.96	0.99	0.49	2.1%	5.3%	−47.9%
Other crops	0.17	0.17	0.17	0.17	0.0%	0.0%	0.0%
Colorado basin hay	0.35	0.28	0.28	0.28	−20.0%	−20.0%	−20.0%
South platte dry hay	0	0	0	0.08	0.0%	0.0%	0.0%
South platte irrigated hay	0.03	0	0	0	−100.0%	−100.0%	−100.0%
Arkansas irrigated hay	0.04	0.03	0.03	0.05	−25.0%	−25.0%	25.0%
All hay	0.41	0.31	0.31	0.41	−24.4%	−24.4%	0.0%

Notes: All values are in M hectares. Source: Model Runs from Colorado Economic Displacement Mathematical Programming (CEDMP) Model.

Sales of the main user of feed, fed beef, do not change much from the calibration to the base, even though water supplies have dropped by 10.3%. Significant and numerous changes in the feed sources, however, do occur. Corn sales to local users other than for livestock feeding decline by about 101 k tons from 2007, while a 3% reduction occurs in corn used for feed, or nearly 177 k tons arises, mainly in a shift to other, smaller grains that use less water. Exports decline by about 25 k tons as well. These shifts together release corn from other uses for a quarter of the increased ethanol demand. However, imports decrease by about 355 k tons, so overall supply is lower from these shifts and cannot fully support growth in corn demand for ethanol, as the variation in exports and imports just offset each other. Thus, production growth is the main source of supply for the increased demand for corn.

Table 4. Production, sales, exports, imports, and prices in the base and simulation scenarios.

Commodity	Variable	Base	Simulation			Percentage change from Base		
			S1	S2-WET	S3-DRY	S1	S2-WET	S3-DRY
Corn	Production	4567	4468	4265	2428	−2.2%	−6.6%	−46.8%
	Sales in Colorado	84	84	84	79	0.0%	0.0%	−6.1%
	Exports	699	693	678	549	−0.7%	−2.9%	−21.5%
	Imports	4128	4194	4326	5532	1.6%	4.8%	34.0%
	Prices [a]	145	145	145	160	0.0%	0.0%	10.0%
Wheat	Production	2251	2294	2722	1402	1.9%	20.9%	−37.7%
	Sales in Colorado	269	269	278	253	0.0%	3.0%	−6.1%
	Exports	2243	2281	2675	1461	1.7%	19.3%	−34.8%
	Imports	261	259	231	313	−1.0%	−11.5%	19.8%
	Prices [a]	220	228	220	243	3.3%	0.0%	10.0%
Fed Beef	Production	1389	1384	1375	1285	−0.3%	−1.0%	−7.5%
	Sales in Colorado	340	340	340	336	0.0%	0.0%	−1.3%
	Exports	1049	1044	1035	949	−0.4%	−1.3%	−9.5%
	Prices [a]	2644	2724	2729	2773	3.0%	3.2%	4.9%
Dairy	Production	1244	1244	1030	1030	0.0%	−17.2%	−17.2%
	Sales in Colorado	1357	1357	1266	1262	0.0%	−6.7%	−7.0%
	Imports	113	113	232	232	0.0%	104.0%	104.0%
	Prices [a]	429	402	500	500	−6.2%	16.6%	16.6%
Hay	Production	3053	2255	2273	3038	−26.1%	−25.5%	−0.5%
	Sales in Colorado	3830	4967	4032	4312	29.7%	5.3%	12.6%
	Imports	777	2712	1759	1274	249.0%	126.4%	64.0%

Notes: [a] Units of Prices are in $/ton; All other values are in k tons. Source: Model Runs from Colorado Equilibrium Displacement Mathematical Programming (CEDMP) Model.

The growth in production is nearly 1041 k tons, which comes from an increase of close to 152 k hectares in corn. This increase is generally in irrigated land in the Arkansas and South Platte basins, but in the Arkansas basin, a significant proportion of the production growth comes on non-irrigated land. Given that irrigated land is withdrawn from production in the Base scenario, growth in corn production must come from a shift out of other crops. This includes a reduction of area harvested for alfalfa hay by nearly one third, or about 223 k hectares, and a reduction of fallow land in the Arkansas basin. This occurred even though hay area in Other Colorado outside the two basins under consideration remained at about 315 k hectares.

The reduction in hay acreage is logical, based on its high water demand, significant use of irrigated land, and the possibility of using imports as a substitute. The irrigated corn area harvested in the South Platte basin increases by about 17.4% (or about 40.5 k hectares) from the calibrated value of 234 k hectares. Another 40.5 k hectares of alfalfa hay, or about 75% of its total area, is lost in the South Platte basin in response to limited water availability. The decline in irrigated corn and hay production negatively influences fed beef operations because locally produced feeds become more expensive.

Several general comments about the Base scenario are worth noting. First, given that most changes in the Base assumptions affect irrigated land, little reallocation occurs in non-irrigated products, such as wheat. While some shifts are found in wheat location, in the aggregate, its area drops by just under 2.0%. Despite the drop in acreage and production, exports, the main use of wheat, rise by about 2.0% or 35 k tons. This is possible mainly because of a shift of local sales of wheat into exports (71 k tons), a small increase in imports (17 k tons), which together permit a growth in exports despite the reduced production.

A second point is that changes in sales, production and consumption of other crops and livestock products occur relative to the calibrated model representing 2007, but for the remainder of this paper, these are not considered in depth. Our focus will be on cattle feeding and dairy, and their inputs, primarily corn and hay, and on wheat as the major non-irrigated product. These commodities account for 75% of area harvested and in-state sales, and about 85% of exports in the 2007 calibration.

The above scenario is created only by withdrawals of water from agriculture due to greater municipal and industrial uses, along with the presence of a larger ethanol industry. This clearly leaves out many possible changes that will occur in the next several decades, with the main ones being technological change and greater population. To reflect these changes, which some models attempt to do, we would need to make assumptions of a wide range of yields and productivity of livestock and dairy, and the increase in consumption of all products from the larger population. This seems to us to be a relatively non-productive effort for a small region like Colorado. Thus our Base is a mixture of the 2007 setting, with selected future effects made to key variables. The proportions of imports and exports stay roughly the same, even though they are not fixed, because the balance between demand and supply is not forecasted into the future. While it is certainly not an exact representation, the Base case permits us to examine important effects of climate change on yields and water availability, without being confounded by added, perhaps unsupportable, changes. Thus, the following results show additional effects due to water and yield changes coming from climate change.

6. Climate Change Scenario Results

As described earlier, three climate change simulations are included in this study. The following discussion of results is split into two sections, where the first summarizes and explains shifts in area within each simulation, which are presented in Table 3. These area shifts are related to a series of price effects that lead to additional variation in feed use, production levels, and exports and imports. These added effects of climate change are found in Table 4.

Simulated Area Effects. Relative to the Base, the area harvested of Colorado *corn* (about 600 k hectares) only changes slightly in S1. In S2-WET, overall area harvested declines by nearly 8%, but the change is not distributed equally across basins. The largest change in cultivated area occurs in S3-DRY, as total harvested area drops by 38%. This decrease is similar across both regions and for irrigated land, as the percentage decline is nearly identical in both the South Platte and Arkansas basins. The greatest impact in S3-DRY occurs in the Arkansas basin's non-irrigated land (−75%), which drops to only 45 from 151 k hectares. Conversely, South Platte non-irrigated corn expands by 11.8% over S2-WET, responding to higher prices coming from the large reduction in irrigated corn area harvested. The drought-like conditions in S3-DRY with high heat cause a large reduction in irrigated corn harvested as yields decrease by 8% from the Base. These results indicate the high sensitivity of corn area to variations created by climate change.

The total area harvested of *wheat* in Colorado (with a baseline of 0.94 M hectares) changes little between S1 and S2-WET, as the wet year leads to a just 4% increase in non-irrigated wheat in both the South Platte and Arkansas River basins. Similar to corn, the largest changes occur in S3-DRY. Due to the dry year's conditions, nearly a 43% reduction of South Platte non-irrigated wheat area occurs, while the Arkansas River basin non-irrigated wheat disappears completely. The latter basin loses over 283 k hectares of cultivated area. Such large changes in non-irrigated wheat represent expected responses to the drought-like conditions, where yields decline by 26% from the wet year conditions in S2-WET. Therefore, a crop that is dependent on rainfall but not on water via irrigation derived from snowpack and storage will see greater variability in total production as climate changes.

Hay is the third commodity examined in Table 3. The initial decrease in irrigation water in S1 causes a 70 k hectare decline in hay acreage outside the two main basins. After that initial decrease, the hay cultivated in Other Colorado remains constant in S2-WET and S-3-DRY, as that region has sufficient irrigation water, compared to its land resource, and cannot produce other crops competitively. In S3-DRY, irrigated hay increases by 20 k hectares in the Arkansas basin. Overall, the reduction in corn area, due to a substitution into imports, leaves irrigated land available for hay in Arkansas and hence some expansion in hay acreage occurs. In the South Platte, non-irrigated corn and hay, to a lesser extent become competitive on land previously in wheat.

Evaluating production, price and trade effects across climate change simulations. In this section, several important market effects are explained, including the scenarios' effects on total production, trade revenues and prices for major commodities produced in the state. The focus is on climate change effects in S2-WET and S3-DRY, but we consider uncertainties in outcomes and possible alternative scenarios as well.

W*heat.* Wheat consists primarily of non-irrigated production, and is generally exported, with local use equivalent to the level of imports. Production increases in S2-WET by about 436 k tons, or 21%, as more rainfall reaches the crop during its early spring growing season and yields improve by 13%. In S3-DRY, with lower rainfall, non-irrigated wheat area is cut nearly in half, with about 485 k hectares going out of production. The shift towards irrigated corn in the South Platte River basin, noted above, occurs because of a price increase of 10% in S3-DRY. However, the same percentage price increase in wheat does not lead to an increase in non-irrigated production in the Arkansas Valley.

These differing responses between corn and wheat come from varying dependence on imports and the fact that there is no irrigated wheat for the Arkansas River basin in the calibrated model, so that commodity cannot enter even with higher prices.

Thus, the wheat crop is extremely sensitive to how climate change affects rainfall, with the variation in exports between S2-WET and S3-DRY being nearly 1.2 M tons. The actual outcomes will also be affected by the performance of other regions, and, indeed, international supply and demand, as much of Colorado's wheat crop leaves the country. As the Northern Plains outside of Colorado should see greater production of wheat with climate change, downward pressure may be exerted on prices in Colorado, although rising international demand could offset that effect [6]. Higher national and international prices, of course, would reverse some of the decline, as Colorado wheat would remain more competitive than in the scenarios presented here.

In sum, this crop's potential outcomes depend importantly on rainfall variation, as well as the international setting, which affects wheat to a greater degree than other crops. The variability in outlook, however, does not affect other commodities critically, such as corn, hay or cattle, as those are more dependent on irrigation from snowpack and statewide precipitation to a greater extent than the timing and amount of local rainfall.

Cattle Feeding. Cattle feeding is the largest industry in Colorado agriculture and is dependent on selling fattened cattle for slaughter out of the state, although little goes to the international market. In simulations S1 and S2-WET, production declines only slightly from the Base, which is related to an increased cost of feed. However, a higher price exists in the output market, which leads to sales revenues nearly the same as in S1, even though water declines and feed becomes more expensive. On the other hand, in S3-DRY, fed beef production declines by nearly 90 k tons, or 8.4%, due to the significantly higher prices of feed and thus fed beef, which is great enough to dampen demand. The small effect in S2-WET is related to the fact that a quarter of fed beef is sold to consumers in Colorado, where a lower own price elasticity is assumed. Thus, the industry can benefit from increased prices in certain ranges, but higher cost feed eventually makes fed beef less competitive with producers outside the state, particularly in S3-DRY.

Several conflicting trends are not modeled in this research. The first is that increased costs might be incurred for feedlots to adapt to higher temperatures, such as adding sheds and mechanical spraying to protect cattle from heat. Also, the lower quality of hay may require increased quantity in rations. On the other hand, temperatures may increase more in other cattle feeding states, such as Texas, giving Colorado a cost advantage over time. Without knowing which effect will dominate, these variations are left for future work.

Feed sources. Examining changes in feed production highlights overall linkages between products and variations across simulations. From Table 5, it is apparent that corn comprises 85% of overall feed use in the state. That source stays roughly the same until S3-DRY, when irrigated hectares drop due to water shortages, but with high temperatures, yields decline from high heat during pollination. Thus, the quantity of corn used as feed drops by nearly 9% compared to S2-WET.

The use of hay grows from the Base in all three simulations, but source of the forage varies considerably between local production and imports, as is shown in Table 4. The use of hay increases in S1 the most, where the overall water reduction occurs from municipal and industrial uses, rather than due to climatic factors. This is because hay can be imported most easily among the forages, and so there is a swell in imports (which grow by nearly 2.5 times over the Base value).

Production drops by 26.1% at the same time, to release irrigation water to be used in other, higher valued crops. In S2-WET, water is less scarce, and yields of non-irrigated pasture and range increase, as do yields of irrigated hay, so less hay is imported and produced.

Table 5. Feed consumed in Base and Climate Change Simulations. Source: Model Runs from Colorado EDMP.

Feed	Base	Simulation			Percentage Change from Base		
		S1 (K tons)	S2-WET (K tons)	S3-DRY (K tons)	S1 (% Change)	S2-WET (% Change)	S3-DRY (% Change)
Hay	3.8	5.0	4.0	4.3	29.7%	5.3%	12.6%
Corn	202.6	201.7	199.8	182.1	−0.5%	−1.4%	−10.1%
Barley	13.8	13.9	14.2	16.7	1.0%	3.0%	20.8%
Oats	2.9	2.9	2.9	2.9	0.0%	0.1%	0.9%
Sorghum	10.4	10.4	10.4	10.4	0.0%	0.0%	0.0%

Production of hay recovers in the third simulation because yield growth of 18% above the Base makes it a profitable user of water. Imports decline because of the general drop in both dairy and cattle feeding seen in S3-DRY. As noted earlier, area is reallocated between the Arkansas and South Platte basins, and the growth occurs due to Colorado feed prices rising in general. In that simulation, corn acreage declines, so irrigated land can shift into hay production. Notably, 283 k hectares are produced in Other Colorado throughout all simulations because there is excess water relative to land in that part of the state.

Corn is the main feed crop that is provided through imports but also has exports. Table 4 showed before that corn is in a net import position, and the internal price does not rise substantially in the first three simulations due to the significance of the import market, where external prices are governed by demand and supply conditions outside Colorado. However, the corn for grain price rises by 10% in S3-DRY due to the general shortage of feed and lower yields of corn in hot and dry conditions. The combination of a water shortage and reduced yields is enough to raise prices to levels where sales of fed beef are affected. This is especially so for exports, which dropped by 9.5% as the industry becomes less competitive. This change leads to lower demand and thus production of corn. Moreover, the ratio of fed beef prices to corn prices declines from about 30 in the first two simulations to 28.7 in S3-DRY, suggesting this change in competitive position.

Effects of Climate Change and Induced Water Loss on Colorado Agricultural Trade. Exports of corn decline by about 22% in S3-DRY relative to the Base scenario, while exports of wheat increase about 19% in S2-WET, due to favorable rainfall and temperature conditions, but decline about 35% in S3-DRY. This leads to a 1.2 M ton swing in exports, which is nearly 60% of average production of wheat in the climate change affected simulations. Beef exports decline about 1.3% and 9.5% in S2-WET and S3-DRY respectively. S2-WET shows 11% decline in wheat imports, while S3-DRY results show that imports of corn, wheat, and dairy increase by 34%, 20% and 104% respectively.

The above changes are all associated with increases in prices, which alter the competitive position of Colorado relative to out of state producers. So, for example, in S3-DRY, wheat prices rise by 10.2% and corn prices increase similarly. For both commodities, exports drop and imports climb as Colorado production becomes more expensive relative to outside sources. Imports of Hay increase in the simulations, with hay imports more than tripling in value in S1 relative to the Base. In S3-DRY, less corn is grown with the reduction in cattle feeding, and thus irrigated land becomes available for hay, which expands from higher prices. This latter outcome is related to the assumption that yields increase for hay from the longer growing season, but decrease in corn from heat and rainfall variation.

Table 6 gives an important perspective on model outcomes provided above. The import and export elasticities for major commodities are first presented, which were constructed to reflect differing external positions. These are key assumptions, of course, because they have a large effect on quantity and price changes in a given simulation. The values are all high, so a "5", for example, indicates that a 1% change in price will lead to a 5% change in quantity, implying quite a large response. Thus, the exports of wheat and fed beef are very responsive to how the internal price changes with respect to the import or export price, which is consistent with a small open economy where local industries face much competition from external sources of supply.

Table 6. Export and import elasticities in the Colorado EDMP, and trade proportions for key commodities.

Commodity	Elasticities	Export or import percent of production
Corn exports	2	15.40%
Wheat exports	5	99.60%
Fed beef exports	5	75.50%
Corn imports	3	90.80%
Hay imports	2	62.70%
Wheat imports	3	11.60%

Corn and wheat's import and export elasticities are worthy of specific mention. The wheat export elasticities exceed its import elasticities, capturing the reality that marketing and distribution systems are export oriented, and there will be a tendency to export wheat output. Wheat production is less likely to develop domestic uses that require more imports, and thus that elasticity is somewhat lower. The reverse is true for corn, where imports support a large feeding industry and a projected ethanol industry, so the import elasticity is higher than the export elasticity.

The wheat import elasticity is lower than the export elasticity to take into account the fact that Colorado is a surplus producer, and, therefore, most infrastructure and institutional relationships focus on exports rather than increasing imports. However, both wheat and corn imports are still elastic relations, as many users of corn and wheat in the Eastern Plains, especially, can purchase needed quantities from nearby locations in Kansas and Nebraska, so it is easy to obtain imports and thus these relationships should be elastic.

The hay elasticity for imports is lower due to an assumption of significant transport costs and therefore tighter regional markets. To bring in more imports to Colorado, therefore, prices must rise faster than in the more widely traded corn and wheat markets. This has a fairly large effect on the local market in S3-DRY, where prices rise internally, forage use is cut, and dairy production decreases. The higher internal prices, driven partly by this elasticity assumption, leads to growth in hay production on irrigated hectares in Arkansas in S3-DRY, especially as corn production declines due to lower demand.

Imports and exports play an important role in describing climate change impacts on Colorado. Exports of wheat and beef, and imports of corn, are all greater than 90% of domestic production, so these products are clearly dependent on external economic performance and trends. We noted earlier that almost all wheat produced in Colorado leaves the state, often for international destinations. The large beef feeding industry is export-oriented, with about three quarters of production leaving the state. Hay is also a commodity where the import market is used quite variably across the simulations.

Welfare Effects. Because the model captures changes in prices and quantities, and has demand and supply functions embedded in the objective function, it is possible to determine changes in producer and consumer surpluses under the different simulations. In this fashion, the model shows how costs of climate change are borne, and could be employed to assess the value of various mitigation strategies in a future study. These results are presented in Figure 2. The measures of economic surplus show approximately a $10.7 M reduction in the S3-DRY scenario, compared to about $2.7 M in the wet year in S2-WET. In other words, the agricultural economy in Colorado loses nearly five times as much in a dry year climate relative to a wet year. The S1 climate scenario is predicted to produce economic net welfare impact that fits in the middle between S2-WET and S3-DRY (at about $6.2 M).

Figure 2. Changes in Producer Surplus (PS) and Consumer Surplus (CS), Million of Dollars.

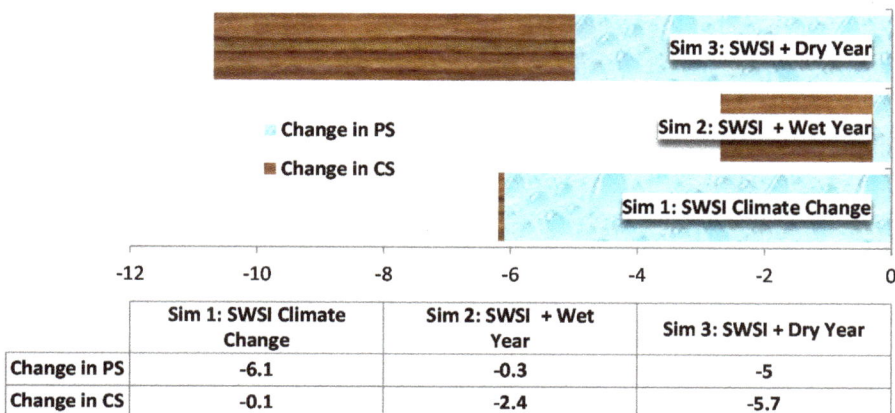

	Sim 1: SWSI Climate Change	Sim 2: SWSI + Wet Year	Sim 3: SWSI + Dry Year
Change in PS	-6.1	-0.3	-5
Change in CS	-0.1	-2.4	-5.7

In S1, most impacts fall on producers through reduced hay area, which has the greatest effect due to its water use, and which is made up by added imports and reduced dairy production. The largest effects naturally come in the dry year simulation, where cultivated area is reduced by up to 60% for some crops and yields can decline by over 10%. The total losses in S3 of more than $10 M are split about evenly between consumers and producers. Even though prices for livestock and major crops often increase by up to 10%, the decline in quantities offsets those better prices, and there is a net loss in producer surplus, which occurs because of the openness of the agricultural economy. The consumers lose in S3-DRY due to the higher overall prices.

Conclusions

Using an Economic Displacement Mathematical Programming (EDMP) model, derived from Harrington and Dubman [34] of the USDA's Economic Research Service. This study examines the effects of climate change on agriculture in Colorado taking into account of selected features projected several decades into the future. Initially, an overview of agriculture in the state and its dependence on water, a critical input, is described. The overview shows that the agricultural economy in Colorado is dominated by livestock, which accounts for 67% of total receipts. Crops, including feed grains and forages, account for 33% of production. Most of agriculture is based on irrigated production, which depends on both groundwater, especially from the Ogallala aquifer, and surface water that comes from runoff derived from snowpack in the Rocky Mountains. Climate studies point to decline in runoff from 6% to 20% by 2050. The timing of runoff is projected to begin and peak earlier in the spring and late-summer, and overall flows may be reduced.

The climate change scenarios evaluated in this paper include three simulations relative to a Base scenario that reflects some key characteristics with regard to future water and yield effects of climate change. Following SWSI projections, the base reflects demographics and economic changes from the calibrated model for 2007. The Base scenario models a 10.3% reduction in agricultural water from increased municipal and industrial water demand, and assumes a 75% increase in corn extracted-ethanol production. The first simulation reduces agricultural water availability by a further 14.0%, for a combined decrease of 24.3%, due to climatic factors and related groundwater depletion. The second simulation describes a year with warmer than historical average temperatures and wetter conditions, which negatively affect yields of irrigated corn and milking cows, but it improves yields for non-irrigated wheat, corn silage, irrigated hay, rangeland and pasture. In contrast, the last simulation describes a drought year, which leads to reduced harvested hectares for corn and wheat, and negatively affects yields for dry land wheat, irrigated corn, pasture and rangeland, while irrigated corn silage and hay output increase.

Three commodities examined in this paper account for a large percent of production in the Colorado agricultural sector: fed beef, wheat and dairy; two others are major sources of feed, including hay and corn. All are strongly affected by the S3-DRY scenario. Cattle feeding is dependent on exports out of the state, and in S3-DRY, fed beef production declines by 7.5% due to the significantly higher prices of feed and the resulting effect on output price. For corn, the hectares decrease by about 38% on irrigated land in both regions, while in the Arkansas basin, non-irrigated land declines by 75%. Due to the dry year's conditions, nearly a 50% reduction of South Platte non-irrigated wheat area occurs, while the Arkansas River basin non-irrigated wheat disappears completely. The wheat crop is extremely sensitive to how rainfall is affected by climate change, with the variation in exports being nearly 1.5 M tons.

The dairy sector reacts strongly to climate variation, given that production decreases by 18% in both warmer scenarios. Dairy is the second largest user of hay, after cow calf producers, and it is the second largest user of grain, after cattle feeding, as its rations require more of each basic feedstuff. Therefore, as feed shortages develop, dairy declines first and frees up significant proportions of grain and forage. The reduction in corn area leaves irrigated land available for hay production in the Arkansas basin, and expansion in irrigated hay occurs in the same basin in drought scenario. In the South Platte, non-irrigated corn becomes competitive on the land that was previously in wheat. Notably, 280 k hectares are in hay production in other parts of Colorado throughout all simulations because excess water relative to land exists in that part of the state.

This model has not taken into account farmers' adaptation strategies, which would reduce the climate impact on yields. Such strategies might include changing planting schedules, production practices or technologies, and the introduction of drought-tolerant varieties. Also, the model has not reflected climate-induced shifts in planting decisions and production practices that lead to various environmental impacts and higher costs. There could be soil and water quality effects through nutrient loss and soil erosion, and a greater use of pesticides to combat a higher prevalence of pests.

These environmental dimensions can be fruitful areas to examine in future research, as would be the development of a wider range of conditions in the analysis of climate change effects in the future. Some of the latter areas could be to look at various productivity growth scenarios before adding the effects of climate change, and also broader alternatives in performance of different commodities. This paper assumes certain large effects, such as the increase in yields for hay and the decrease in dairy output, but others, such as using the current set of relative prices and import and export positions as starting points, may seem to understate the climate change impacts on the agricultural economy of Colorado. A more extensive examination of these settings could provide additional insights.

Acknowledgment

The authors thank David Harrington and Robert Dubman at the Economic Research Service, U.S. Department of Agriculture, for providing the U.S. Equilibrium Displacement Mathematical Programming Model, which the authors modified to reflect Colorado's agricultural economy and water specifications.

Appendix

Positive Mathematical Programming

Returning to the matrix notation of Equation (1), Z is subject to the following constraints:

A$_{11}$x ≤ free Indicator accounts, necessary for analytical purposes, not shown in the Tableau;
A$_{21}$x ≤ b Resource constraints;
A$_{31}$x ≤ 0 Commodity balance equations;
I31x = c Calibration constraints, dropped after calibration;
U11 ≤ 0 Input accounts.

The additional notation is:

A is the matrix of technical coefficients;

I is an identity matrix of calibration constraints;

U is the matrix of inputs in dollar value to sector's activities;

b is a vector of right hand sides of resource constraints;

c is a vector of calibration quantity targets used only in calibration phase.

The resource constraints involve land and water for crop activities. Cropland, pasture, range land and land in the conservation reserve programs are quantified and include land fallowed as part of crop rotations including wheat-fallow. The supply of water available to agriculture is fixed, while the demand for water is exogenously determined for each crop by the State of Colorado's Consumptive Use Model (StateCU) component of the CDSS, which is based on a modified Blaney-Criddle method. (Other constraints include livestock facilities for livestock and labor for both crop and livestock activities).

The block of commodity balance equations runs across the production, demand and trade sections of the model. These are accounting constraints that distribute production across its uses. Corn, wheat and hay production are separated by location for the South Platte, Arkansas, and St. Luis Valley, and the Upper Colorado basins, and are identified by whether they are irrigated or non-irrigated production. Within this block, the two rows for corn and ethanol/distilled grain are highlighted. The corn balance equation allocates crop production from each basin and type of farming activity (irrigated *versus* non-irrigated) across basins and imports to ethanol production, domestic non-farm sales, and exports. In addition feed use of corn is calculated as a residual and transferred to the grain ration equation.

For example, the following is the corn commodity balance equation with the variable acronyms:

$$-60.818 \text{ SPCRND} - 176.868 \text{ IRSPCRN} - 179.625 \text{ ARCRNIR} - 45.75 \text{ DCRNAR} -134.134 \\ \text{CORNCO} + \text{CRNTUS} + 0.357 \text{ ETHCO} + \text{SELCRNCO} + \text{EXPCRNCO} - \text{IMPCRNCO} \leq 0 \tag{A1}$$

where, SPCRND, IRSPCRN, ARCRNIR, DCRNAR and CORNCO are corn production activities (harvested hectares) in South Platte non-irrigated, irrigated land, Arkansas non-irrigated, Arkansas irrigated land and the rest of Colorado; CRNTUS is the production allocated to feed; ETHCO is the ethanol production in M gallons; SELCRNCO, EXPCRNCO and IMPCRNCO are the levels of non-farm domestic sales, exports and imports in tons. The coefficients on the hectares are yields (tons/hectare), while the coefficient with ethanol production is the conversion ratio (liters of ethanol/ton of corn).

The feed requirements are calculated as intermediate inputs and are not priced in CDEMP. The model includes two rations. The grain ration equation is formulated as follows:

$$\sum_i \sum_b \alpha'' g - .064 \text{ eth} + \sum \beta'' K \leq 0 \tag{A2}$$

where, α is the vector of coefficients converting crops into feed ration components; and g is the vector of grain feed crops (corn, barley, oat and sorghum); eth is the level of ethanol production; and β is the vector of ration requirements in as fed form by livestock types; and k is the vector of livestock activity levels. Note that both g and K are subsets of cl_{ib}, and α and β are subsets of A_{31}, the matrix of technical coefficients.

The forage ration equation has similar structure:

$$\sum_i \sum_b \mu'''h + \sum \theta'''k \leq 0 \tag{A3}$$

where, μ is the vector of coefficients converting hay and pasture forage into feed ration components and h is the vector of forage activities (silage, cropped hay and pastures, permanent pastures and rangeland), θ is the vector of ration requirements in *as fed* form identified by livestock types k. Here h and k are both subsets of cl_{ib}, and μ and θ are subsets of A_{31}.

Harrington and Dubman [35] suggested changing one or more of the following EDMP model's parameter(s) to calibrate a base scenario:

1- Modify the scenario intercept for parallel shift of supply or demand function;
2- Modify the Hessian for rotation of the supply or demand function;
3- Modify the Right Hand Side (RHS) coefficients to change the resource availability;
4- Change the crop's yield, livestock productivity, or change the transfer from primary to semi or finished product coefficients.

Table A1. Area and production of crops and livestock activities, actual, and calibrated values for the Colorado EDMP.

Crop or Commodity	Units	Historical 2007 quantity	Calibrated quantity	Hessian element	Intercept
Ethanol	Million Liters	648.97	660.96	0.00	1.51
South platte dry corn	Million Hectares	0.10	0.11	−76.73	909.07
South platte irrigated corn	Million Hectares	0.21	0.23	−38.93	1173.13
Arkansas dry corn	Million Hectares	0.04	0.13	−81.46	436.30
Total corn	**Million Hectares**	**0.36**	**0.47**		
South Platte dry wheat	Million Hectares	0.04	0.00	−183.25	1160.79
South Platte irrigated wheat	Million Hectares	0.55	0.63	−11.25	1549.58
Arkansas dry wheat	Million Hectares	0.33	0.31	−22.30	703.28
Wheat, other [a]	Million Hectares	0.04	0.00	−157.75	2262.93
Total wheat	**Million Hectares**	**0.87**	**0.93**		
Sorghum	Million Hectares	0.07	0.07	−44.27	1683.21
Potatoes	Million Hectares	0.02	0.03	−13871.39	4041.04
South Platte irrigated. hay, all	Million Hectares	0.15	0.12	−10.04	1077.62
Arkansas dry hay, all	Million Hectares	0.09	0.11	−121.73	1082.96
Hay all, other	Million Hectares	0.31	0.34	−9.71	3251.96
Hay all, total	**Million Hectares**	**0.55**	**0.57**		
Fed beef	Thousand Ton	1235.6	1241.9	−0.2	182.1
Hogs,	Thousand Ton	161.6	165.7	−57.3	76.4
Dairy	Thousand Ton	1228.3	1236.5	−1.1	42.3
Broiler	Thousand Ton	157.5	173.4	−6.1	92.8
Eggs, independent	Million dozens	8.83	9.72	0	1.9
Eggs, contracted	Million dozens	79.5	87.45	0	2.5
Turkey, independent	Thousand Ton	13.6	16.8	−108.3	108
Turkey, contracted	Thousand Ton	20.9	21.3	−141.3	135.4

Note: [a] Other basins include San Luis Valley and Colorado River basin.

86

Conflicts of Interest

The authors declare no conflict of interest.

References

1. Colorado Water Conservation Board. *Statewide Water Supply Initiative Report Overview*; Colorado Department of Natural Resources: Denver, CO, USA, 2004.
2. Ray, A.; Barsugli, J.; Averyt, K.; Wolter, K.; Hoerling, M.; Doesken, N.; Udall, B.; Webb, R.S. *Climate Change in Colorado: A Synthesis to Support Water Resources Management and Adaptation*; University of Colorado Boulder: Boulder, CO, USA, 2008.
3. Parry, M.L. *Climate Change 2007: Impacts, Adaptation and Vulnerability: Contribution of Working Group II to the Fourth Assessment Report of the Intergovernmental Panel on Climate Change*; Cambridge University Press: Cambridge, UK, 2007; Volume 4.
4. Solomon, S.D.; Qin, M.; Manning, Z.; Chen, M.; Marquis, K.B. *Contribution of Working Group I to the Fourth Assessment Report of the Intergovernmental Panel on Climate Change*; Cambridge University Press: Cambridge, UK, 2007.
5. Thorvaldson, J.; Pritchett, J. *Economic Impact Analysis of Irrigated in Four River Basins in Colorado*; Colorado Water Resources Research Institute: Fort Collins, CO, USA, 2006.
6. Gunter, A.; Goemans, C.; Pritchett, J.G.; Thilmany, D.D. Linking an Equilibrium Displacement Mathematical Programming Model and an Input-Output Model to Estimate the Impacts of Drought: An Application to Southeast Colorado. In Proceedings of Agricultural & applied Economics Association's 2012 AAEA Annual Meeting, 12–14 August 2012; Agricultural and Applied Economics Association: Seattle, WA, USA, 2012.
7. Malcolm, S.; Marshall, E.; Aillery, M.; Heisey, P.; Livingston, M.; Day-Rubenstein, K. *Agricultural Adaptation to a Changing Climate: Economic and Environmental Implications Vary by US Region*; USDA-ERS Economic Research Report No. 136; United States Department of Agriculture-Economic Research Service (USDA-ERS): Washington, DC, USA, 2012.
8. Islam, A.; Ahuja, L.R.; Garcia, L.A.; Ma, L.; Saseendran, A.S.; Trout, T.J., Modeling the impacts of climate change on irrigated corn production in the Central Great Plains. *Agric. Water Manag.* **2012**, *110*, 94–108.
9. Herrero, M.P.; Johnson, R. High temperature stress and pollen viability of maize. *Crop. Sci.* **1980**, *20*, 796–800.
10. Kelly, E.Z.; Tunc-Ozdemir, M.; Harper, J.F. Temperature stress and plant sexual reproduction: Uncovering the weakest links. *J. Exp. Bot.* **2010**, *61*, 1959–1968.
11. Izaurralde, R.C.; Thomson, A.M.; Morgan, J.; Fay, P.; Polley, H.; Hatfield, J.L. Climate impacts on agriculture: Implications for forage and rangeland production. *Agron. J.* **2011**, *103*, 371–381.
12. Morgan, J.A.; LeCain, D.R.; Pendall, E.; Blumenthal, D.M.; Kimball, B.A.; Carrillo, Y.; Williams, D.G.; Heisler-White, J.; Dijkstra, F.A.; West, M. C4 grasses prosper as carbon dioxide eliminates desiccation in warmed semi-arid grassland. *Nature* **2011**, *476*, 202–205.

13. Backlund, P.; Janetos, A.; Schimel, D.; Walsh, M. The effects of climate change on agriculture, land resources, water resources, and biodiversity in the United States. In *The Effects of Climate Change on Agriculture, Land Resources, Water Resources, and Biodiversity in the United States*; Synthesis and Assessment Report 4.3; U.S. Department of Agriculture: Washington, DC, USA, 2008.

14. Preston, B.L.; Jones, R. *Climate Change Impacts on Australia and the Benefits of Early Action to Reduce Global Greenhouse Gas Emissions*; The Commonwealth Scientific and Industrial Research Organisation (CSIRO): Clayton South, Australia, 2006.

15. Moons, C.P. H.; Sonck, B.; Tuyttens, F.A.M. Importance of outdoor shelter for cattle in temperate climates. *Livest. Sci.* **2014**, *159*, 87–101.

16. Lambertz, C.; Sanker, C.; Gauly, M. Climatic effects on milk production traits and somatic cell score in lactating Holstein-Friesian cows in different housing systems. *J. Dairy Sci.* **2014**, *97*, 319–329.

17. National Agricultural Statistics Service (NASS). Department of Agriculture. Census of Agriculture. Available online: http://www.agcensus.usda.gov/ (accessed on 15 June 2012).

18. Colorado Water Conservation Board. *Conservation Levels Analysis Final Report*; Colorado Department of Natural Resources: Denver, CO, USA, 2010.

19. Hardling, B.L; Wood, A.W.; Prairie, J.R. The implications of climate change scenario selection for future stream flow projection in the Upper Colorado River Basin. *Hydrol. Earth Syst. Sci. Discuss.* **2012**, *9*, 847–894.

20. Howitt, R.E.; MacEwan, D.; Medellín-Azuara, J.; Lund, J.R. *Economic Modeling of Agriculture and Water in California Using the Statewide Agricultural Production Model*; Department of Agricultural and Resource Economics, Department of Civil and Environmental Engineering, Center for Watershed Sciences: Davis, CA, USA, 2010.

21. Preckel, P.V.; Harrington, D.; Dubman, R. Primal/dual positive math programming: Illustrated through an evaluation of the impacts of market resistance to genetically modified grains. *Am. J. Agric. Econ.* **2002**, *84*, 679–690.

22. Schmid, E.; Sinabell, F. *Using the Positive Mathematical Programming Method to Calibrate Linear Programming Models*; University für Bodenkultur Wien, Department für Wirtschafts-u. Sozialwiss, Inst. für NachhaltigeWirtschaftsentwicklung: Vienna, Austria, 2005.

23. Heckelei, T.; Britz, W. Models Based on Positive Mathematical Programming: State of the Art and Further Extensions. In *Modeling Agricultural Policies: State of the Art and New Challenges*; Monte Università Parma: Parma, Italy, 2005; pp. 48–73.

24. De Frahan, B.H.; Buysse, J.; Polomé, P.; Fernagut, B.; Harmignie, O.; Lauwers, L.; Van Huylenbroeck, G.; Van Meensel, J. Positive Mathematical Programming for Agricultural and Environmental Policy Analysis: Review and Practice. In *Handbook of Operations Research in Natural Resources*; Springer: New York, NY, USA, 2007; pp. 129–154.

25. Buysse, J.; Van Huylenbroeck, G.; Lauwers, L. Normative, positive and econometric mathematical programming as tools for incorporation of multifunctionality in agricultural policy modeling. *Agric. Ecosyst. Environ. Ecosyst. Environ.* **2007**, *120*, 70–81.

26. Heckelei, T.; Wolff, H. Estimation of constrained optimisation models for agricultural supply analysis based on generalised maximum entropy. *Eur. Review Agric. Econ.* **2003**, *30*, 27–50.

27. Osterburg, B.; Offermann, F.; Kleinhanss, W. A Sector Consistent Farm Group Model for German Agriculture. In *Agricultural Sector Modeling and Policy Information Systems*; *Vauk Verlag*: Kiel, Germay, 2001; pp. 152–160.

28. Judez, L.; De Miguel, J.; Mas, J.; Bru, R. Modeling crop regional production using positive mathematical programming. *Math. Comput. Model.* **2002**, *35*, 77–86.

29. Baskaqui, A.; Butault, J.; Rousselle, J. Positive Mathematical Programming and Agricultural Supply within EU under Agenda 2000. In Proceedings of the 65th European Seminar of the European Association of Agricultural Economists (EAAE), Bonn, Germany, 29–31 March 2000, Wissenschaftsverlag Vauk Kiel KG: Kiel, Germany, 2001; p. 200.

30. Paris, Q.; Montresor, E.; Arfini, F.; Mazzocchi, M.; Heckelei, T.; Witzke, H.; Henrichsmeyer, W. An Integrated Multi-Phase Model for Evaluating Agricultural Policies through Positive Information, Agricultural Sector Modelling and Policy Information Systems. In Proceedings of the 65th European Seminar of the European Association of Agricultural Economists (EAAE), Bonn, Germany, 29–31 March 2000, Wissenschaftsverlag Vauk Kiel KG: Kiel, Germany, 2001; pp. 100–110.

31. Sinabell, F.; Streicher, G. Programme Evaluation with Micro-Data: The Use of FADN Data to Evaluate Effects on the Market Situation of Programme Participants. In Proceedings of 87th EAAE-Seminar: Assessing Rural Development Policies of the CAP, Vienna, Austria, 21–23 April 2004.

32. Howitt, R.E. Positive mathematical programming. *Am. J. Agric. Econ.* **1995**, *77*, 329–342.

33. Muth, R.F. The derived demand curve for a productive factor and the industry supply curve. *Oxf. Econ. Pap.* **1964**, *16*, 221–234.

34. Piggott, R.R.; Piggott, N.E.; Wright, V.E. Approximating farm-level returns to incremental advertising expenditure: Methods and an application to the Australian meat industry. *Am. J. Agric. Econ.* **1995**, *77*, 497–511.

35. Samuelson, P.A. Spatial price equilibrium and linear programming. *Am. Econ. Review* **1952**, *42*, 282–303.

36. Harrington, D.H.; Dubman, R. *Equilibrium Displacement Mathematical Programming Models: Methodology and a Model of the U.S. Agricultural Sector*; Technical Bulletin No. (TB-1918); United States Department of Agriculture: Washington, DC, USA, 2008.

37. Johnson, G.; Quance, C.L. *The Overproduction Trap in US Agriculture: A Study of Resource Allocation from World War I to the Late 1960's*; The John Hopkins University Press: Baltimore, MD, USA, 2011

38. Connor, J.D.; Schwabe, K.; King, D.; Knapp, K. Irrigated agriculture and climate change: The influence of water supply variability and salinity on adaptation. *Ecol. Econ.* **2012**, *77*, 149–157.

39. Henseler, M.; Wirsig, A.; Herrmann, S.; Krimly, T.; Dabbert, S. Modeling the impact of global change on regional agricultural land use through an activity-based non-linear programming approach. *Agric. Syst.* **2009**, *100*, 31–42.

40. Qureshi, M.E.; Ahmad, M.-U.-D.; Whitten, S.M.; Kirby, M. A Multi-Period Positive Mathematical Programming Approach for Assessing Economic Impact of Drought in the Murray-Darling Basin, Australia. In Proceedings of Australian Agricultural and Resource Economics Society, 56th Conference, Freemantle, Australia, 7–10 February, 2012.

41. Withey, P.; van Kooten, G.C. *The Effect of Climate Change on Land Use and Wetlands Conservation in Western Canada: An. Application of Positive Mathematical Programming*; Working Paper 2011–04; Resource Economics and Policy Analysis Research Group (REPA), Department of Economics, University of Victoria: Victoria, Canada, 2011.

42. Schlenker, W.; Hanemann, W.M.; Fisher, A.C. Water availability, degree days, and the potential impact of climate change on irrigated agriculture in California. *Clim. Change* **2007**, *81*, 19–38.

43. Bauman, A.; Goemans, C.; Pritchett, J.; McFadden, D.T. Estimating the economic and social impacts from the drought in Southern Colorado. *J. Contemp. Water Res. Educ.* **2013**, *151*, 61–69.

44. Schaible, G.D.; Kim, C.; Aillery, M.P. Dynamic adjustment of irrigation technology/water management in western US agriculture: Toward a sustainable future. *Can. J. Agric. Econ.* **2010**, *58*, 433–461.

45. Pritchett, J.G.; Davies, S.P.; Fathelrahman, E.; Davies, A. Welfare Impacts of Rural to Urban Water Transfers: An Equilibrium Displacement Approach. In Proceedings of Agricultural & Applied Economics Association (AAEA), Canadian Agricultural Economics Society (CAES), & Western Agricultural Economics Association (WAEA) Joint Annual Meeting, Denver, CO, USA, 25–27 July 2010.

46. Colorado Decision Support System (CDSS). Colorado Division of Water Resources. Databases and software. Available online: http://cdss.state.co.us/Pages/CDSSHome.aspx (accessed on 20 March 2012).

47. Takayama, T.; Judge, G.G. Equilibrium among spatially separated markets: A reformulation. *Econom. J. Econ. Soc.* **1964**, *32*, 510–524.

48. Helming, J.F. A Model of Dutch Agriculture Based on Positive Mathematical Programming with Regional and Environmental Applications. Ph.D. Thesis. Wageningen University, Wageningen, the Netherlands, 11 February 2005.

49. Colorado Water Conservation Board. *Colorado River Water Availability Study, March, 2010. Phase I. Draft*; Colorado Department of Natural Resources: Denver, CO, USA, 2010.

Rainfall Enhances Vegetation Growth but Does the Reverse Hold?

John Boland

Abstract: In the literature, there is substantial evidence presented of enhancement of vegetation growth and regrowth with rainfall. There is also much research presented on the decline in rainfall with land clearance. This article deals with the well documented decline in rainfall in southwest Western Australia and discusses the literature that has been presented as to the rationale for the decline. The original view was that it was the result of climate change. More recent research points to the compounding effect of land use change. In particular, one study estimated, through simulation work with atmospheric models, that up to 50% of the decline could be attributed to land use change. For South Australia, there is an examination the pattern of rainfall decline in one particular region, using Cummins on the Eyre Peninsula as an example location. There is a statistically significant decrease in annual rainfall over time in that location. This is mirrored for the vast majority of locations studied in South Australia, most probably having the dual drivers of climate and land use change. Conversely, it is found that for two locations, Murray Bridge and Callington, southeast of Adelaide, there is marginal evidence for an increase in annual rainfall over the last two decades, during which, incidentally, Australia experienced the most severe drought in recorded history. The one feature common to these two locations is the proximity to the Monarto plateau, which lies between them. It was the site of extensive revegetation in the 1970s. It is conjectured that there could be a connection between the increase in rainfall and the revegetation, and there is evidence presented from a number of studies for such a connection, though not specifically relating to this location.

Reprinted from *Water*. Cite as: Boland, J. Rainfall Enhances Vegetation Growth but Does the Reverse Hold? *Water* **2014**, *6*, 2127–2143.

1. Introduction

There have been numerous studies detailing the deleterious effects of vegetation clearance on rainfall totals. Junkermann *et al.* [1] report on trends in rainfall in Western Australia. They state that the western tip of the continent has experienced a reduction of precipitation by about 30% (from an average of 325 mm/a) since the 1970s, attributed to a change in the large scale surface pressure patterns of the southern ocean ([2,3]) with a concurrent reduction in surface water fluxes [4].

This article discusses this reduction and focusses also on two areas in South Australia, starting first with details of an investigation of rainfall trends in Cummins, on Eyre Peninsula, 34.26°S, 135.73°E. Despite some perceptions of increased rainfall in the region, a significant decrease in rainfall is found if you compared before and after 1975, roughly mirroring the Western Australian experience, even if not as extreme. Noteworthy is the fact that the land clearance in this area took place principally from the 1950s [5], substantially later than in some other parts of South Australia.

The perception is that extensive plantings on the Monarto plateau in the 1970s has led to increased rainfall on the plateau and close by. Unfortunately, the only official weather station on the plateau is at the Monarto Zoo, and has only been in operation in recent times and even then only sporadically. However there are official stations at Murray Bridge (35.12° S, 139.27° E) and Callington (35.12° S, 139.04° E) adjacent. Gallant *et al.* [6] studied rainfall trends from 1910-2005 in six regions of Australia. In their Southeast region, which includes this study area, their conclusion was that there has been *a significant decrease in annual total rainfall of 20 mm per decade since 1950 (that) stems mainly from decreases during autumn*. It is in this context that results from rainfall trend analysis are presented for these stations and two further from the Monarto plateau. It would appear that these two locations at the edge of the plateau are going against the trend, with increasing rainfall if one compares the period pre 1989 with that after, though not at a significant level. The year 1989 was chosen as this is one decade after the end of the Monarto plantings. On the other hand, the trends at Mt Barker and Tailem Bend are consistent with the findings of Gallant *et al.* [6] of diminishing rainfall in this area of the country, though not at as great a rate as they report.

Thus, there appears to be some evidence in Australia of the influence of introducing vegetation on rainfall as well as the more easily supported evidence of lowering of rainfall with land clearance. One must be guarded in this conclusion as there are always confounding factors, for instance changing weather patterns with climate change being one. But it does influence us to conduct more research into the topic.

2. Rainfall Trends in Southwest Western Australia

To illustrate the change in rainfall over time in Western Australia, see Figure 1. The mean annual rainfall for Perth Airport up to 1975 was 836 mm. and from 1975 on was 721 mm. A two sample *t*-test was run to see if the mean before 1975 was significantly higher than after. To perform such a test, the null hypothesis, H_0, is the one that is tested, since it is an exact statement-that the means before and after are equal. The alternative hypothesis, H_a is that the mean before is greater than the mean after 1975. One selects a level of significance, normally $\alpha = 0.05$. One calculates what is called the *p*-value. This is essentially the probability of the test data being consistent with the null hypothesis. If the *p*-value is less than α, then the probability of the test data being consistent with the H_0 is low and H_0 is rejected. The test was performed and it was found to be significant at a *p*-value of <0.01. Thus, one rejects the H_0, and concludes that the mean rainfall before 1975 was significantly higher than after.

While this large scale phenomenon has been going on, there would seem to have been more local changes as well. It is necessary to give some background on land use changes in the region. Note that early research [7] concluded that the demonstrable rainfall decline was most likely due to a combination of climate change and climate variability. Increasingly, though, the more the phenomenon was investigated, the greater the case for the influence of land use change as well. Pitman *et al.* [8] found that up to 50% of the reduction in rainfall can be explained through land use change. One pertinent piece of evidence from their atmospheric circulation model simulations is that the reduction of rainfall on the coast and the increase inland (matching different land use patterns as shall be seen below) match the observations.

Figure 1. Annual rainfall totals for Perth Airport from 1944 to 2011 with smoothed trend.

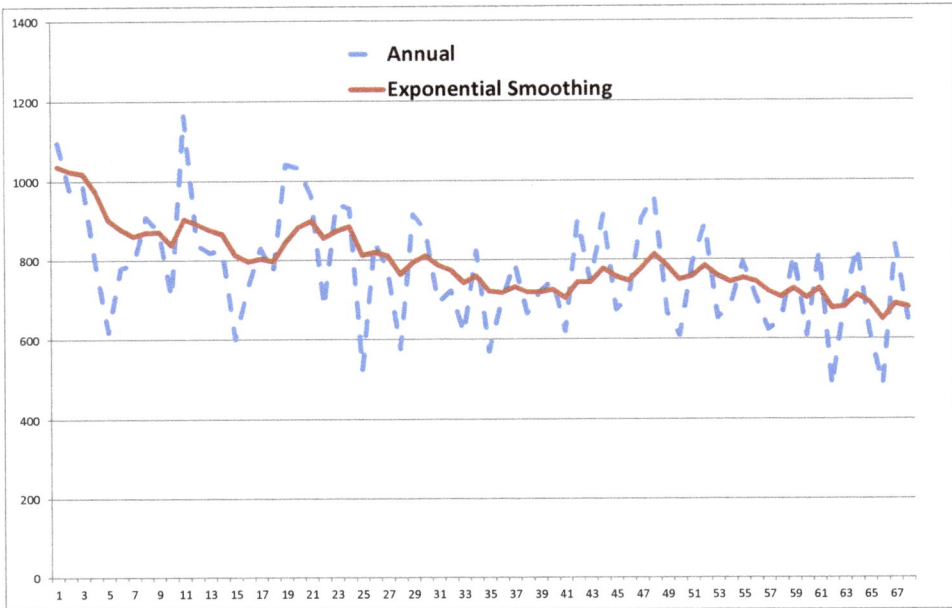

At essentially the state boundary between Western Australia and South Australia, there is the approximately 1500 km Vermin Proof Fence (formerly referred to as the Rabbit Proof Fence) erected in the early part of the 20th century. On the western side the land was predominantly cleared for agriculture, whereas the eastern side is natural vegetation. Sadler [7] found that there has been a shift in precipitation from the agricultural regions to the regions of natural vegetation with an increase of about 20%. Junkermann *et al.* [1] conducted atmospheric experiments above both the agricultural regions and the area of natural vegetation, attempting to understand the mechanisms behind the differential changes in rainfall. They concluded that extra cloud condensation nuclei (CCN) were generated over the agricultural areas due to salt lake geochemistry. The extra aerosols mean that there is an increase in cloud droplets competing for condensable water, thus depressing precipitation. This phenomenon does not occur over the areas of natural vegetation, resulting in a decoupling of the water cycle on a regional basis. Similarly, Pielke *et al.* [9] discussed the possible mechanisms for such a phenomenon. They give results from the Bunny Fence Experiments of 2005–2007 where they compare the latent heat fluxes (low and not very variable for the forested areas and for the agricultural areas low after harvest but high during the growing season) and also that the sensible heat flux is higher over the native vegetation during August to December. This, they say, results in vigorous boundary layer development and higher planetary boundary layer over the vegetated area. Figure 2 gives an illustration of the effects of this in a satellite photograph in which there is substantial cloud formation over the native vegetation and none over the agricultural land. The edge of the cloud formation appears to coincide very well with the western edge of the native vegetation. The deforestation is thus linked to a reduction in rainfall. This significant result gives credence to the premise that land use change in the form of clearance reduces rainfall, but can afforestation enhance rainfall?

Figure 2. Cloud formation over Western Australia-Satellite image originally processed by the Bureau of Meteorology from the geostationary meteorological satellite MTSAT-1R operated by the Japan Meteorological Agency.

Chikoore and Jury [10] present an interesting discussion of this topic. They focus on understanding the role of vegetation in the African climate system. They quote studies that show that vegetation growth and distribution are largely determined by climate [11,12]. They also refer to vegetation and land use feedbacks on climate [11,13]. Their specific tasks were twofold. One was to determine if greening caused by one rainfall event produced evapotranspiration that affects the next wet spell. The second was to understand if vegetation could help, in their terms, "anchor" cloud bands regardless of external forcing by large scale circulation and heating anomalies. They found positive answers using means such as principal component analysis to support their conjectures in both cases. Sprack *et al.* [14] observed that there is an enhancement of rainfall in tropical regions when the rain bearing clouds passed over forested areas.

Other recent research supports the conjecture that there is a connection between vegetation and maintenance or even enhancement of precipitation. Makarieva *et al.* [15] provides interesting evidence to support this theory. They state that the dependence of annual precipitation on distance from the ocean differs markedly between the world's forested and non-forested continent-scale regions. In the non-forested regions, precipitation declines exponentially with distance from the ocean. In contrast, in the forest-covered regions precipitation does not decrease or even grows along several thousand kilometers inland.

The implication is that there may even be an increase in rainfall in forested areas as one progresses inland. Beltran-Przekurat *et al.* [16] performed some interesting simulations based on South America with their scenarios of land use, agriculture, grasslands and afforested areas. They were performed for three separate sets of El Nino - Southern Oscillation conditions. The general result vis-a-vis rainfall was that while there was insignificant difference between the grassland and agricultural settings, there was an increase in rainfall with afforestation.

Thus, both observational evidence and also simulations using atmospheric models lends support to the concept that there can be enhancement of rainfall because of afforestation. Our task is to examine the interplay between vegetation and rainfall in South Australia, evidence will be given to support the depletion of rainfall through land use change, and also that increased vegetation can enhance rainfall, if placed advantageously.

3. Rainfall Trends on Eyre Peninsula

In this section, details are given of an investigation of rainfall trends in a particular region of South Australia, Cummins, on Eyre Peninsula, 34.26° S, 135.73° E. A subsequent section will deal with trends in and around the Monarto region, 35.08° S, 139.13° E, of South Australia. Note that in all the analysis in this paper, annual rainfall totals are used. Seasonal analysis was also performed and the results by season did not show any significant difference from the annual analysis. Also, in this study, no account was taken of any extraneous variables such as temperature trends or ENSO for example and their possible connection with rainfall. This will be examined in future work.

There is a history of land clearance on Eyre Peninsula, primarily for agricultural activities. In the Cummins-Wanilla Basin Catchment Management Plan [5], the authors point out that native vegetation covers only 6% or 50 km^2 of the Basin with the majority of the land utilised for various agricultural purposes. The Cummins-Wanilla Basin, as is the Eyre Peninsula as a whole, is of high botanical significance. There occur a number of species and associations of species of high conservation significance. South of Cummins was once dominated by Low Open Forest, including *Eucalyptus cladocalyx* (sugar gum) with an understorey of *Xanthorrhoea sp.* (yacca), and various *Acacia sp.* (wattle). Closer to Cummins the community was predominantly Open Scrub with various Mallee type *Eucalyptus sp.* as well as *Melaleuca lanceolata* (dryland tea tree) [5].

As part of an Australian Research Council Discovery Indigenous Researcher Development grant, one of the prime areas of activity has been to identify specific locations where revegetation projects may be able to enhance rainfall, particularly where there is evidence of rainfall decline. Interestingly, two members of the project team interviewed some people in the Cummins region as to their views on the project and what they would think about attempting to increase the rainfall in the region. The reaction of two of the people interviewed are described below:

One chap said, "We have lots of water lying on the ground so we don't need any more rain."
Another woman said, "We don't need any more rain, we have enough now"

The question arises then as to whether, as these statements seem to imply, the rainfall has actually increased in the Cummins region over time. To begin the analysis, Figure 3 shows the annual rainfall totals for Cummins from 1915 to 2004. Later years were excluded because of missing months of data. Coincidentally, many of the latter years coincided with a severe drought and they may have only increased the idea of decline. Superimposed on the figure is the exponentially smoothed version of the rainfall totals constructed using

$$Y_t = \alpha X_t + (1 - \alpha)Y_t, \qquad Y_0 = X_0, 0 < \alpha < 1 \tag{1}$$

X_t is the original data series and if $\alpha = 1$ we get the original series. So, higher values of α decrease the level of smoothing. From the figure it appears that there may be a downward trend to the totals. To determine whether there is, comparison of the annual totals before and after 1975 is performed, consistent with the analysis for Western Australia, with the mean rainfall before being 440 mm. and after being 395 mm. This does not seem as dramatic a shift as in Western Australia, but still may be a significant change. A two sample t-test was performed with the following hypotheses:

$$H_0 \quad : \quad \mu_1 = \mu_2$$
$$H_1 \quad : \quad \mu_1 > \mu_2$$
$$\alpha \quad = \quad 0.05$$

In this case, H_0 was rejected with a p-value of 0.014. Thus, evidence of a statistically significant reduction in rainfall exists. This appears at odds with the views expressed above by the two people interviewed. How can we have such a contradiction-an appearance of too much, or at least sufficient, rain and scientific evidence of depletion in rainfall? One can hypothesize that, as in the Western Australian example given in the Introduction, there is a link between the rainfall decline and land clearance. Even more significantly, conjectures can be formed as to the disjunction between the reality of depletion of rainfall and the perception of waterlogging.

It is easy to imagine that with the clearance of land there will be less interception of the rain that falls, that is more runoff [17]. This runoff can tend to pool in regions where the soil structure is such that there is little soakage, that is where you have clay soils or indeed, silting up over time because of runoff of material as well as the rainfall. What is happening is that the people who believe there is sufficient or even an increase in rainfall may be seeing only the superficial markers, the pooling. They are not seeing the decline in rainfall as it is happening at longer time scales than their levels of perception allow for. If they were truly perceptive, they would notice the decline in rainfall and then ask the truly pertinent question, why is there more pooling with less rain! Common sense perceptions that are couched within cultural understandings don't necessarily include an analysis of environmental change. Noteworthy is the fact that the land clearance in this area took place principally from the 1950s [5], substantially later than in some other parts of South Australia.

Figure 3. Annual rainfall totals for Cummins for 1945–2004 with smoothed trend.

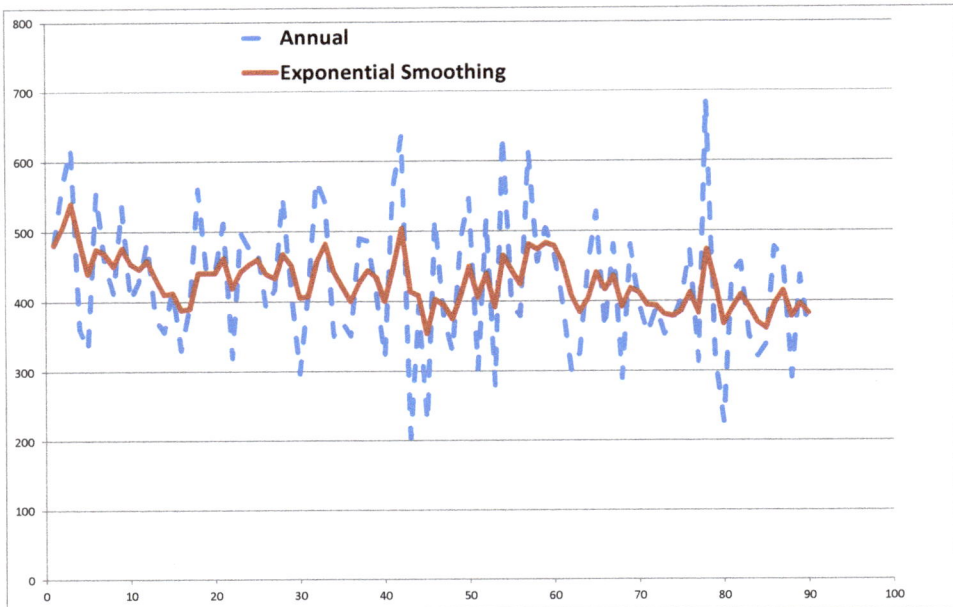

3.1. Evidence of Beneficial Effects of Upland Plantings

When driving around the countryside near Cummins in springtime, the pooling of water on the ground was noticed, but there was also the lack of vegetation on nearby hilltops, invoking the question of is it sensible to draw some conclusions about connections. Recently, George Monbiot wrote an article in The Guardian [17] about the vagaries of agricultural subsidies in England and how they are geared to promote flooding because they promote the clearing of upland vegetation. He referred to an opposite approach taken by a group of farmers in Wales, by the Pontbren River. The Pontbren Project [18] *is an innovative approach to using woodland management and tree planting to improve the efficiency of upland livestock farming, led by a group of neighbouring farmers in mid-Wales.* They found that not only had they been able to develop markets for some of the wooden products derived from their new plantings, but also they noticed less runoff into lowland areas in regions where they were revegetating. They then invited scientists to investigate. *Supported by government funding, this internationally important research has revealed why strategically located belts of trees are so effective at reducing the amount of water running off improved upland grasslands* [18]. This research team [19] set up three experimentation sites where surface runoff was measured, supplemented by measurements of soil infiltration rates and soil and vegetation physical properties. On one site they excluded sheep, at another they both excluded sheep and planted native broadleaf tree species, and the third was used as a control. Five years after planting, they found that median soil infiltration rates were 67 times greater in plots planted with trees compared to grazed pasture.

As well as the general results of greatly aiding retention of the rain that falls in the places where it is needed rather than having it run off and cause flooding elsewhere, there is a key point they emphasize. They stress that it is the strategic location of belts of vegetation that is important, not just the random plantings that tend to happen with some revegetation schemes. In [20], there is a comprehensive discussion of the many benefits of integrating bands of native vegetation in agricultural practice. They give specific details on the mechanisms whereby a cleared landscape creates a hotter and drier landscape and how the integrated bands of native vegetation restore water recycling into the landscape, through making it cooler and returning more water to the lower atmosphere through turbulent exchanges with wind and centres of forced convection. Additionally, they describe, referring to [21], how vegetation bands increase the retention of overland flow and leak this water to hillslopes and adjacent streams slowly over time, therefore, helping to buffer streams from large peak runoff rates. One can see from this latter point how the strategic planting of hilltops can be beneficial in minimising flooding and waterlogging. Let's now pass on to another potentially good result not from strategic placement but from happenstance.

4. Rainfall Trends at Monarto and Surrounding Areas

In the 1970s, the then government of South Australia decided to build a satellite city at Monarto, 70 km to the south east of Adelaide. They were concerned that the burgeoning urban sprawl of the city of Adelaide would be at the expense of highly productive agricultural land. To circumvent this, it was decided that a new development on marginal land at Monarto, linked to Adelaide by high speed rail, would prove a better option. As part of beautifying this site, heavily cleared previously for grazing and cropping, millions of seedlings of Australian native plant species were planted from 1973 to 1979. Many years on, local anecdotal "evidence" infers an increase in rainfall on the plateau on which Monarto lies. This conjecture is tested.

The perception is that the extensive plantings on the Monarto plateau in the 1970s has led to increased rainfall on the plateau and close by. The only official weather station on the plateau is at the Monarto Zoo, and has only been in operation in recent times and even then only sporadically. However there are official stations at Murray Bridge (35.12° S, 139.27° E) and Callington (35.12° S, 139.04° E) adjacent to the plateau. The structure of the analysis in this section will be to examine the annual trend in rainfall before and after the intervention, that is the planting of the vegetation. This is not precisely what will be done, since some time after the plantings is allowed for establishment. Since the last of the organised plantings was in 1979, the analysis uses 1989 as the change year, the year it is assumed that there has been sufficient establishment. A comparison is made of what is happening at these two stations with two others, one further away to the East (Tailem Bend 35.26° S, 139.46° E) and one to the West (Mt. Barker 35.07° S, 138.86° E). These are in areas where there wasn't the same level of interference during this time. It should be noted that all of the area in which these four stations are located underwent substantial clearing in the century prior to the 1970s.

Gallant *et al.* [6] studied rainfall trends from 1910–2005 in six regions of Australia. In their Southeast region, which includes this study area, their conclusion was that there has been a significant decrease in annual total rainfall of 20 mm per decade since 1950 (that) stems mainly from decreases

98

during autumn. It is in this context that results are presented from rainfall trend analysis for these four stations.

Let's examine Murray Bridge rainfall over the whole period of record first. This is displayed in Figure 4. From this diagram, it appears that contrary to the conclusions in [6], the rainfall in latter years at Murray Bridge has either maintained the same level or indeed slightly risen. To test this premise divide the data into two sets, pre 1989 and from 1989 onwards. The mean rainfall before 1989 was 341 mm and 373 after. Then test the hypotheses.

$$H_0 \quad : \quad \mu_1 = \mu_2$$
$$H_1 \quad : \quad \mu_1 < \mu_2$$
$$\alpha \quad = \quad 0.05$$

For this test, the p-value $= 0.073$. In essence the null hypothesis of equal rainfall before and after 1989 cannot be rejected, but the p-value is quite close to the level of significance, and so the conclusion can be regarded as tentative. Indeed, if one changed the hypothesis test where one were evaluating the conjecture from [6], that the annual rainfall was 20 mm greater before 1989 *versus* the rainfall difference was less than that, the *p*-value for that test is 0.01. In any case, one can conclude that the rainfall is either similar before and after 1989 or greater after, rather than the expectation of diminishing rainfall.

Figure 4. Annual rainfall totals for Murray Bridge from 1886 to 2011.

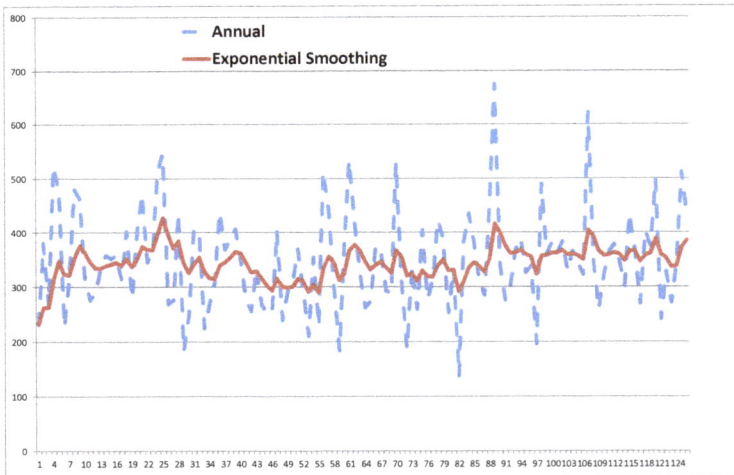

Let's now turn to the other station in close proximity to the Monarto plateau, that of Callington. A similar set of tests shows not disparate results, with there being little statistical likelihood that the rainfall could have decreased after 1989 and a not significant but indicative chance of an increase, at least when compared to the previous few decades. See Figure 5 for an illustration of this. One might surmise that after land clearance the rainfall decreased and that it is now being restored to historical levels. The mean rainfall before 1989 was 369 mm and 400 after.

Figure 5. Annual rainfall totals for Callington from 1900 to 2011.

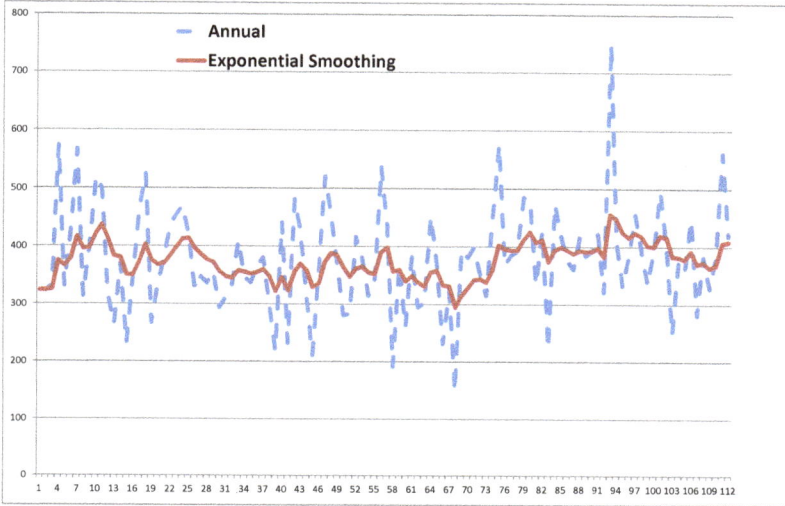

It would appear that these two locations at the edge of the plateau are going against the trend identified by [6]. It still could be that the entire region is following the same trend over time. In this case, there would be no argument for the conjecture that the increase in vegetation has aided the total rainfall. To check on this, examine the trends over time for the two locations more removed from the plateau. First, let's look at the time series of annual rainfall at Mt Barker, as shown in Figure 6. An hypothesis test for equal rainfall before 1989 *versus* less after has a p-value of 0.103. So there is a lower average rainfall after of 727 mm versus 775 before but it is not significantly lower.

Figure 6. Annual rainfall totals for Mt Barker from 1861 to 2006.

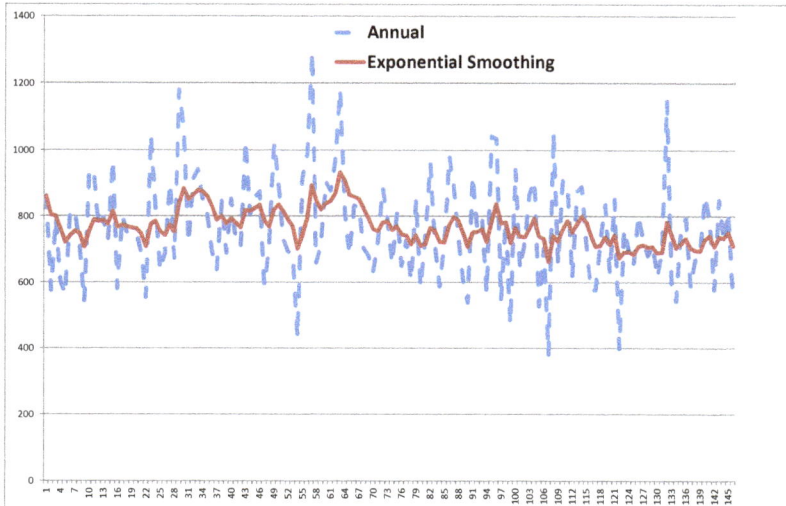

Now move to examining a town east of the plateau, Tailem Bend, for rainfall trends. The mean rainfall before 1989 was 380 mm and after was 348, as illustrated in Figure 7. An hypothesis test of the mean rainfall being the same before and after 1989 *versus* being less after was not rejected with a *p*-value of 0.069. This was in a way similar to the test for Murray Bridge but with opposite alternate hypothesis, being greater after 1989 at Murray Bridge and less after 1989 at Tailem Bend, even though the two locations are only 20 km apart.

Figure 7. Annual rainfall totals for Tailem Bend from 1908 to 1998.

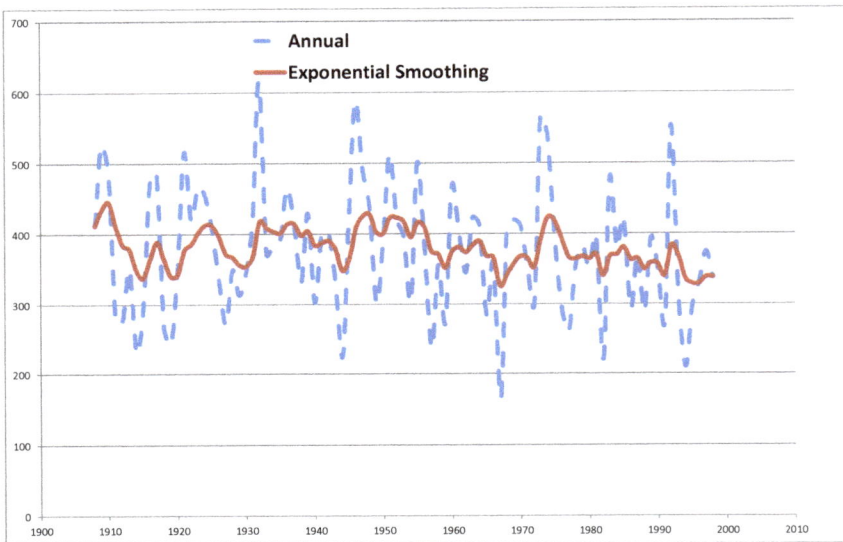

To summarise the rainfall analysis at the four locations, there is insufficient statistical evidence to conclude any difference in the mean rainfall before and after 1989 at any station. There is however, a definite pattern in the results. The trends at Mt Barker and Tailem Bend are consistent with the findings of Gallant *et al.* [6] of a diminishing rainfall in this area of the country, though not at as great a rate. Estimates of the change per decade in rainfall at the four stations are Murray Bridge (+9.5), Callington (+9.6), Mt Barker (−6.8) and Tailem Bend (−8.2). Note that, as expected, the slopes are not statistically significant, but are on the other hand actual estimates of change. Also, the changes are positive near the plateau and negative further away, consistent with an influence there. And, as has been emphasized, if one tests that the changes are consistent with the findings of Gallant *et al.* [6] for the stations near the plateau, this hypothesis is rejected.

5. Cross Checking with Other Locations in South Australia

It is important to make sure that the analysis has not been selective. Thus, yearly rainfall totals were accessed from a number of other sites across southern South Australia and examined whether one could reasonably infer that there was any trend upwards or downwards in any of them. From Figures 8–13, it can be reliably concluded that in none of these locations is there any evidence of any long term increase in rainfall. Nor can one conclude that there is any evidence of recent increases as in Murray Bridge and Callington. It appears to be some evidence of recent decrease in four cases (Tumby Bay, Minlaton, Glen Osmond and Keith) and essentially level over time in Port Wakefield and Loxton. These inferences will have to be monitored as more data comes to hand. It will be part of a future exercise to use satellite images, aerial maps and so on to try an understand the land use and land cover changes over time in these areas and see how they may relate to the rainfall trends.

Figure 8. Annual rainfall totals for Tumby Bay, 34.38° S, 136.10° E, for 1906–2011.

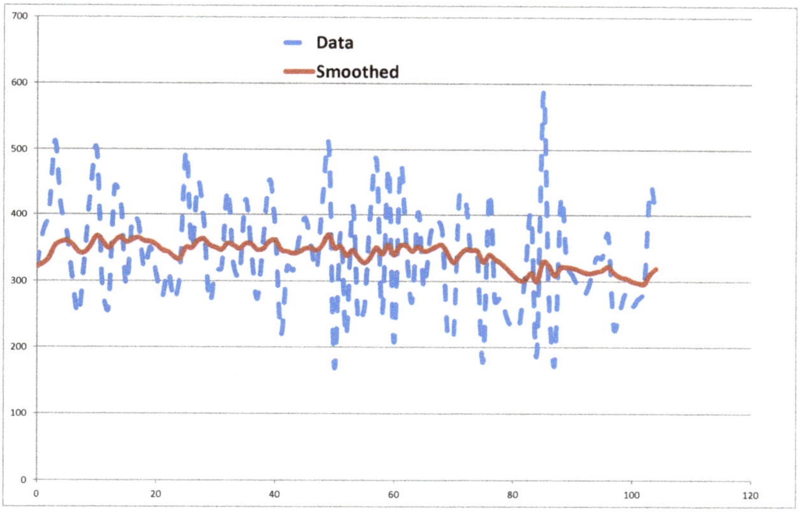

Figure 9. Annual rainfall totals for Minlaton, 34.77° S, 137.60° E, for 1880–2013. Note that exponential smoothing was not done for this location as there are some missing data.

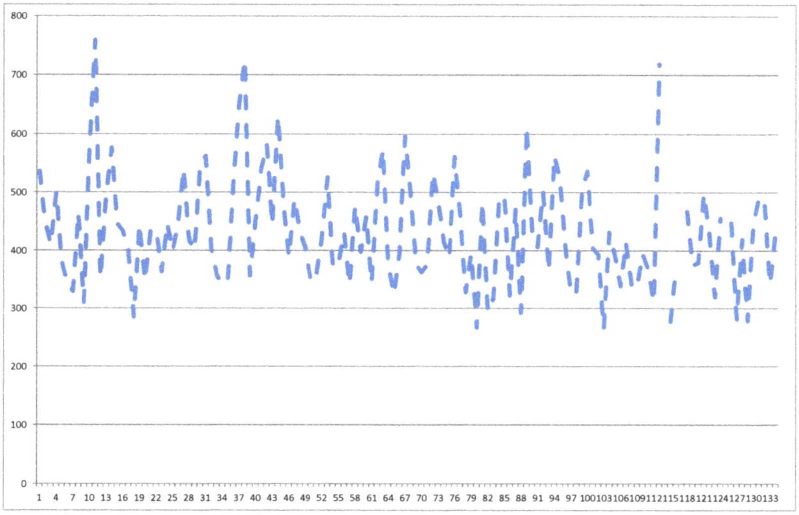

102

Figure 10. Annual rainfall totals for Port Wakefield, 34.18° S, 138.15° E, 1874–2013.

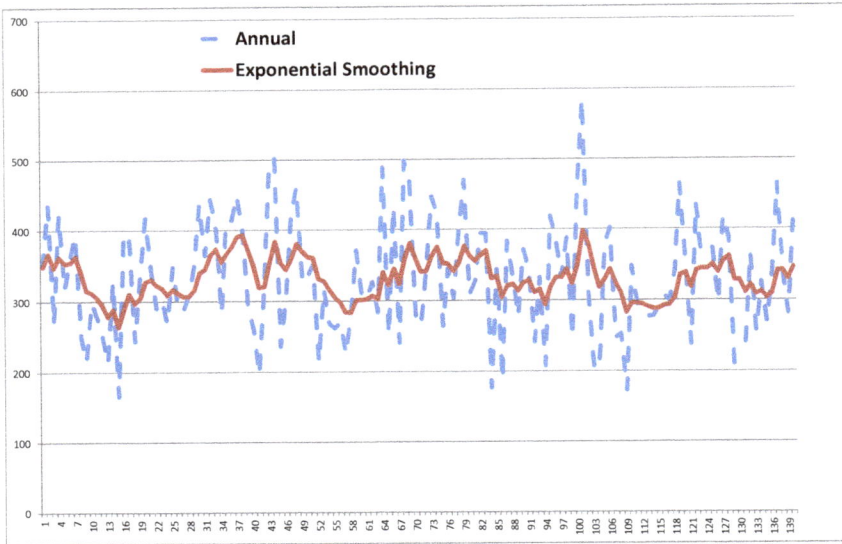

Figure 11. Annual rainfall totals for Glen Osmond, 34.95° S, 138.65° E, for 1884–2013. Note that exponential smoothing was not done for this location as there are some missing data.

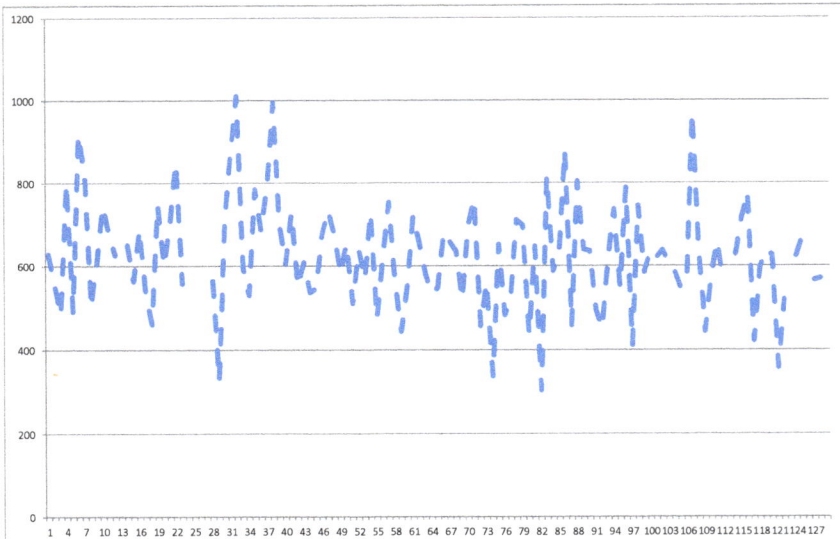

Figure 12. Annual rainfall totals for Loxton, 34.44° S, 140.50° E, for 1897–2002.

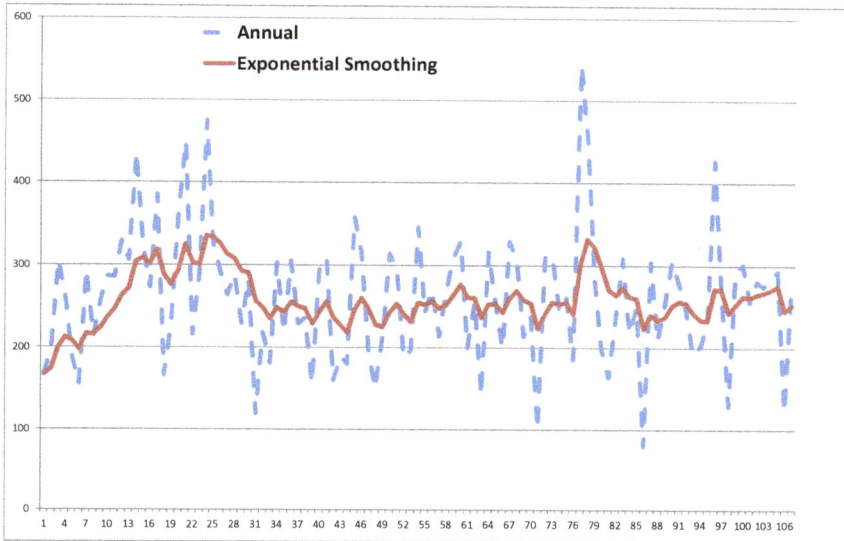

Figure 13. Annual rainfall totals for Keith, 36.10° S, 140.36° E, for 1907–2013.

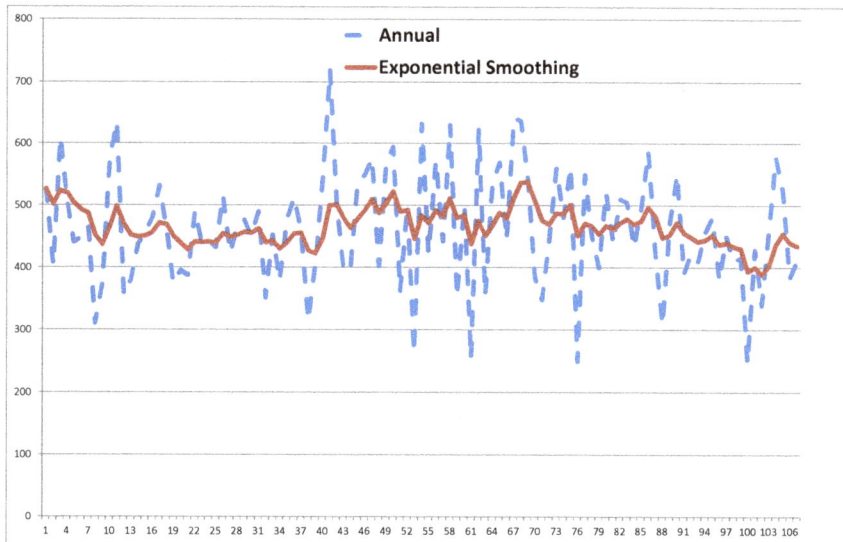

6. Possible Complicating Factors

There are some issues that must be considered in this study that may affect the conclusions:

- It may be surmised by some that when one sees rainfall trending down in any of the locations, it is due to the impacts of climate change on rainfall - see cf. Gallant *et al.* [6]. That may well be, but in fact if that were the case, then to actually have rainfall trending up, it is even easier to accredit land use change as a factor.

- Similarly, if one can attribute changes in rainfall to climate change plus land use change, the conclusion can still be made that what is happening at Murray Bridge and Callington is most probably due to positive land use change.
- What if one has differing things happening in different seasons, and it must be noted that often this type of analysis is performed season by season? Any investigations done on seasonal totals mirror in general what has been reported on an annual basis. This has not been checked for every location, but where it is, this has been the result.

7. Conclusions

This has been an investigation to add some knowledge to the interplay between vegetation and rainfall. As was stated, there has been a significant amount of work on how rainfall can diminish after clearance of large areas of vegetation. This continues to be substantiated. In a recent comprehensive analysis of the depletion of rainfall in Western Australia, Andrich and Imberger [22] state that the reduction of native vegetation from 60% to 30% of the land area in the wheatbelt, between 1950 and 1970, coincided with an average 21% reduction in inland rainfall relative to coastal rainfall. The coastal part did not experience land-use change and the rainfall remained stationary over the same period. It was found that for the forested coastal strip region south of Perth, land clearing that removed 50% of the native forests between 1960 and 1980 coincided with a 16% reduction in rainfall relative to stationary coastal rainfall.

There has been less work reported on how rainfall might be enhanced by strategic planting schemes. What is meant by strategic is selection of sites that will aid in interference with weather patterns in order to add to the rainfall totals where they are needed. Interestingly, Andrich and Imberger [22] go on in their conclusions to state that there is a pressing need to undertake large scale reforestation with native trees to mitigate the long term changes in climate. They even coin the phrase **reforestation for water production**, which should be taken up as a catchcry.

Fortuitously, the plantings that were made in the Monarto area in the 1970s seem to have been by chance of great enough extent and in a sensible area to be able to enhance the rainfall in the surrounding towns of Murray Bridge and Callington. In August, 2013 strategic planting schemes near Cummins on Eyre Peninsula have begun. It is hoped that benefits of this will accrue in the future.

Acknowledgments

This work is supported by an Australian Research Council (ARC) Discovery Indigenous Researcher Development Grant DI110100028-Indigenous knowledge: water sustainability and wild fire mitigation, and the Goyder Institute for Water Research.

Conflicts of Interest

The author declares no conflict of interest.

References

1. Junkermann, W.; Hacker, J.; Lyons, T.; Nair, U. Land use change suppresses precipitation. *Atmos. Chem. Phys.* **2009**, *9*, 6531–6539.
2. Allan, R.J.; Haylock, M.R. Circulation features associated with the winter rainfall decrease in Southwestern Australia. *J. Clim.* **1993**, *6*, 1356–1367.
3. Smith, I.N.; McIntosh, P.; Ansell, T.J.; Reason, C.J.C.; McInnes, K. Southwest Western Australian winter rainfall and its association with Indian Ocean climate variability. *Int. J. Climatol.* **2000**, *20*, 1913–1930.
4. Climate Change and Water. In *Technical Paper of the Intergovernmental Panel on Climate Change*; Bates, B.C., Kundzewicz, Z.W., Wu, S., Palutikof, J.P., Eds.; IPCC Secretariat: Geneva, Switzerland, 2008; p. 210.
5. Sindicic, M.; STREAMCARE Working Group. *Cummins-Wanilla Basin Catchment Management Plan*; Department of Environment, Water and Natural Resources: Port Lincoln, Australia, 2002.
6. Gallant, A.; Hennessy, K.; Risbey, J. Trends in rainfall indices for six Australian regions: 1910–2005. *Aust. Met. Mag.* **2007**, *56*, 223–239.
7. Sadler, B. Climate variability and change in south west Western Australia, Indian Ocean Climate Initiative Panel, c/-Department of Environment, Water and Catchment Protection, WA. Available online: http://www.waterandclimateinformationcentre.org/resources/8012007 IOCI2002.pdf (accessed on 25 February 2014).
8. Pitman, A.J.; Narisma, G.T.; Pielke, R.A.; Holbrook, N.J. Impact of land cover change on the climate of southwest Western Australia. *J. Geophys. Res.* **2004**, *109*, doi:10.10292003JD004347.
9. Pielke, R.A.; Pitman, A.; Niyogi, D.; Mahmood, R.; Mcalpine, C.; Hossain, F.; Goldewijk, K.K.; Nair, U.; Betts, R.; Souleymane, F.; *et al.* Land use/land cover changes and climate: Modeling analysis and observational evidence. *Wiley Interdiscip. Rev. Climate Change* **2011**, *2*, 828–850.
10. Chikoore, H.; Jury, M.R. Intraseasonal variability of satellite-derived rainfall and vegetation over Southern Africa. *Earth Interact.* **2010**, *14*, 1–26.
11. Wang, G. A conceptual modeling study on biosphere-atmosphere interactions and its implications for physically based climate modeling. *J. Clim.* **2004**, *17*, 2572–2583.
12. Woodward, F.I. *Climate and Plant Distribution*; Cambridge University Press: Cambridge, UK, 1987.
13. Zeng, N.; Neelin, J.D.; Lau, K.-M.; Tucker, C.J. Enhancement of interdecadal climate variability in the Sahel by vegetation interaction. *Science* **1999**, *286*, 1537–1540.

14. Spracken, D.V.; Arnold, S.R.; Taylor, C.M. Observations of increased tropical rainfall preceded by air passge over forests. *Nature* **2012**, *49*, 282–285.

15. Makarieva, A.M.; Gorshkov, V.G.; Li, B.-L. Precipitation on land versus distance from the ocean: Evidence for a forest pump of atmospheric moisture. *Ecol. Complex.* **2009**, *6*, 302–307.

16. Beltran-Przekurat, A.; Pielke, R.A., Sr.; Eastman, J.L.; Coughenour, M.B. Modelling the effects of land-use and land-cover changes on the near-surface atmosphere in southern South America. *Int. J. Climatol.* **2012**, *32*, 1206–1225.

17. Monbiot, G. Drowning in Money. *The Guardian*, 13 January 2014.

18. The Pontbren Project. Available online: http://www.pontbrenfarmers.co.uk/ (accessed on 25 February 2014).

19. Marshall, M.R.; Ballard, C.E.; Frogbrook, Z.L.; Solloway, I.; McIntyre, N.; Reynolds, B.; Wheater, H.S. The impact of rural land management changes on soil hydraulic properties and runoff processes: Results from experimental plots in upland UK. *Hydrol. Process* **2014**, *28*, 2617–2629.

20. Ryan, J.G.; McAlpine, C.A.; Ludwig, J.A. Integrated vegetation designs for enhancing water retention and recycling in agroecosystems. *Landsc. Ecol.* **2010**, *25*, 1277–1288.

21. O'Loughlin, E.M.; Nambiar, E.K.S. *Plantations, Farm Forestry and Water–A Discussion Paper*; Rural Industries Research and Development Corporation: Canberra, Australia, 2001

22. Andrich, M.A.; Imberger, J. The effect of land clearing on rainfall and freshwater resources in Western Australia: A multifunctional sustainability analysis. *Int. J. Sustain. Dev. World Ecol.* **2013**, *20*, 549–563.

Climatic Characteristics of Reference Evapotranspiration in the Hai River Basin and Their Attribution

Lingling Zhao, Jun Xia, Leszek Sobkowiak and Zongli Li

Abstract: Based on the meteorological data from 46 stations in the Hai River Basin (HRB) from 1961–2010, the annual and seasonal variation of reference evapotranspiration was analyzed. The sensitivity coefficients combined with the detrend method were used to discuss the dominant factor affecting the reference evapotranspiration (ET_o). The obtained results indicate that the annual reference evapotranspiration is dominated by the decreasing trends at the confidence level of 95% in the southern and eastern parts of the HRB. The sensitivity order of climatic variables to ET_o from strong to weak is: relativity humidity, temperature, shortwave radiation and wind speed, respectively. However, comprehensively considering the sensitivity and its variation strength, the detrend analysis indicates that the decreasing trends of ET_o in eastern and southern HRB may be caused mainly by the decreasing wind speed and shortwave radiation. As for the relationship between human activity and the trend of ET_o, we found that ET_o decreased more significantly on the plains than in the mountains. By contrast, the population density increased more considerably from 2000 to 2010 on the plains than in the mountains. Therefore, in this paper, the correlation of the spatial variation pattern between ET_o and population was further analyzed. The spatial correlation coefficient between population and the trend of ET_o is -0.132, while the spatial correlation coefficient between the trend of ET_o and elevation, temperature, shortwave radiation and wind speed is 0.667, 0.668, 0.749 and 0.416, respectively. This suggests that human activity has a certain influence on the spatial variation of ET_o, while natural factors play a decisive role in the spatial variation of reference evapotranspiration in this area.

Reprinted from *Water*. Cite as: Zhao, L.; Xia, J.; Sobkowiak, L.; Li, Z. Climatic Characteristics of Reference Evapotranspiration in the Hai River Basin and Their Attribution. *Water* **2014**, *6*, 1482-1499.

1. Introduction

Hydrologists have found that climate change has resulted in some changes in the water cycle [1–3]. One major challenge of recent hydrological modeling activities is the assessment of the effects of climate change on the terrestrial water cycle [4]. Hydrological models are usually based on the calculation of reference evapotranspiration and reducing it to the actual evapotranspiration by considering the soil moisture status [5] or the number of days since the last rainfall event [6]. Therefore, analyzing how climate change affects reference evapotranspiration (ET_o) is critical for understanding the impact of climate change on the hydrological cycle. According to Allen *et al.* [7], ET_o is the evapotranspiration from the reference surface, which is a hypothetical grass reference crop with an assumed crop height of 0.12 m, a fixed surface resistance of 70 s m^{-1} and an albedo of 0.23.

The Hai River Basin (HRB) is one of seven largest river basins and also one of the most developed areas in China, with the population accounting for about 10% of the nation's total. The middle and lower reaches of the basin are important wheat production regions in China. This region has a semi-humid and semi-arid climate and has been strongly influenced by human activities. The annual precipitation is 539 mm, while the annual pan evaporation is 1100 mm, making the basin vulnerable to climatic variations [7]. In recent decades, several eco-environmental problems in that area have come to the fore under the combined impacts of climate change and intensified human activities. Water resources in the HRB are currently used for irrigation, aquaculture and industries. Due to the very limited available water resources in the basin, water has been diverted from other basins to supply it for agriculture and to maintain essential ecosystem functions [8].

In order to understand how climatic variables affect ET_o, some studies have been carried out to evaluate evapotranspiration in the context of climate change. Zheng *et al.* [9] analyzed the cause of the decreased pan evaporation during 1957–2001 in the HRB, and found the reason to be the declining wind speed. Xu *et al.* [10] proved that the decreasing wind speed and net radiation were responsible for the ET_o changes in the Changjiang River Basin of China. Liu [11], who analyzed the pan evaporation from 1955 to 2000 in China, found that the decrease in solar radiance was most likely the driving force of the reduced pan evaporation in China. Furthermore, sensitivity analysis has also been performed on the impacts of climate change [12–18]. However, the temporal pattern of ET_o is not only influenced by the sensitivity of climatic variables, but also by their variation patterns.

The spatial pattern of ET_o in HRB has not been addressed in the literature, yet. In this study, we calculated ET_o using the FAO-56 Penman–Monteith equation and analyzed the temporal-spatial pattern in ET_o and its driving variables. Attribution analysis was then performed to quantify the contribution of each input variable to ET_o variation. The objective of this paper is to exhibit the temporal-spatial variation pattern of ET_o over the past 50 years in the HRB, then to detect the reason for these characteristics and to quantify the contribution of the climatic variation to ET_o.

2. Study Area and Data

The Hai River Basin is located in north China and is one of seven largest river basins in the country. The basin is bounded in the north by Mount Tangshan, in the west by Mount Taihang and in the east by the Bohai Sea. Land surface elevation in the mountainous north and west of the study area is generally above 2000 m a.s.l. On the floodplains, however, surface elevation hardly exceeds 100 m a.s.l. The basin occupies an area of 3.2×10^5 km^2 (34.9–42.8° N, 112.0–119.8° E) and includes five provinces and the two megacities of Beijing and Tianjin (Figure 1). Climatically, the HRB belongs to the East Asian monsoon region. The annual mean temperature varies from 8 °C to 12 °C, while annual precipitation is about 539 mm; relative humidity varies from 50% to 70%.

Data from 46 National Meteorological Observatory stations included daily observations of the maximum, minimum and mean air temperatures (T_{max}, T_{min}, T_a), wind speed (U), relative humidity (Rh) and sunshine duration (n) for the period from 1960 to 2010 and pan evaporation (E_{pan}) for 1960–2001. E_{pan} was measured using a metal pan, 20 cm in diameter and 10 cm high, installed 70 cm above the ground. The data have been provided by the National Climatic Center of China

Meteorological Administration. The locations of the stations are shown in Figure 1, while the details of the stations are listed in Table 1. In Table 1, the annual ET_o is the average value from 1960 to 2010 calculated using the FAO-56 Penman–Monteith method.

Figure 1. The location of the Hai River Basin (HRB).

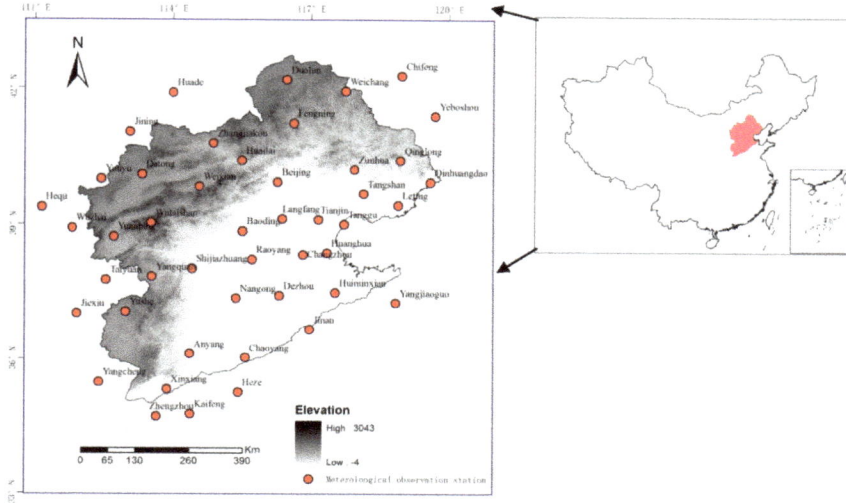

Table 1. Basic data on the investigated stations in HRB.

Name	No.	Longitude (°)	Latitude (°)	Elevation (m)	Name	No.	Longitude (°)	Latitude (°)	Elevation (m)
Wutaishan	53588	113.53	39.03	2896	Changzhou	54616	116.83	38.33	10
Weixian	53593	114.57	39.83	910	Tanggu	54623	117.72	39.00	3
Yuanping	53673	112.72	38.73	828	Huanghua	54624	117.35	38.37	7
Shijiazhuang	53698	114.42	38.03	81	Nangong	54705	115.38	37.37	27
Yangquan	53782	113.55	37.85	742	Dezhou	54714	116.32	37.43	21
Yushe	53787	112.98	37.07	1041	Huiminxian	54725	117.53	37.50	12
Anyang	53898	114.37	36.12	76	Chaoyang	54808	115.58	36.03	43
Xinxiang	53986	113.88	35.32	73	Huade	53391	114.00	41.90	1483
Duolun	54208	116.47	42.18	1245	Shiyu	53478	112.45	40.00	1346
Fengning	54308	116.63	41.22	660	Jiying	53480	113.07	41.03	1419
Weichang	54311	117.75	41.93	843	Hequ	53564	111.15	39.38	862
Zhuangjiakou	54401	114.88	40.78	724	Wuzhai	53663	111.82	38.92	1401
Huailai	54405	115.50	40.40	537	Taiyuan	53772	112.55	37.78	778
Zunhua	54429	117.95	40.20	55	Jiexiu	53863	111.92	37.03	744
Qinglong	54436	118.95	40.40	227	Yangcheng	53975	112.40	35.48	660
Qinhuangdao	54449	119.60	39.93	2	Chifeng	54218	118.97	42.27	568
Beijing	54511	116.28	39.93	54	Yeboshou	54326	119.70	41.38	662
Langfang	54518	116.38	39.12	9	Yangjiaogou	54736	118.85	37.27	6
Tianjin	54527	117.17	39.10	3	Jinan	54823	116.98	36.68	52
Tangshan	54534	118.15	39.67	28	Heze	54906	115.43	35.25	50
Leting	54539	118.90	39.42	11	Zhengzhou	57083	113.65	34.72	110
Baoding	54602	115.52	38.85	17	Kaifeng	57091	114.38	34.77	73
Raoyang	54606	115.73	38.23	19	Datong	53487	113.33	40.10	1067

110

3. Methodologies

3.1. Penman–Monteith Method

The Penman–Monteith method recommended by FAO (Food and Agriculture Organization) [19] as the standard method for determining reference evapotranspiration was used in this study. The method was selected because it is physically based and explicitly incorporates both physiological and aerodynamic parameters.

$$ET_0 = \frac{0.408\Delta(R_n - G) + \gamma\frac{900}{(T+273)}U_2(e_s - e_a)}{\Delta + \gamma(1 + 0.34U_2)} \tag{1}$$

where, ET_0 is the reference evapotranspiration (mm/day); R_n is net radiation at the crop surface (MJ/m^2/day); G is soil heat flux density (MJ/m^2/day); T is mean daily air temperature (°C), U_2 is wind speed at 2 m height (m/s); e_s is saturation vapor pressure (kPa); ($e_s - e_a$) is the saturation vapor pressure deficit (kPa); Δ is the slope of vapor pressure (kPa/°C) and γ is the psychometric constant (kPa/°C). The computation of all data required for the calculation and relevant procedures are given in Chapter 3 of the FAO Paper 56 [19].

3.2. Trend Detection and Sensitivity Analysis Method

The rank-based nonparametric Mann–Kendall statistical test [20,21] is commonly used for trend detection, because of its robustness for non-normally distributed and censored data, which are frequently encountered in hydroclimatic time series. In this method, the test statistic, Z, is as follows:

$$Z = S / \sigma_s^2 \tag{2}$$

$$\text{with } S = \sum_{i=1}^{n-1}\sum_{j=i+1}^{n} \text{sgn}(x_j - x_i) \tag{3}$$

$$\sigma_s^2 = \frac{n(n-1)(2n+5) - \sum_{i-1}^{n} e_i i(i-1)(2i+5)}{18} \tag{4}$$

$$\text{sgn}(x) = \begin{cases} 1 & \text{if } x > 0 \\ 0 & \text{if } x = 0 \\ -1 & \text{if } x < 0 \end{cases} \tag{5}$$

Equation (2) gives the standard deviation of S with correction for ties in the data, with e_i denoting the number of ties of extent i. The upward or downward trend in the data is statistically significant if $|Z| > \mu_{1-\alpha/2}$, where $\mu_{1-\alpha/2}$ is the $(1-\alpha/2)$ quantity of the standard normal distribution and when $\alpha = 0.05, u_{1-\alpha/2} = 1.96$. Positive Z indicates an increasing trend in the time series, while negative Z, a decreasing one.

Original measurements of air temperature (T_a), wind speed (U) and relative humidity (Rh) were chosen for the sensitivity analyses. The fourth applied variable is shortwave radiation (R_s). This is because shortwave radiation is one of the input variables in a number of semi-physical and semi-empirical equations that are used to derive the net energy flux required by the Penman method [22]. Following the procedure described by Allen [19], R_s can be estimated with the following formula that relates surface shortwave radiation to extraterrestrial radiation and daily sunshine duration:

$$R_s = (a_s + b_s \frac{n}{N})R_a \qquad (6)$$

where R_s is shortwave radiation, n is daily sunshine duration (h); N is maximum possible duration of sunshine or daylight hours (h); n/N is relative sunshine duration; R_a is extraterrestrial radiation and a_s and b_s are the regression constants. The recommended values $a_s = 0.25$ and $b_s = 0.75$ were used in this study.

In multivariate models, different variables have different dimensions and different ranges of values, which make it difficult to compare the sensitivity by partial derivatives. Consequently, the partial derivative is transformed into a non-dimensional form:

$$Sx = \lim_{\Delta x/x} \left(\frac{\frac{\Delta ETo}{ETo}}{\frac{\Delta x}{x}} \right) = \frac{\partial ETo}{\partial x} \cdot \frac{x}{ETo} \qquad (7)$$

Basically, a positive/negative sensitivity coefficient of a variable indicates that ET_o increases/decreases as the variable increases; the larger the sensitivity coefficient, the larger the effect a given variable has on ET_o.

3.3. Spatial Correlation Coefficient

Correlation coefficients depict the spatial relationship between two datasets. The correlation between two variables is a measure of dependency between these variables. It is the ratio of the covariance between the two datasets divided by the product of their standard deviations. Because it is a ratio, it is a unit-less number. The equation to calculate the correlation is [23]:

$$Corr_{ij} = \frac{Cov_{ij}}{\delta_i \delta_j} \qquad (8)$$

where, Cov_{ij} is the covariance; δ_i, δ_j are the standard deviations of dataset i and j, respectively.

The calculated covariance matrix in this paper contains values of variances and covariances. The variance is a statistical measure showing how much variance there is from the mean. The remaining entries within the covariance matrix are the covariances between all pairs of the input datasets. The following formula is used to determine the covariance between datasets i and j:

$$Cov_{ij} = \frac{\sum_{k=1}^{N}(Z_{ik} - \mu_i)(Z_{jk} - \mu_j)}{N-1} \qquad (9)$$

where, Z_{ik}, Z_{jk} are the values of dataset i and dataset j, respectively in location k; μ_i, μ_j are the average values of datasets i and j, respectively; i, j is the order of dataset; N is the number of dataset; K denotes a particular location.

Correlation ranges from +1 to −1. A positive correlation indicates a direct relationship between two datasets, such as when the cell values of one datasets increase, the cell values of another datasets are also likely to increase. A negative correlation means that one variable changes inversely to the other. A correlation of zero means that two datasets are independent of one another.

3.4. Detrend Method

The variation pattern of ETo is determined by multi-climatic variables, including their sensitivity to ETo and variation fluctuations. The detrend method is a combination method that considers both the sensitivity coefficient and the fluctuation of the climatic variables.

The detrend method is a way of quantifying the contribution of climatic variables to the annual variation of ET_o. This method shows the contribution in graphs and vividly describes how the climatic variables influence ET_o. In this study, the following steps were performed: (1) use of the simple linear regression method to detect the changing slope of the main climatic variables; (2) detection of the significance of the slope by the *t*-test; (3) removal of the significant slope of main climatic variables to make them stationary time series; (4) recalculation of reference evapotranspiration using each time the original series of three variables and the detrend data of one variable; (5) comparison of the results with the original reference evapotranspiration; the observed difference is considered as the influence of those variables on the trend [22].

4. Results and Discussion

4.1. Correlation between ET_o and E_{pan}

Figure 2 shows the monthly correlation coefficients (R^2) between ET_o and E_{pan} in HRB. As can be seen, the lowest correlation coefficient is 0.93, while the highest is 0.97. Spatially, mountain areas in the northwest and coastal areas in the southeast have higher correlation coefficients than plains in the central part of the study area. Mountain and coastal areas are more humid than plains, which is consistent with the research conclusion of Brutsaert [24] that in humid areas, reference evaporation has a better relationship with pan evaporation.

4.2. Spatial-Temporal Variation of ET_o in HRB

To study the spatial distribution of the trend of ET_o from 1960 to 2010 in the HRB, the Mann–Kendall test was used for each station to establish the ET_o trends. The trends of annual and seasonal ET_o were tested at the 95% confidence level. Decreasing trends in annual ET_o were observed at 28 stations located mostly on the eastern and southern plains of the HRB. However, increasing trends in the annual ET_o were observed at three sites (Datong, Wutaishan, Weixian) located in the western mountain region of the HRB (Figure 3).

Figure 2. Correlation between ET_o and E_{pan} in the HRB.

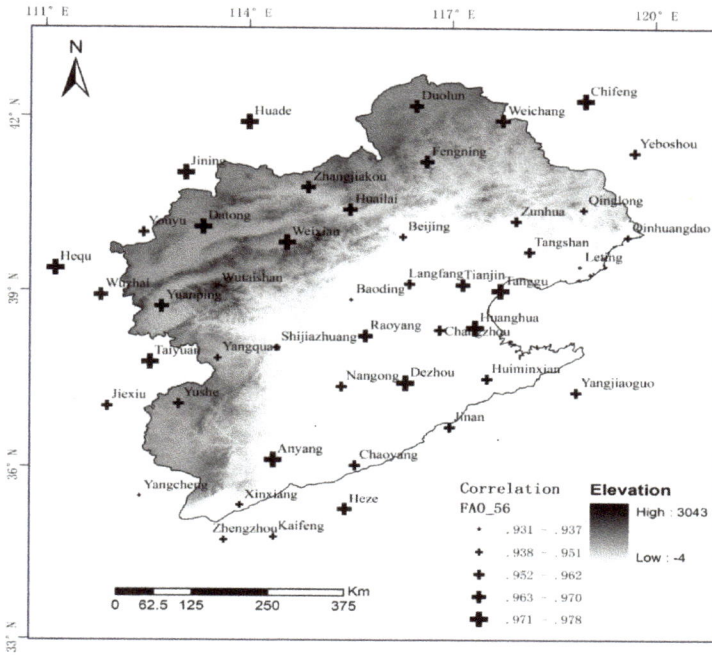

Figure 3. The annual trend of ET_o in the HRB.

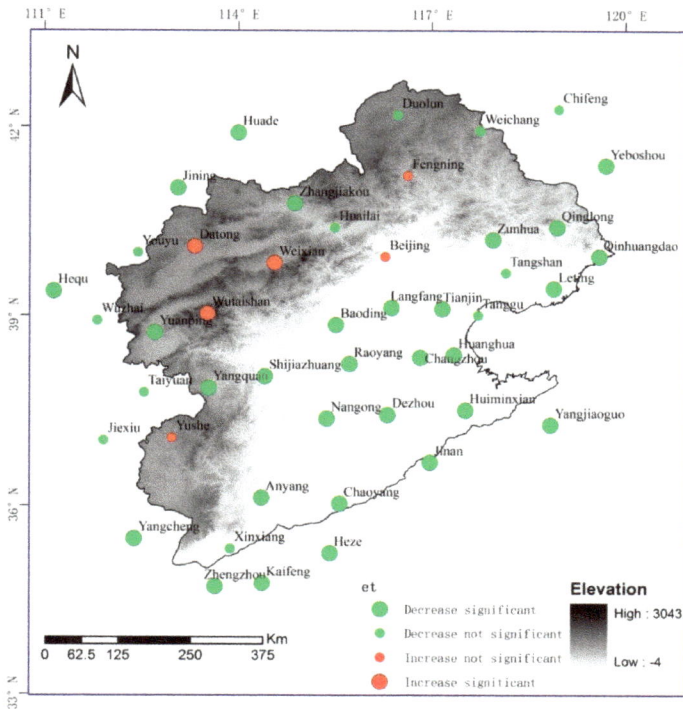

As for the variation pattern of seasonal ET_0, Figure 4 shows large differences in the spatial distribution of trends in seasonal ET_0 in the HRB. As can be seen, stations with decreasing trends in spring are mainly distributed in the southeastern plain region of the study area. The significantly increasing trend of ET_0 was found only at Wutaishan station in the western mountain region of the HRB. Twenty-nine stations characterized by decreasing trends of ET_0 in summer are concentrated mainly in the eastern and southern regions of the HRB. Only one station is dominated by the increasing trend of ET_0 in summer. In autumn, only nine out of 46 stations show a trend in ET_0. Among them, eight stations present decreasing trends, while one station shows an increasing trend. The changing trend patterns in winter are similar to those in autumn: 16 out of 46 stations are dominated by the decreasing winter ET_0, while two stations in the western plateau region display the increasing trend.

4.3. Variation Pattern of Climatic Variables

Shortwave radiation decreases in the whole basin, and most trends are significant at the 0.05 significance level (Figure 5). The maximum and minimum temperatures increased in the whole HRB (Figure 5). The maximum temperatures in the southern mountain area increase more significantly than on the northern plains; the average p-value of the Mann–Kendall test for the whole basin is 2.9 (Table 2). The minimum temperatures increase more obviously than the maximum ones; the average p-value of the Mann–Kendall test for the whole basin is five. Relative humidity decreases in most of the basin, and this trend is significant in most sites of the mountain areas. As to the wind speed, the decreasing trends are significant in most parts of the basin at the 0.05 significance level.

Figure 4. The seasonal trend of ET_0 in the HRB. (**a**) Spring; (**b**) Summer; (**c**) Winter; (**d**) Autumn.

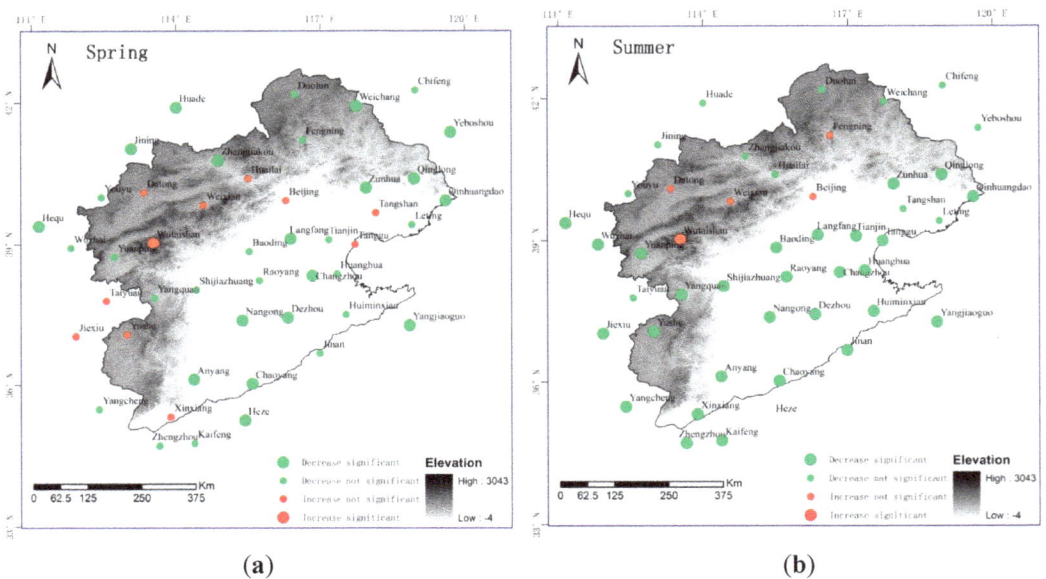

(a)

(b)

Figure 4. *Cont.*

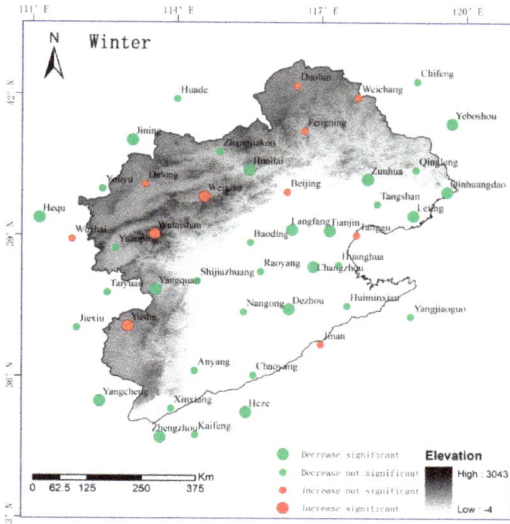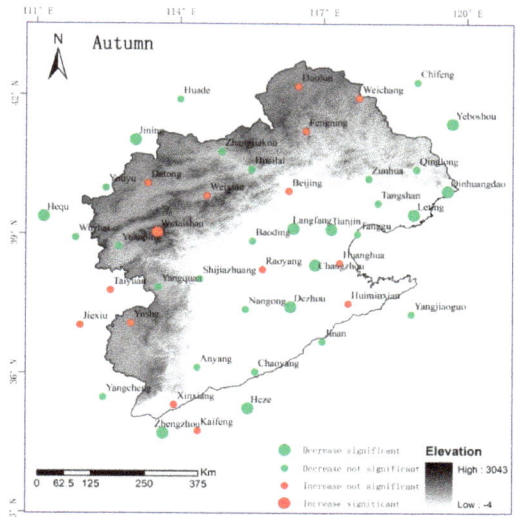

(c)   (d)

Figure 5. The trend of main climatic variables in the HRB. (**a**) The trend of T_a; (**b**) The trend of R_s; (**c**) The trend of U; (**d**) The trend of Rh.

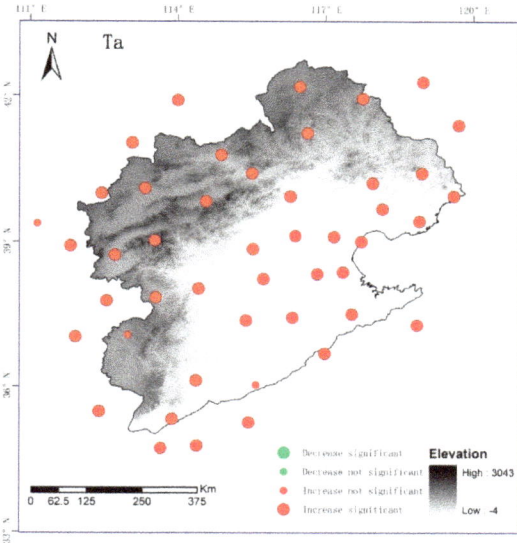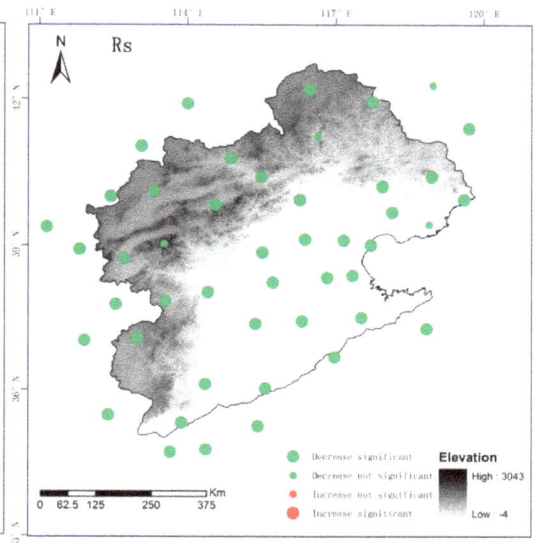

(a)   (b)

Figure 5. *Cont.*

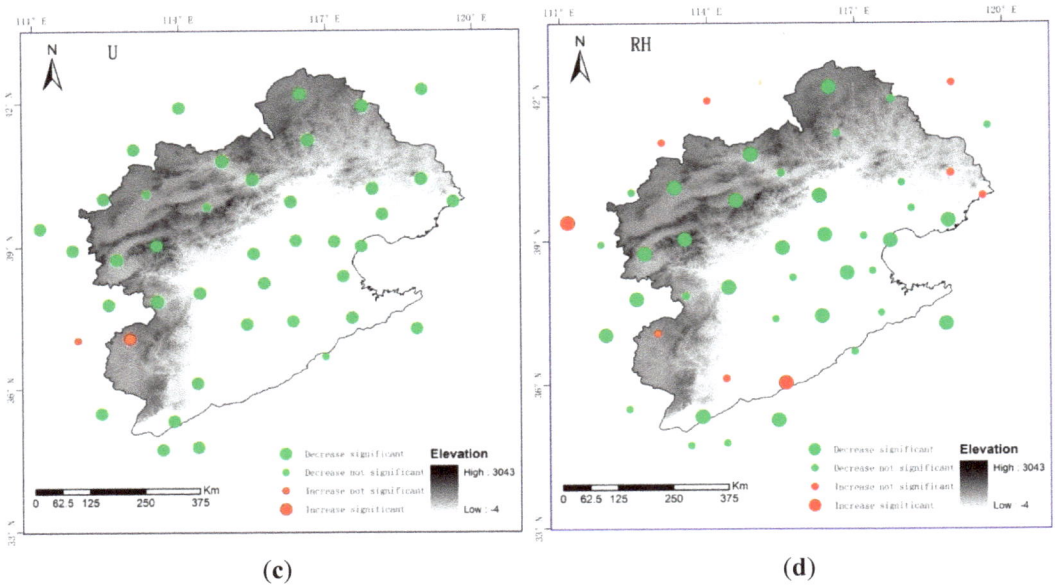

(c)

(d)

Table 2. The Mann–Kendall test value of the main climate variables at the three stations.

Station name	T_a	U	Rh	R_s
Datong	4.78	−1.55	−1.81	−2.54
Wutaishan	4.30	−5.21	−3.57	−0.54
Weixian	4.87	−0.92	−3.07	−2.04
Mean value of the HRB	4.16	−4.12	−1.41	−3.94

The increasing temperature and decreasing relative humidity will make ET_o increase, while the decreasing short wave radiation and wind speed will make ET_o decrease. Therefore, the decreasing trends of ET_o in the HRB suggest that the increasing temperature and decreasing relative humidity slightly discount the decreasing trend, but do not change its direction; the decreasing shortwave radiation and wind speed commonly result in the decreasing ET_o in the southeastern coastal area of the HRB.

In order to detect the reasons for the increasing trend of ET_o in Datong, Wutaishan and Weixian stations, this paper analyzed the climate variables at these three locations. We found a Mann–Kendall test value of temperature as large as at the other investigated stations. However, relative humidity decreased faster, while shortwave radiation and wind speed (except Wutaishan) were slower compared to the mean value calculated for the whole HRB. Therefore, it can be concluded that the increasing temperature, decreasing faster relative humidity and decreasing slower shortwave radiation and wind speed result in the increasing trend of ET_o at these three stations.

4.4. Sensitivity of Climatic Variables

Since ET$_o$ is an important indicator of climatic changes, the sensitivity coefficient was used in this study to analyze how climatic variables affect ET$_o$. The non-dimensional form of the sensitivity coefficient was employed to estimate the sensitivity of climatic variables in the HRB. Figure 6 gives the annual sensitivity coefficient of four climatic variables (T_a, Rh, R_s and U) to ET$_o$ from 1960 to 2010 estimated by the FAO-56 Penman–Monteith method. Figure 6 suggests that temperature, wind speed and short wave radiation are less sensitive in the mountain areas than on the coastal plains, while relative humidity is less sensitive in the area closest to the coast than on the plains and in the mountains.

As for the trend of sensitivity to the climatic variables, the decreasing trends are found in temperature, short wave radiation and relative humidity, while the increasing trend is in wind speed. As variations in sensitivity coefficients to main climatic variables are detected before and after 1990, so the data series of sensitivity coefficients were divided into two groups of 1960–1990 and 1991–2010, respectively, to calculate mean sensitivity coefficients to climatic variables. The obtained results show that before 1990, the sensitivity coefficients to T_a, Rh, U and R_s were 0.500, −0.641, 0.205 and −0.357, respectively, while after 1990, they were 0.477, −0.586, 0.221 and 0.349, respectively.

Figure 6. The sensitivity coefficients of climatic variables in the HRB. (**a**) The sensitivity coefficients of T$_a$; (**b**) The sensitivity coefficients of Rh; (**c**) The sensitivity coefficients of U; (**d**) The sensitivity coefficients of R$_s$.

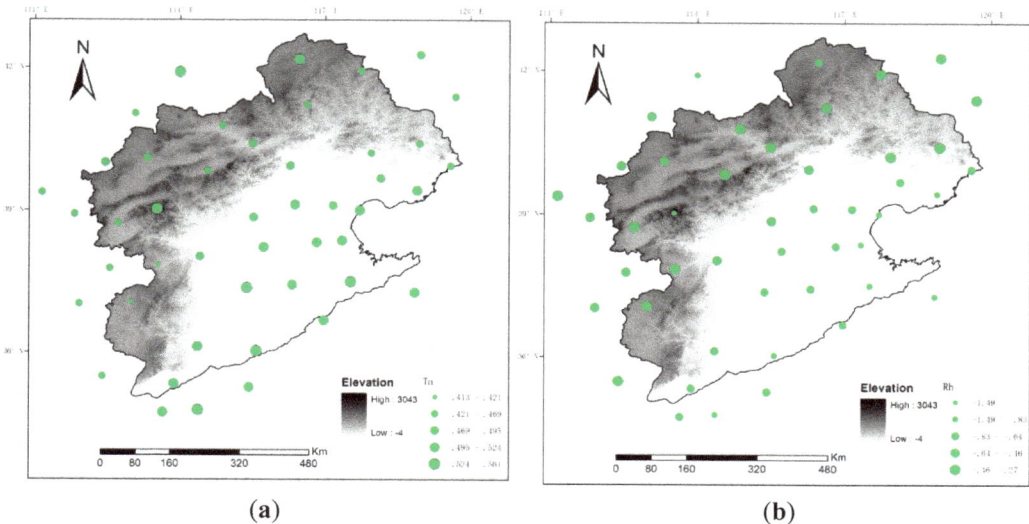

(a)   (b)

Figure 6. *Cont.*

(c)

(d)

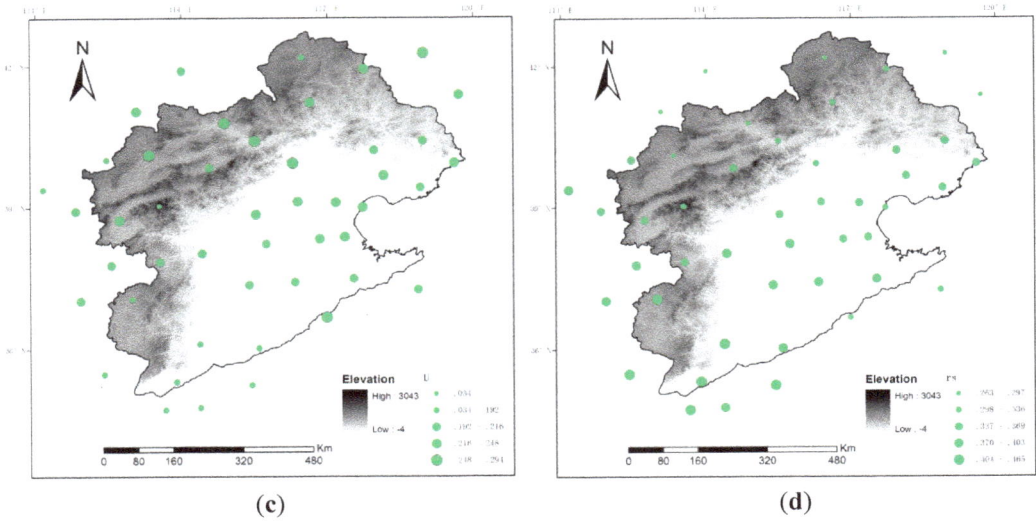

4.5. ET₀ with Detrend Climatic Variables

Figure 7 gives the trend of ET_o estimated with the detrend data series. The trends of ET_o with detrend temperature are decreasing, except one station, while the variation pattern of ET_o with detrend relative humidity is similar to the original ET_o. This shows that there are larger differences between the original ET_o and the recalculated one with detrend wind speed or detrend shortwave radiation than that with the detrend temperature or relative humidity. This suggests that the decreasing wind speed and shortwave radiation may be the main causes of the decreasing ET_o in the HRB. As for the decreasing trend of shortwave radiation, previous studies have shown that the decrease in global radiation is the most likely the cause, which is a regional phenomenon. By examining the regional total radiation in eastern China, Zhang [25] concluded that the regional total radiation is decreasing due to the increased air pollution in that area. Another study by Liu [26] also proved that air pollution may result in the decrease of R_s in HRB. Therefore, they speculate that aerosols may play a critical role in the decrease of solar radiation in China.

Figure 7. The annual trend of ET_o with detrend climatic variables in the HRB. (**a**) The annual trend of ET_o in the HRB; (**b**) The trend of ET_o with detrend U; (**c**) The trend of ET_o with detrend T_a; (**d**) The trend of ET_o with detrend R_s; (**e**) The trend of ET_o with detrend Rh.

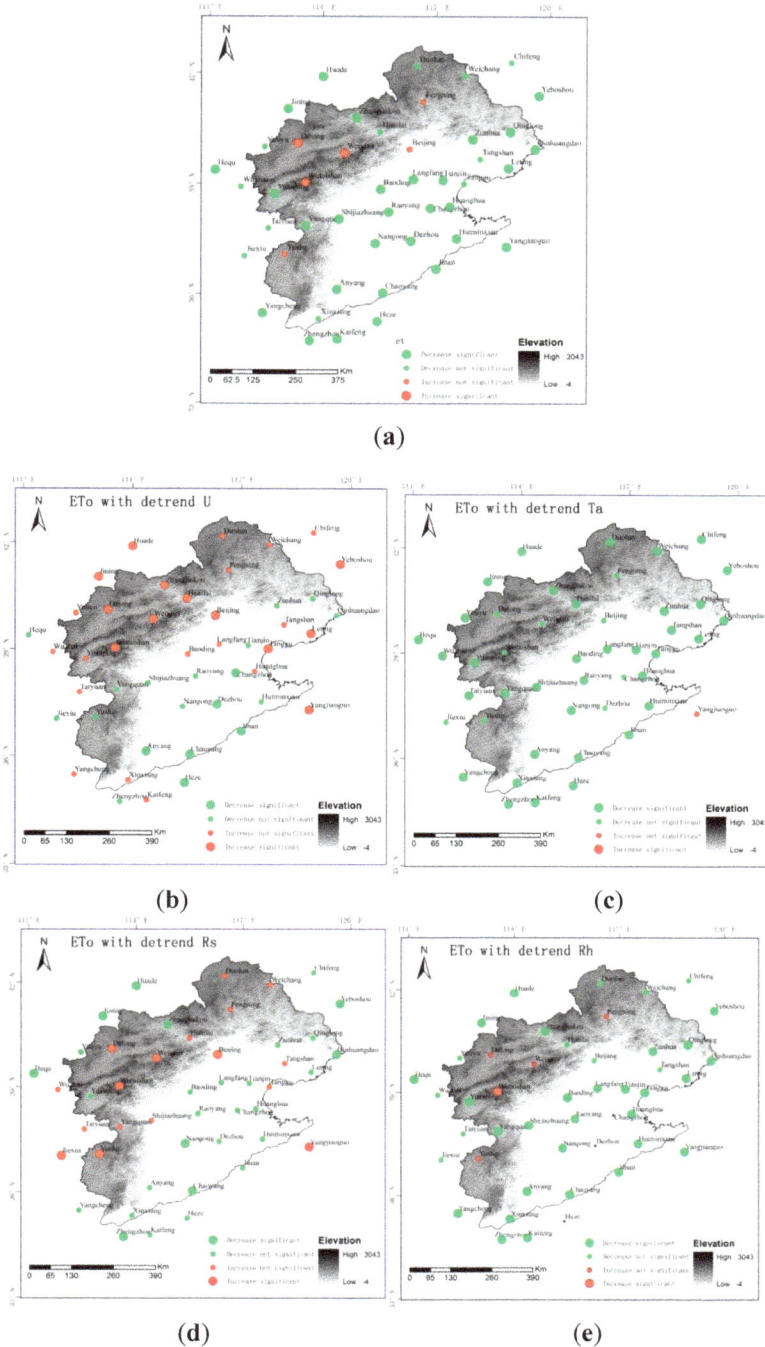

(**a**)

(**b**)

(**c**)

(**d**)

(**e**)

4.6. Relationship between ET₀ and Human Activity

The relationship between the variation pattern of ET_o and human activity refers to how human activity affects climatic variables. In order to detect the relationship between human activity and the variation of climatic variables, the investigated sites were divided into two groups; elevation was taken as the sole criterion of that division. Areas with an elevation higher than 500 m a.s.l. are usually defined as mountains, while below 500 m a.s.l. as plains. In general, the population density is higher and human activity is more intense on the plains than in the mountain areas. Figure 8 gives the GDP (Gross Domestic Product) increase from 2000 to 2010 and the distribution of plains in the HRB, respectively. As can be seen, the GDP increased noticeably from 2000 to 2010 on the plains of the HRB. This may suggest that human activity is concentrated on the plains. At the same time, the decrease of ET_o is more visible on the plains than in the mountains.

Figure 8. The GDP variation and distribution of plain areas in the HRB. (**a**) GDP variation during 2010 to 2000 in HRB; (**b**) DEM below 500 m a.s.l. in HRB.

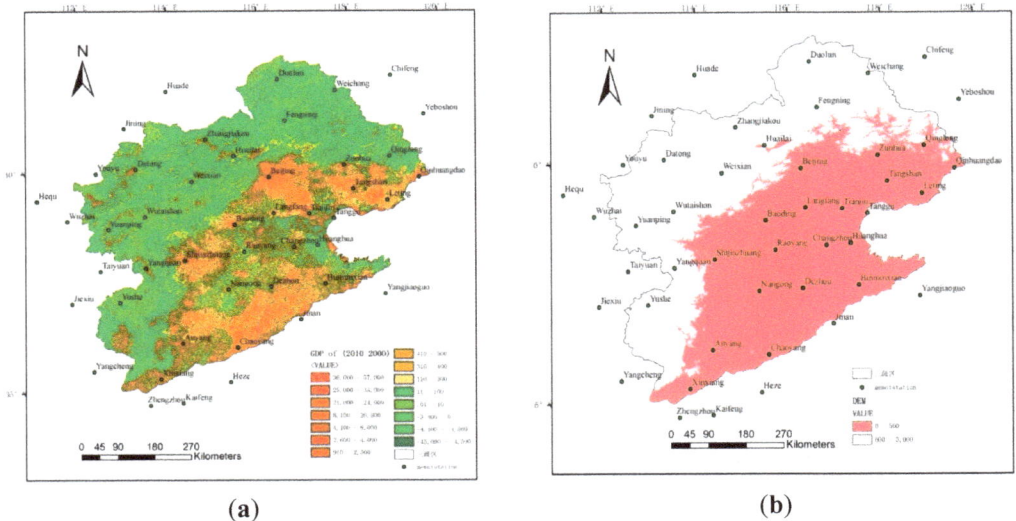

(a)   (b)

Table 2 gives the trend of the investigated climatic variables in these two elevation groups. It can be seen that the trend of the shortwave radiation, wind speed and minimum temperature on the plains is more obvious than that in the mountain areas. The detrend results show that the decreasing shortwave radiation and wind speed are the main causes of the decreasing ET_o on the plains of the HRB. Therefore, we deduce that there is a relationship between the human activity and the decreasing shortwave radiation and wind speed. Liu *et al.* [26] found that during the period from 1957 to 2008, the solar radiation decreased significantly in the HRB and that the trend was more significant in the densely populated areas than in the sparsely populated ones. The spatial distribution of the aerosol index increase is consistent with the solar radiation decrease. The aerosol increase resulting from human activities was an important reason for the decrease in solar radiation. This phenomenon was also found by some other studies [27–29]. As for the wind speed,

Jiang *et al.* [30] analyzed the change of wind speed over China from 1956 to 2004. They found that the effect of urbanization on annual MWS (Mean Wind Speed) is more and more obvious with time, especially in the 1980s. It corresponds with the substantial development of urbanization. At the same time, the declining trend of mean wind speed at the urban stations is more serious than that at the rural ones over the last 50 years. This is coincident with the results given in Table 3.

Table 3. The trend of climatic variables in different elevation groups.

Elevation	R_s	T_{max}	T_{min}	Rh	U
>500 m a.s.l.	−3.689	3.698	4.727	−1.157	−3.773
<500 m a.s.l.	−4.601	2.170	5.198	−1.348	−4.739

Furthermore, in order to determine the relationship between ET_o and human activities, this paper analyzed with the use of the spatial correlation coefficient method the spatial correlation between ET_o, population and meteorological factors, which have a definite effect on ET_o. The results show that the spatial correlation coefficient between population and ET_o is −0.132, while the spatial correlation coefficients between population, R_s and Rh are −0.307 and −0.144, respectively. Moreover, the spatial correlation coefficients between elevation and ET_o is 0.667, while the spatial correlation coefficients between ET_o, R_s, U and T_a are 0.749, 0.416 and 0.668, respectively. This proves that human activity has a certain influence on the spatial variation of ET_o, while natural factors play a decisive role in the spatial variation character of reference evapotranspiration (Table 4).

Table 4. The spatial correlation coefficients between ET_o and different factors.

R_s	U	T_a	Rh	Population	Elevation
0.749	0.416	0.668	−0.267	−0.132	0.667

With increasing altitude and decreasing pressure, the atmosphere is getting thinner and the heat radiation losses become faster. Therefore, the temperature decreases with increasing altitude. At the same time, when the sunlight scattering and reflection decrease, the solar radiation increases. While the proportion of water vapor in the air is relatively large, it is usually concentrated in the lower parts of the atmosphere. Relatively small air moisture content at high altitudes results in low relative humidity there. Friction caused by the uneven Earth surface retards the air flow, so that the wind speed is relatively smaller near the ground. With higher elevation (up to 6000 m a.s.l.), the wind speed becomes less impacted by the uneven ground.

5. Conclusions

In this paper, the spatial and temporal characteristics of annual and seasonal ET_o (1960–2010) in the Hai River Basin (HRB) were examined, and the possible causes of changes in ET_o were detected. The following conclusions may be drawn from the study:

(1) Most stations in the HRB have decreasing trends in the annual ET_o at a confidence level of 95%. These stations are distributed mainly in the southern and eastern coastal areas of HRB. Three stations (Datong, Wutaishan and Weixian) in the western area of HRB show significant increasing trends in the annual ET_o. As for the seasonal changes, similar characteristics with respect to the annual ET_o were identified only in summer, while during the other three seasons (spring, autumn and winter), the trends were less obvious.

(2) The spatial patterns of the Mann–Kendall trends of the annual meteorological variables show that the maximum and minimum temperatures increase significantly at the 0.05 significance level. However, the increase of the minimum temperature is more apparent than that of the maximum ones all over the basin. Wind speed and shortwave radiation show decreasing trends in the whole basin, and the trends are significant in the eastern and southern parts of the HRB. The sensitivity analysis shows that relativity humidity is the most sensitive variable to ET_o, followed by temperature, shortwave radiation and wind speed as the least sensitive to ET_o in the whole HRB.

(3) Comprehensively considering the sensitivity and variation strength of the meteorological variables, the detrend analysis indicates the decreasing trends in ET_o dominant in the eastern and southern area of HRB. These may be caused mainly by the behavior of wind speed and shortwave radiation. Meanwhile, the obtained detrend results suggest that the increasing temperature is the main cause of the increasing trend of ET_o in Datong, Wutaishan and Weixian stations.

(4) The spatial correlation coefficient between population and the trend of ET_o is −0.132, and the correlation coefficient between the trend of ET_o and natural factors is even higher. This suggests that human activity has a certain influence on the spatial variation of ET_o, while natural factors play a decisive role in the spatial variation character of reference evapotranspiration in this area.

Acknowledgments

The authors are grateful to the National Climate Center of China for providing climatic data. Creative Talents Fund of Guangzhou Institute of Geography, Young Talents Fund of Guangdong Academy of Sciences (rcjj201303) and the Natural Science Foundation of China (No. 51279140) contributed to this research.

Author Contributions

Ling-ling Zhao is the first person to complete the paper, is responsible for the calculation of the paper, written and modified. Leszek Sobkowiak responsible for checking the language and spelling, Jun Xia and Zongli Li give recommendations in the paper forming process.

Conflicts of Interest

The authors declare no conflict of interest.

References

1. Okechukwu, A.; Luc, D.; Souley, Y.K.; le, B.E.; Ibrahim, M.; Abdou, A.; Théo, V.; Bader, J.; Moussa, I.B.; Gautier, E.; *et al.* Increasing river flows in the Sahel? *Water* **2010**, *2*, 170–199.
2. Huntington, T.G. Evidence for intensification of the global water cycle: Review and synthesis. *J. Hydrol.* **2006**, *319*, 83–95.
3. Changchun, X.; Yaning, C.; Yuhui, Y.; Hao, X.M.; Shen, Y.P. Hydrology and water resources variation and its response to regional climate change in Xinjiang. *J. Geogr. Sci.* **2010**, *20*, 599–612.
4. Zhao, L.L.; Xia, J.; Xu, C.Y.; Wang, Z.G.; Sobkowiak, L.; Long, C. Evapotranspiration estimated methods in hydrological simulation. *J. Geogr. Sci.* **2013**, *23*, 359–369.
5. Feddes, R.A.; Kowalik, P.J.; Zaradny, H. *Simulation of Field Water Use and Crop Yield*; Centre for Agricultural Publishing and Documentation: Wageningen, Netherlands 1978.
6. Ritchie, J.T. Model for predicting evaporation from a row crop with incomplete cover. *Water Resourc. Res.* **1972**, *8*, 1204–1213.
7. Bo, T.; Ling, T.; Shaozhong, K.; Lu, Z. Impacts of climate variability on reference evapotranspiration over 58 years in the Haihe river basin of North China. *Agric. Water Manag.* **2011**, *98*, 1660–1670.
8. Jun, X.; Huali, F.; Chesheng, Z.; Cunwen, N. Determination of a reasonable percentage for ecological water-use in the Haihe River Basin, China. *Pedosphere* **2006**, *16*, 33–42.
9. Zheng, H.X.; Liu, X.M.; Liu, C.M.; Dai, X.Q.; Zhu, R.R. Assessing contributions to panevaporation trends in Haihe River Basin, China. *J. Geophys. Res. Atmos.* **2009**, *114*, doi:10.1029/2009JD012203.
10. Xu, C.Y.; Gong, L.B.; Jiang, T.; Chen, D.L. Decreasing reference evapotranspiration in a warming climatea case of Changjiang (Yangtze) River catchment during 1970–2000. *Adv. Atmos. Sci.* **2006**, *23*, 513–520.
11. Liu, B.H.; Xu, M.; Henderson, M.; Gong, W.G. A spatial analysis of pan evaporation trends in China, 1955–2000. *J. Geophys. Res.* **2004**, *109*, doi: 10.1029/2004JD004511.
12. Bormann, H. Sensitivity analysis of 18 different potential evapotranspiration models to observed climatic change at German climate stations. *Clim. Chang.* **2011**, *104*, 729–753.
13. Coleman, G..; DeCoursey, D.G. Sensitivity and model variance analysis applied to some evaporation and evapotranspiration models. *Water Resourc. Res.* **1976**, *12*, 873–879.
14. Gao, G.; Chen, D.L.; Ren, G.Y.; Chen, Y.; Liao, Y.M. Spatial and temporal variations and controlling factors of potential evapotranspiration in China: 1956–2000. *J. Geogr. Sci.* **2006**, *16*, 3–12.
15. Gong, L.B.; Xu, C.Y.; Chen, D.L.; Halldin, S.; Chen, Y.Q. David. Sensitivity of the Penman–Monteith reference evapotranspiration to key climatic variables in the Changjiang (Yangtze River) basin. *J. Hydrol.* **2006**, *329*, 620–629.
16. Goyal, R.K. Sensitivity of evapotranspiration to global warming: A case study of arid zone of Rajasthan (India). *Agric. Water Manag.* **2004**, *69*, 1–11.

17. Hupet, F.; Vanclooster, M. Effect of the sampling frequency of meteorological variables on the estimation of the reference evapotranspiration. *J. Hydrol.* **2001**, *243*, 192–204.

18. McCuen, R.H. A sensitivity and error analysis of procedures used for estimating evaporation1. *JAWRA J. Am. Water Resourc. Assoc.* **1974**, *10*, 486–497.

19. Allen, R.G.; Pereira, L.S.; Raes, D.; Smith, M. *Crop Evapotranspiration-Guidelines for Computing Crop Water Requirements-FAO Irrigation and Drainage Paper 56*; Food and Agriculture Organization: Rome, Italy, 1998; 1–15.

20. Kendall, M.G. *Rank Correlation Measures*; Charles Griffin: London, UK, 1975; Volume 202.

21. Mann, H.B. Nonparametric tests against trend. *Econ. J. Econ. Soc.* **1945**, *13*, 245–259.

22. Xu, C.Y.; Gong, L.B.; Jiang, T.; Chen, D.L.; Singh, V.P. Analysis of spatial distribution and temporal trend of reference evapotranspiration and pan evaporation in Changjiang (Yangtze River) catchment. *J. Hydrol.* **2006**, *327*, 81–93.

23. Chong, S.Z. *The Probabilistic Approach in Water Science and Technology*; Science Press: Beijing, China, 2010. (In Chinese)

24. Brutsaert, W. Land-surface water vapor and sensible heat flux: Spatial variability, homogeneity, and measurement scales. *Water Resourc. Res.* **1998**, *34*, 2433–2442.

25. Zhang, Y.Q.; Liu, C.M.; Tang, Y.H.; Yang, Y.H. Trends in pan evaporation and reference and actual evapotranspiration across the Tibetan Plateau. *J. Geophys. Res. Atmos.* **2007**, *112*, doi: 10.1029/2006JD008161.

26. Liu, C.M.; Liu, X.M.; Zheng, H.X.; Zeng, Y. Change of the solar radiation and its causes in the Haihe River Basin and surrounding areas. *J. Geogr. Sci.* **2010**, *20*, 569–580.

27. Grimenes, A.A.; Thue-Hansen, V. The reduction of global radiation in south-eastern Norway during the last 50 years. *Theor. Appl. Climatol.* **2006**, *85*, 37–40.

28. Qian, Y.; Kaiser, D.P.; Leung, L.R.; Xu, M. More frequent cloud-free sky and less surface solar radiation in China from 1955 to 2000. *Geophys. Res. Lett.* **2006**, *33*, doi: 10.1029/2005GL024586.

29. Qian, Y.; Wang, W.G.; Leung, L.R.; Kaiser, D.P. Variability of solar radiation under cloud-free skies in China: The role of aerosols. *Geophys. Res. Lett.* **2007**, *34*, doi: 10.1029/2006GL028800.

30. Jiang, Y.; Luo, Y.; Zhao, Z.C.; Tao, S.W. Changes in wind speed over China during 1956–2004. *Theor. Appl. Climatol.* **2010**, *99*, 421–430.

Scenario-Based Impacts of Land Use and Climate Change on Land and Water Degradation from the Meso to Regional Scale

Aymar Y. Bossa, Bernd Diekkrüger and Euloge K. Agbossou

Abstract: Scale-dependent parameter models were developed and nested to the Soil and Water Assessment Tool-SWAT to simulate climate and land use change impacts on water-sediment-nutrient yields in Benin at a regional scale (49,256 km²). Weighted contributions of relevant landscape attributes characterizing the spatial pattern of ongoing hydrological processes were used to constrain the model parameters to acceptable physical meanings. Climate change projections (describing a rainfall reduction of up to 25%) simulated throughout the Regional Model-REMO, very sensitive to a prescribed degradation of land cover, were considered. Land use change scenarios in which the population growth was translated into a specific demand for settlements and croplands (cropland increase of up to 40%) according to the development of the national framework, were also considered. The results were consistent with simulations performed at the meso-scale (586 km²) where local management operations were incorporated. Surface runoff, groundwater flow, sediment and organic N and P yields were affected by land use change (as major effects) of -8% to $+50\%$, while water yield and evapotranspiration were dominantly affected by climate change of -31% to $+2\%$. This tendency was more marked at the regional scale as response to higher scale-dependent rates of natural vegetations with higher conversions to croplands.

Reprinted from *Water*. Cite as: Bossa, A.Y.; Diekkrüger, B.; Agbossou, E.K. Scenario-Based Impacts of Land Use and Climate Change on Land and Water Degradation from the Meso to Regional Scale. *Water* **2014**, *6*, 3152–3181.

1. Introduction

Unsustainable land use is driving land degradation, which in the form of soil erosion, nutrient depletion, water scarcity, salinity and disruption of biological cycles is a fundamental and persistent problem, diminishing productivity, biodiversity, other ecosystem services, and contributing to climate change [1]. A global survey suggests that 40% of agricultural land is already degraded to the point that yields are greatly reduced, and a further 9% is degraded to the point that it cannot be reclaimed for productive use by farm level measures [2]. According to the Global Assessment of Human-induced Soil Degradation (GLASOD) [1,3,4] estimates, degradation of cropland appears to be most prevalent in Africa, affecting already in the 1990s 65% of cropland areas, compared with 51% in Latin America and 38% in Asia. Many studies have been conducted in parts of Africa to understand the processes as well as determinant and promoting causes, related to the specific climatological, meteorological and soil conditions [5–16], and have been capitalized into this current work.

Vulnerability to change, whether climate-induced or related to anthropogenic-induced changes in land use/land cover, is a major threat, consisting at the same time of a water dimension (reduced water availability), an agrarian dimension (falling yields) and an environmental dimension (weakening of the soil and increasing erosion) [7]. Several investigations on the global change processes and its impact on the hydrological cycle [9,17–26] have shown that global climate change has a significant influence on the regional water and soil resources. Neumann *et al.* [23] investigated climate trends of temperature, precipitation and river discharge in the Volta Basin (West Africa) and have concluded on weak trends towards a decrease in rainfall with no clear trend on discharge as the anthropogenic influences (e.g., building of dams, intensified irrigation) were not quantified. Legesse *et al.* [21] evaluated the hydrological response of a catchment to climate and land use changes in Tropical Africa and found that a 10% decrease in rainfall produced a 30% reduction on the simulated hydrologic response of the catchment, while a 1.5 °C increase in air temperature would result in a decrease in the simulated discharge of about 15%. Moreover, they indicated that a conversion of the present day dominantly cultivated/grazing land by woodland would decrease the discharge at the outlet by about 8%. Chaplot [20] examined the effects of increasing CO_2 concentrations and rainfall changes associated with changes in average daily rainfall intensity, and surface air temperature on loads of water, NO_3-N and sediments from watersheds exhibiting different environmental conditions. He found over a 100-year simulated period: (1) flow and sediment discharges affected by precipitation changes while temperature and changes in atmospheric CO_2 concentration had a smaller effect; (2) CO_2 concentration was the main controlling factor of NO_3-N loads; and (3) global changes in the humid watershed had a greater effect on the water and soil resources. Ward *et al.* [25] evaluated the impact of land use and climate change on future suspended sediment yields and found an increase in all simulations due to conversion of forest to agricultural land. Conjoint sensitivity analyses have shown that although land use change acts as the primary control on long-term changes in sediment yield, the sensitivity of sediment yield to changes in climate increases as the percentage of deforested land increases. Mahe *et al.* [24] modeled the impact of land use change on soil water holding capacity and river flow in West Africa and found that the total reduction in water holding capacity is estimated to range from 33% to 62% between 1965 and 1995. This was explained by the decline in the extent of natural vegetation from 43% to 13% of the total basin area, whilst the cultivated areas increased from 53% to 76% and the area of bare soil nearly tripled from 4% to 11%. Li *et al.* [22] modeled the hydrological impact of land use change (West Africa) and pointed out that total deforestation (clearcutting) increases the simulated runoff ratio from 0.15 to 0.44, and the annual streamflow by 35%–65%, depending on location in the basin, although forests occupy only a small portion (<5%) of the total basin area. They mentioned that there is no significant impact on the water yield and river discharge when the deforestation (thinning) percentage is below 50% or the overgrazing percentage below 70% for savanna and 80% for grassland areas; however, the water yield is increased dramatically when land cover change exceeds these thresholds. Faramarzi *et al.* [19] modeled the impacts of climate change on freshwater availability in Africa and the results indicated that the mean total quantity of water resources is likely to increase, but for individual catchments and countries, variations are substantial. Cornelissen *et al.* [18] assessed the suitability of different hydrological model types for simulating scenarios of future discharge behavior in West Africa in the context of climate and land use change. They found that all models simulate an increase in surface runoff due to land use change. The application of climate change scenarios resulted in considerable variation between the models and points not only to uncertainties in climate change scenarios. The conclusions drawn out from the above-presented studies are concordant with the different methodologies rolled out. They demonstrated significant sensitivities of soil and water resources to increase CO_2, temperature, rainfall and land use depending on the study locations, the catchment characteristics, the modeling approaches, the land use change drivers, the structure of the atmosphere-ocean global climate models as well as the regional climate models behind the climate projections used. None of them has addressed the chemical dimension of soil degradation, notably the soil organic N and P loads and delivered together with sediment at catchment outlet. Organic N and P loads highly depend on landscape heterogeneity and spatial patterns of hydrological processes which are well known to smooth out with increasing catchment size resulting in more uncertain model parameters (without physical meaning) and more uncertain impact calculations at large scale, but almost none of the available impact studies have addressed this scaling problem, what is also offered in the present study.

It is widely accepted that the complexity of hydrological processes as well as land and water degradation depend on the environmental heterogeneity such as soil pattern, topography, geology, vegetation and anthropogenic impacts. Thus, the process-dependent hierarchization of landscape elements from local to regional scales [27] is recently reflected in the development of several computationally efficient conceptual and distributed physical-based models (e.g., SWAT), attempting to quantify the hydrological variability occurring at a range of scales. Therefore, catchments may be subdivided into a number of smaller units such as sub-catchments, hillslopes, hydrological response units, contour-based elements, and square grid elements [6,28–31], but opened discussions on the modeling uncertainty issue, including scaling-effects in model internal aggregation [32,33] for large scale applications. This often affects the magnitudes of model parameters which may finally have no consistent physical meanings, carrying too poor information [33]. A scale dependent parameterization approach may significantly reduce these sources of uncertainty and the problem of lack or non-accurate measurement data (e.g., stream water-sediment-nutrient measurements) at large catchment scale. This is crucial and of high interest for impact assessment of climate and land use change at large-scale in a data-poor environment like Benin.

Previous integrated modeling works in West Africa and in the upper Ouémé catchment of roughly 15,000 km^2 in size in Benin [9,16], have largely contributed to improving knowledge of recent land degradation processes from local to regional scales. The results have shown that amongst others silt and clay particle loads totaling 0.5 ton ha^{-1} a^{-1} with an associated organic nitrogen load of 0.8 $kg \cdot ha^{-1}$ a^{-1} [16]. These results clearly indicate that the study area is impacted by land and water degradation processes, primary seen as human-induced or natural processes that negatively affect ecosystem as for resources storing and recycling. Scenario-based land use and climate change [34–37] may impact the degradation process to a level of +50% of the observations [15,16]. The current work attempts to expand all above-mentioned dimensions of erosion-related degradation under different land use and climate scenarios to large catchment scales (up to 49,256 km^2), while minimizing biases due to scaling-up processes such as model internal aggregation processes in the SWAT model. These thorough modeling exercises are still

128

challenging the world scientific community and have never been parts of previous studies in Benin, as transferability of results through spatial scales stays an important underlying question.

This study aims to investigate how global change impacts on water/land degradation at different spatial scales in Benin. It specifically means to investigate the degradation trends at different catchment scales: (1) a meso-scale catchment (586 km²) investigation incorporating local-scale farming practices [16]; and (2) a large-scale (49,256 km²) investigation based on a regionalized model parameterization [38] based on scale-dependent model parameters for simulating water-sediment-nutrient fluxes. This latter approach makes use of physical catchment properties depending on the spatial scale as explanatory variables for model parameters using regression techniques.

2. Materials and Methods

2.1. Study Area

Located at about 90% in the Republic of Benin between 6°48' and 10°12' N of latitude and as part of the stable margin of the West African Craton, the Ouémé catchment (49,256 km², cf. Figure 1) is mainly characterized by a Precambrian basement, consists predominantly of complex migmatites granulites and gneisses, including less abundant mica shists, quarzites and amphibolites [9]. Syn-and post-tectonic intrusions of mainly granites, diorites, gabbros and volcanic rocks are present [39]. With a topographic relief generally low (highest elevation point of 617 meter) the land surface is slightly ondulating (granitic-gneissic plateau), strongly fractured (granitic peneplain) with typical seasonally waterlogged linear depressions (inland valleys) [9].

At a regional scale, fersialitic soils (ferruginous tropical sols) are predominant, characterized by clay translocation and iron segregation (ferruginous tropical sols with iron segregation), which lead to a clear horizon differentiation [40]. A local scale description has shown a typical catena with lixisols/acrisols on the upper and middle slopes, following by plinthosols on the downslopes, gleysols in the inland valleys and fluvisols on the fluvial plain [41].

Situated in a wet (Guinean coast) and a dry (Northern Soudanian zone) tropical climate, the Ouémé catchment records annual mean temperatures of 26 °C to 30 °C, annual mean rainfalls of 1280 mm (from 1950 to 1969) and 1150 mm (from 1970 to 2004) at a climatic station close to 9° N latitude [9]. As shown in the Figure 1, the Soudanian zone has a unimodal rainfall season that peaks in August whereas the Guinean zone exhibits a bimodal rainfall season that peaks in June and October.

The catchment landscape is characterized by forest islands, gallery forest, savannah, woodlands, agricultural lands and pastures. Agriculture and other human activities have led to large-scale deforestation and fragmentations leaving only small relics of the natural vegetation types within a matrix of degraded secondary habitats [9].

With a length of about 510 km and with two most important tributaries, Zou (150 km) and Okpara (200 km), the Ouémé river drains into Lake Nokoué (150 km²) and flows through the coastal lagoon system into the sea. Rainfall-runoff variability is high in the catchment, leading to runoff coefficients varying from 0.10 to 0.26 (of the total annual rainfall), with the lowest values for the savannahs and forest landscapes [9].

Figure 1. Location and climate condition of the study area, after Speth *et al.* [9]. The investigated catchments are Donga-Pont (586 km²), Vossa (1935 km²), Térou-Igbomakoro (2344 km²), Zou-Atchérigbé (6978 km²), Kaboua (9459 km²), Bétérou (10,072 km²), Savè (23,488 km²), Ouémé-Bonou (49,256 km²).

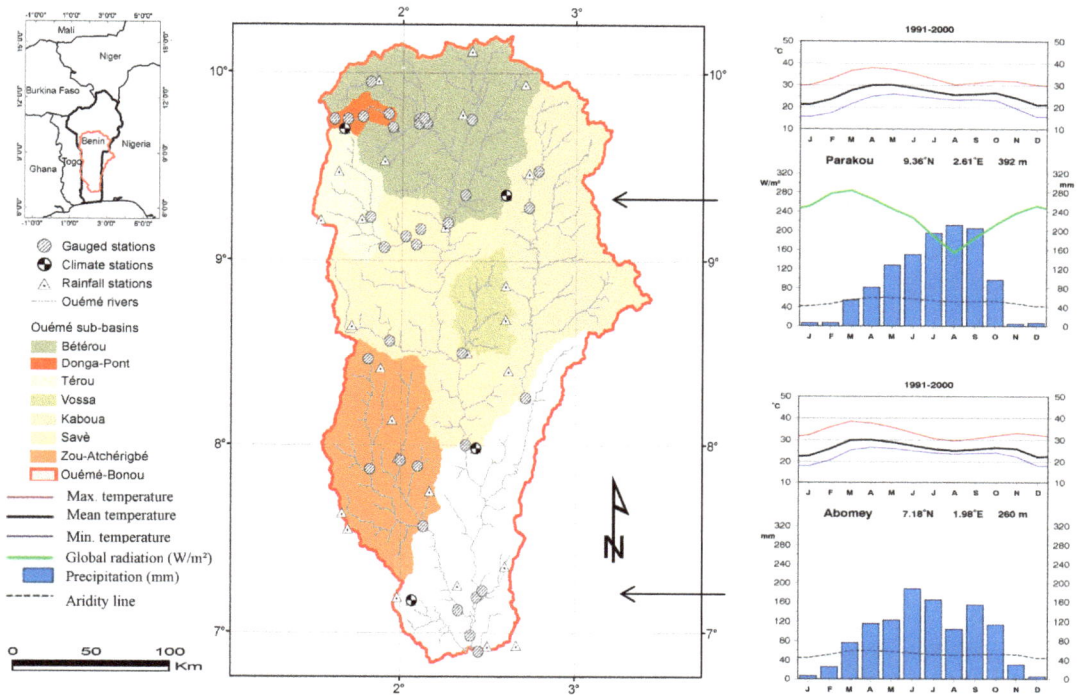

2.2. Modeling Approach

The SWAT (Soil and Water Assessment Tool) model is an eco-hydrological model developed by the United States Department of Agricultural-Research-Service (USDA-ARS) [31]. It is a continuous-time model that operates at a daily time-step. It allows the assessment of various subsurface flows and storages and related sediment and nutrient loads, taking into account the feedback between plant growth, water, and nutrient cycle, and helps to understand land management practice effects on water, sediment, and nutrient dynamics. It is a catchment scale model which can be applied from small (km²) to regional (100,000 km²) scale. SWAT subdivides the catchment into sub-catchments based on a Digital Elevation Model (DEM). Each sub-catchment consists of a number of Hydrological Response Units (HRUs) which are homogeneous concerning soil, relief, and vegetation. The HRUs are not georeferenced and not linked to each other within the sub-catchment.

In SWAT surface runoff is simulated using a modified version of the SCS CN method [42]. Lateral flow is simulated using the kinematic method of Sloan and Moore [43]. Percolation occurs when the soil field capacity is exceeded, recharging two aquifer systems: an unconfined aquifer generating base flow to the catchment streams, and a confined (deep) aquifer generating base flow to streams outside the catchment. The mass of nitrate lost from the soil horizons is determined using the nitrate concentration in the mobile water multiplied by the water volume flowing in each pathway. Rainfall-runoff erosion is estimated using the Modified Universal Soil Loss Equation (MUSLE) [44]. Organic N attached to sediments is estimated based on the loading function of McElroy et al. [45] and modified by Williams and Hann [46] to consider each runoff events. The model computes evaporation from soils according to Ritchie [47]. Actual soil water evaporation is estimated using exponential functions of soil depth and water content. Plant transpiration is computed as a linear function of potential evapotranspiration and leaf area index.

The overall modeling approach is summarized in Figure 2, showing the nature and source of the different data layers, their scales and types of parameters and investigations [48,49], in the structure as required for applying the SWAT model. A 90 m resolution Digital Elevation Model from the Shuttle Radar Topography Mission-SRTM was used. A SOil and TERrain (SOTER) digital database established at the scale 1:200.000 for the whole Ouémé catchment, in corporation with INRAB (Institut National de la Recherche Agricole du Bénin) is considered in this study (*cf.* Bossa *et al.* [50] for more details and an overview of the map and soil properties). This database includes different soil properties that were determinant for the model setup and parameterization: saturated hydraulic conductivity, organic CNP, bulk density, texture, erodibility factor, available water content, hydrology group, *etc.* The land use/cover map considered in this study has been established at 250 m resolution from 3 scenes satellite images LANDSAT ETM+ of 2003 [37] with an overall accuracy of 87%. More than 650 observation points were checked during the ground checks and 17 land use/cover classes were defined. Agricultural calendars depending on rainy season onsets, rainfall rhythms, and crop growth cycles/management over the period 2004–2009 (activity reports from the Regional Center of Agricultural Promotion-CeRPA and Ministry of Agriculture, Livestock and Fisheries-MAEP) have been used for the baseline agricultural practices introduced in SWAT (cf. Tables 1 and 2 for an example). Climate data (rainfall, temperature, solar radiation, wind speed and air humidity) were collected from 35 stations managed by the German Research Project IMPETUS, IRD (Institut de Recherche pour le Développement, France), and DMN (Direction de la Météorologie Nationale).

Besides discharge data continuously available for more or less 10 years (1998–2008) at 8 gauging stations, water samples (9 liters per day) were collected in 2004, 2005, 2008, 2009 and 2010 at 4 gauging stations (Donga-Pont, Bétérou, Térou and Zou-Atchérigbé, cf. Figure 1) and filtered in order to calculate daily suspended sediment concentration. Multi-parameter probes YSI 600 OMS (including one turbidity-broom sensor YSI 6136) were installed at the same stations to register turbidity at a high temporal resolution (used to calculate continuous time series of suspended sediment concentrations) to consider the hysteresis effects on the relationship between sediment and discharge. After filtration the obtained sediments were analyzed in the laboratory for organic Nitrogen and non-soluble/organic Phosphorus content. Weekly water samples were collected (2008–2010) for analyzing Nitrate and soluble Phosphorus.

Figure 2. Schematization of the modeling approach. Soil and land use data are from IMPETUS [48] and INRAB (Institut National de la Recherche Agricole du Bénin [49]), Climate data are from IMPETUS, IRD (Institut de Recherche pour le Développement, France), and DMN (Direction de la Météorologie Nationale), Geology data is from OBEMINES (Office Béninoise des MINES). CountryStat: Benin National Statistics (Food and Agriculture data network). CeRPA: Regional Center of Agricultural Promotion. MAEP: Ministry of Agriculture, Livestock and Fisheries. SSC means suspended sediment concentration.

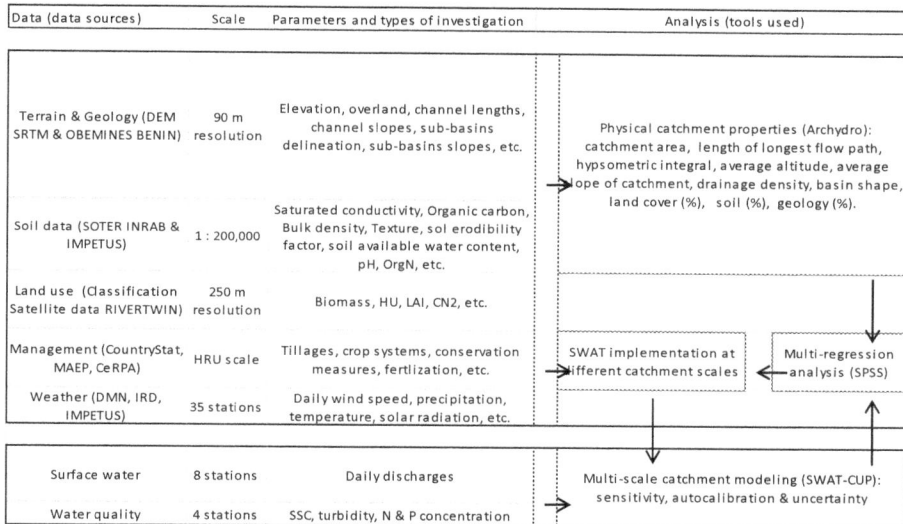

Table 1. Management operations considered for croplands in the Donga-Pont catchment. T1, T2 and T3: Tillage operation; F1 and F2: Fertilization.

Croplands	T1	F1	T2	T3	F2
Elemental N-P (kg ha^{-1})		10.5–6			
Elemental N (kg ha^{-1})					10.5
Tillage depth (cm)	25		10	10	
Mixing efficiency	0.5		0.25	0.25	

Table 2. Management operations considered for pastures in the Donga-Pont catchment. NH3N is the dissolved nitrogen, easily convertible into nitrate.

Pastures	Grazing
Grazing days	300
Biomass eaten: beef/dairy-sheep-goat (kg ha^{-1} d^{-1})	76-24-28
Biomass trampled: beef/dairy-sheep-goat (kg ha^{-1} d^{-1})	15-5-6
Manure beef (1%N-0.4%P-3%ORGN-0.7%ORGP-95%NH3N) (kg ha^{-1} d^{-1})	38
Manure sheep (1%N-0.4%P-3%ORGN-0.7%ORGP-95%NH3N) (kg ha^{-1} d^{-1})	12
Manure goat (1%N-0.4%P-3%ORGN-0.7%ORGP-95%NH3N) (kg ha^{-1} d^{-1})	64

The general input data were used to compute selected physical catchment attributes and beyond six individual Ouémé sub-catchments (Donga-Pont, Térou, Bétérou, Zou-Atchérigbé, Vossa and Kaboua, cf. Figure 1) were considered for SWAT calibration. Multi-scale auto-calibration and uncertainty analysis were performed applying the SUFI-2 procedure (Sequential Uncertainty Fitting version 2, SWAT-CUP interface [51]) so discharge, sediment, nitrate and organic N and P were simultaneously calibrated. SPSS software was used to statistically analyze two different matrixes of calibrated parameter sets and computed catchment attributes. A correlation analysis was performed to identify physical catchment attributes meaningful for each calibrated model parameter. Multiple regression analyses were later on performed to establish the regionalization rules which are in fact assumed to highly capture the catchment heterogeneity as well as the spatial pattern of the hydrological processes. Table 3 shows the calibration and validation periods as well as quality measures of the simulations for the different sub-catchments investigated.

Table 3. Model goodness of fit to measurements for the different sub-catchments involved in the multiple regression analysis, for model calibration. Information concerning validation are provided in brackets.

		Donga	Vossa	Térou	Atchérigbé	Kaboua	Bétérou
Discharge	Period	2006–2008	1998–2000	2002–2005	2007–2008	2004–2006	2006–2009
		(1998–2005)	(1995)	(1998–2001, 2006)	(2001–2006, 2009)	(1995–1998)	(1998–2005)
	R^2	0.72 (0.58)	0.75 (0.63)	0.75 (0.61)	0.89 (0.71)	0.73 (0.55)	0.75 (0.64)
	NS	0.72 (0.51)	0.75 (0.34)	0.74 (0.51)	0.82 (0.62)	0.67 (0.40)	0.60 (0.59)
Sediment	Period	2008	–	2004–2005	2008	–	2008–2009
		(2005)	–	(2006)	(2009)	–	(2004–2005)
	R^2	0.69 (0.58)	–	0.44 (0.33)	0.66 (0.67)	–	0.43 (0.27)
	NS	0.67 (0.55)	–	0.41 (0.32)	0.64 (0.67)	–	0.30 (0.14)
Nitrate	Period	2008	–	–	2008	–	2008–2009
		(2008–2009)	–	–	(2009)	–	(2008)
	R^2	0.99 (0.95)	–	–	0.86 (0.62)	–	0.73 (0.52)
	NS	0.99 (0.78)	–	–	0.81 (0.54)	–	(0.46)

Regarding discharge simulations, poor model efficiency were obtained for validation in Vossa and Kaboua sub-catchments (0.34 and 0.40 respectively) due mainly to peak overestimation caused partly by land use map derived from 2003 Landsat images, which considered more agricultural areas than the reality of the validation period (1995–1998). Critical model performances were obtained for sediment simulation in the Bétérou sub-catchments, where the model efficiency decreased even to 0.14. This may be mainly caused by strong hysteresis effects observed at this station, which was not equipped of turbidity probe as used at the Donga-Pont and Atchérigbé gauging stations to minimize this effect. Nitrate load was in general well represented in the model with coefficients of determination ranging from 0.62 to 0.99 and model efficiencies ranging from 0.54 to 0.99. Higher performances were observed for smaller sub-catchments. Calibrated model parameters are presented in Table 4 for all investigated sub-catchments.

Table 4. Calibrated model parameter matrix involved in the multiple regression analysis. The letters v, r and a mean values, relative change and absolute change, respectively. CN2: SCS Curve Number, ALPHA_BF: base flow recession constant; SOL_K: soil hydraulic conductivity, RCHRG_DP: aquifer percolation coefficient, GWQMN: minimum water level for base flow generation, REVAPMN: threshold water level in a shallow aquifer for capillary rise, ESCO: Soil evaporation compensation factor, GW_DELAY: groundwater delay, Ch_K2: Effective channel hydraulic conductivity, USLE_P: Practice factor, USLE_K: Soil erodibility factor, SPEXP: Exponent for calculating max sediment retrained, SURLAG: Surface runoff lag coefficient, NPERCO: Nitrate percolation coefficient.

Parameter	Description	Donga	Vossa	Térou	Atchérigbé	Kaboua	Bétérou
ESCO (v)	Soil evaporation compensation factor (–)	0.38	0.28	0.49	0.35	0.28	0.43
SOL_Z (r)	Soil depth (mm)	0.27	0.01	0.37	0.16	0.06	0.03
CN2 (r)	Curve Number (–)	6.65	3.86	5.55	6.24	3.97	2.51
GWQMN (v)	Threshold depth for ground water flow to occur (mm)	38.75	7.50	47.50	28.50	30.50	43.50
REVAPMN (v)	Threshold water level in shallow aquifer for revap (mm)	15.25	6.50	45.50	18.50	26.50	26.50
Ch_K2 (v)	Effective channel hydraulic conductivity (mm/hr)	3.95	1.00	12.77	10.65	1.00	12.52
Sol_K (r)	Saturated hydraulic conductivity (mm/hr)	–0.78	–0.65	–0.73	–0.35	–0.82	–0.66
GW_DELAY (v)	Ground water delay (day)	15.08	17.12	23.25	10.87	16.04	24.80
USLE_P (v)	Practice factor (–)	0.13	0.10	0.07	0.15	0.18	0.00
USLE_K (r)	Soil erodibility factor (0.013 t m²h/(m³ t cm))	0.03	0.16	0.08	0.14	0.25	–0.57
SPEXP (v)	Exponent for calculating max sediment retrained (–)	1.35	1.07	1.21	1.20	1.28	1.38
SURLAG (v)	Surface runoff lag coefficient (–)	0.19	0.35	0.24	0.25	0.17	0.10
ALPHA_BF (v)	Base flow recession factor (day)	0.06	0.17	0.07	0.12	0.11	0.15
NPERCO (v)	Nitrate percolation coefficient (–)	0.49	0.88	0.74	0.71	0.32	0.67
RCHRG_DP (v)	Fraction of deep aquifer percolation (–)	0.25	0.20	0.17	0.22	0.15	0.29

In the following paragraphs of this section, details of the calibration and validation issues are presented as follows for the Atchérigbé sub-catchment (6978 km²) to provide a complete overview on the measurements involved into the multi-scale modeling step. The validation of the regionalization rules is presented for the Savè sub-catchment (23,488 km²) in this same section and we should highlight that orders of magnitude of impacts of climate and land use change scenarios are presented and compared for the two targeted spatial scales (Donga-Pont: 586 km² and Ouémé-Bonou: 49,256 km²) in the result section.

Simulated *versus* observed daily water discharge and sediment yield are shown in Figures 3 and 4 for the Atchérigbé sub-catchment (cf. Figure 1). Recession periods were generally well represented. Less accurate predictions of single peaks are also shown in some years, partly due to the measurement errors during exceptional flooding years (2003 and 2007) in which over bank full discharge was observed at the gauging station. Differences are also usually caused by the SWAT structure, since it is a continuous time model with a daily time step and sub-scale processes such as single-event flood routing cannot be efficiently predicted. In addition, the daily measured precipitation for 24 h starts at 6:00 am and may not well match to the daily average discharge values, which were measured for 24 h from midnight on [52]. As it can be seen from the figures in the year 2008, discharge measurement gaps of even more than 10 days can happen due mainly to technical problems.

Figure 3. Simulated *vs.* observed daily discharge for the Atchérigbé sub-catchment (6978 km^2). Calibration period was 2007 to 2008 (R^2 = 0.89 and ME = 0.83), validation period was 2001–2006, and 2009 (R^2 = 0.71 and ME = 0.62).

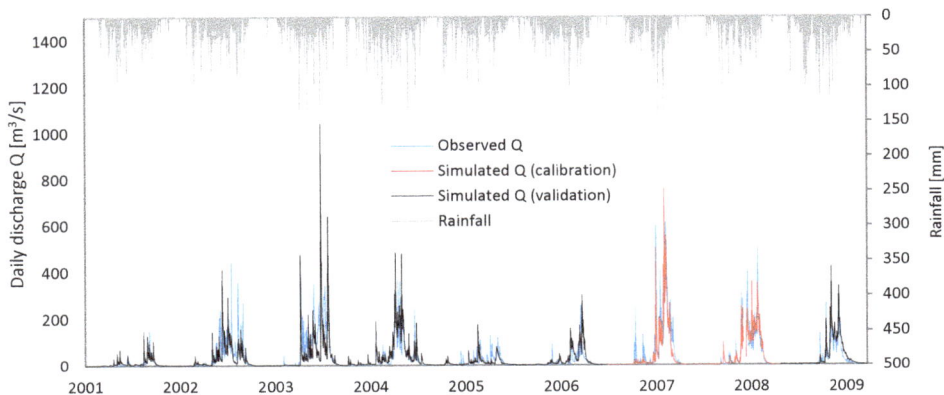

Simulated *versus* observed daily stream water nitrate load are shown in Figure 5 for the same Atchérigbé sub-catchment. Similarly to the sediment yield, nitrate peaks accompanied discharge peaks mainly caused by combined effects of increase nitrate loading and increase in water volume. Due to the sampling time scale (one time a week) several peaks were missed, but did not affect the model calibration.

According to FAO [53], water degradation by sediment has a chemical dimension—the silt and clay fraction, primary carrier of adsorbed chemicals, like nitrogen and phosphorus, which are transported by sediment into the aquatic system. Figure 6 shows weekly simulated *versus* observed organic N and P delivery at the Atchérigbé gauging station. Organic N and P were not calibrated. Since it was assumed that a good adjustment of soil nutrient pools, nitrate and sediment loads would be reflected in their simulations, only a validation was performed. Model goodness-of-fit were acceptable: 0.58 (R^2) and 0.78 (NS) for organic Nitrogen and 0.89 (R^2) and 0.96 (NS) for organic Phosphorus.

Table 5 shows the computed regionalization rules and derived parameter sets for the Savè (23,488 km², cf. Figure 1) sub-catchment and Ouémé-Bonou (49,256 km², cf. Figure 1) catchment. Physical catchment attributes depending on spatial scale were used as explanatory variables of SWAT model parameters. With respect to discharge, validation was performed for the Savè sub-catchment with a goodness-of-fit around 0.7 for model efficiency and R^2 (Figure 7).

Figure 4. Simulated *vs.* Observed daily sediment yield for Atchérigbé sub-catchment (6978 km²). Calibration period was 2008 (R² = 0.66 and ME = 0.64), validation period was 2009 (R² = 0.67 and ME = 0.67). SSC = suspended sediment.

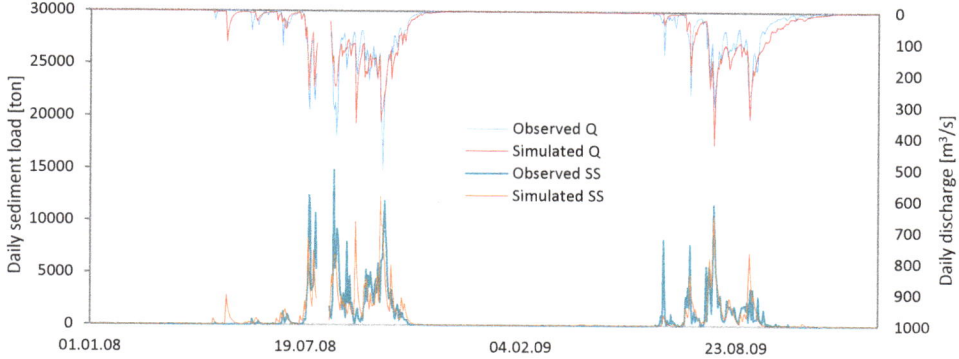

Figure 5. Simulated *vs.* observed daily nitrate load for Atchérigbé sub-catchment (6978 km²). Calibration period was 2008 (R² = 0.86 and ME = 0.81), validation period was 2009 (R² = 0.62 and ME = 0.54).

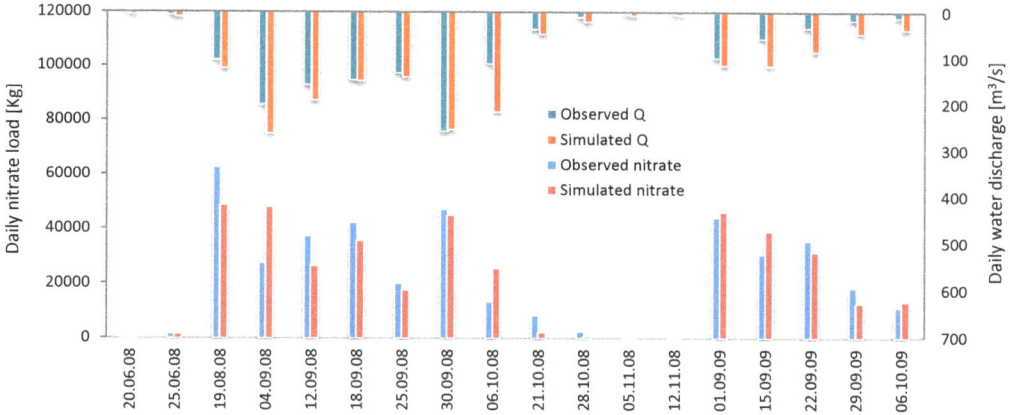

Figure 6. Simulated *vs.* observed weekly organic N and P load for the Atchérigbé sub-catchment (6978 km²). Only validation was performed from 2008 to 2009 with R² = 0.58 and ME = 0.78 for organic Nitrogen and R² = 0.89 and ME = 0.96 for organic Phosphorus.

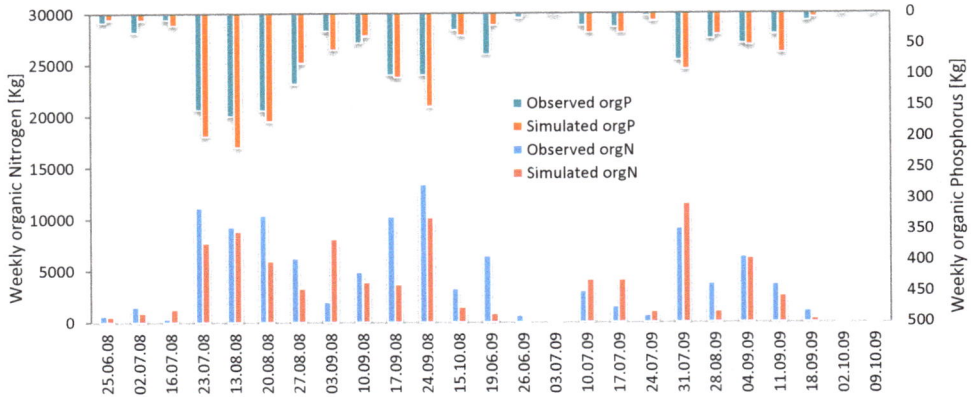

Table 5. Best regression-based parameter model and resulting values for three independent catchments (Savè: 23,488 km², Ouémé-Bonou: 49,256 km²). CN2: SCS Curve Number, ALPHA_BF: base flow recession constant; SOL_K: soil hydraulic conductivity, RCHRG_DP: aquifer percolation coefficient, GWQMN: minimum water level for base flow generation, REVAPMN: threshold water level in a shallow aquifer for capillary rise, ESCO: Soil evaporation compensation factor, GW_DELAY: groundwater delay, Ch_K2: Effective channel hydraulic conductivity (mm/h), USLE_P: Practice factor (–), USLE_K: Soil erodibility factor [0.013 t m²h/(m³ t cm)], SPEXP: Exponent for calculating max sediment retrained (–), SURLAG: Surface runoff lag coefficient (–), NPERCO: Nitrate percolation coefficient (–).

Parameters	Equations	R²	Savè	Ouémé-Bonou
ESCO	= 0.935 − 0.217 (Average slope of catchment) + 0.00327 (% Alterites)	0.92	0.34	0.37
SOL_Z	= 0.758 − 0.01 (% Migmatites)	0.81	0.02	0.08
SOL_K	= 26.991 − 0.278 (% Percentage of level)	0.92	−0.58	−0.76
CN2	= 10.0 − 0.0824 Migmatites (%)	0.49	3.94	4.4
GWQMN	= 185 − 49.2 (Average slope of catchment) − 0.255 (% Migmatites)	0.85	26.89	37.28
REVAPMN	= 16.5 + 0.769 (% Alterites)	0.6	20.37	18.56
Ch_K2	= 56.1 − 16.0 (Average slope of the catchment) − 0.461 (% Granites)	0.98	8.12	11.53
ALPHA_BF	= − 0.0794 + 0.00300 (% Migmatites)	0.87	0.14	0.12
GW_DELAY	= 19.0 − 0.248 (% Crop land) + 0.165 (% Savannah)	0.98	22.17	16.82
USLE_P	= 0.129 − 0.0143 (% Lateritic consolidated soil layer)	0.51	0.06	0.07
USLE_K	= 0.162 − 0.0848 (% Lateritic consolidated soil lay)	0.85	−0.24	−0.18
NPERCO	= 1.72 − 3.80 (% Hypsometric integral) + 0.00779 (% Circularity Index) − 0.033 (% Elongation ratio)	0.85	0.08	0.47
RCHRG_DP	= − 0.758 + 0.462 (Drainage density (km/ km²))	0.55	0.24	0.99
SPEXP	= 1.47 − 0.00454 Lixisol (%) + 0.00011 Migmatites (%)	0.7	1.22	1.47
SURLAG	= 0.109 + 0.003 Lixisol (%) − 0.016 Lateritic consolidated soil layers (%)	0.93	0.19	0.22

2.3. Climate and Land Use Change Scenarios

The climate scenarios used in this study were computed by Paeth *et al.* [36] for a part of Africa from −15° S to 45° N latitude using the regional climate model REMO driven by the IPCC (Intergovernmental Panel on Climate Change) SRES (Special Report on Emission Scenarios) scenarios A1B and B1. The IPCC SRES scenario A1B characterizes a globalized world of rapid economic growth and comparatively low population growth. The SRES scenario B1 also characterizes a future globalized world with a low population growth. REMO is a regional climate model that is nested in the global circulation model ECHAM5/MPI-OM Paeth *et al.* [54]. REMO was forced on a grid of 50 km resolution throughout the first half of the 21st century over West Africa.

Figure 7. Observed *vs.* simulated total discharge (validation) using the regression-based parameters for the Savè catchment (23,488 km²), with 0.71 for R² and 0.67 for model efficiency (ME). Savè was chosen for the validation because measurements at Ouémé-Bonou are not reliable.

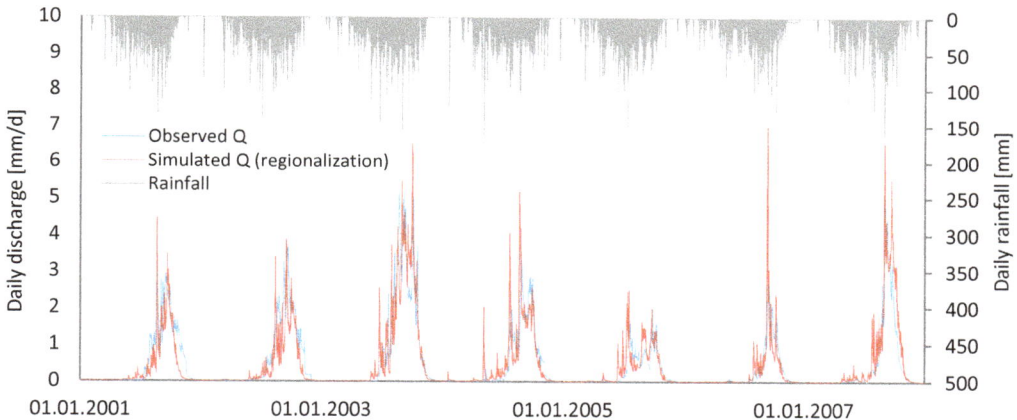

Initial runs of REMO over West Africa have shown systematically underestimated rainfall amounts and variability with a shift in the pattern towards more weak events and fewer extremes. This was addressed by applying the Model Output Statistics—MOS to correct monthly bias using other near-surface parameters such as temperature, sea level pressure and wind. Since the regional-mean (precipitation) strongly differed from the observed spatial patterns of daily rainfall events, a conversion of the MOS-corrected regional-mean from REMO to local rainfall event patterns has been done. Virtual station data, matching the rainfall stations in Benin, were useful to adjust the results to the statistical characteristics of observed daily precipitation at the rainfall stations by probability matching.

Figure 8a shows mean monthly REMO rainfall amounts over 1960–2000 compared with measurements over 1998–2005 for the upper Ouémé catchment (14,500 km² including Donga-Pont (586 km²), Térou-Igbomakoro (2344 km²) and Bétérou (10,072 km²)), while Figure 8b presents mean monthly water discharges simulated with SWAT using REMO outputs over 1960–2000 compared with measurements over 1998–2005. These figures suggested that REMO and SWAT represent correctly the observations.

Figure 8. (a) Mean monthly rainfall from REMO output (period 1960–2000) compared with measurements (period 1998–2005) for the upper Ouémé catchment [15]; **(b)** Simulated mean monthly water discharge with SWAT using REMO output (period 1960–2000) compared measurements (period 1998–2005) for the upper Ouémé catchment [15].

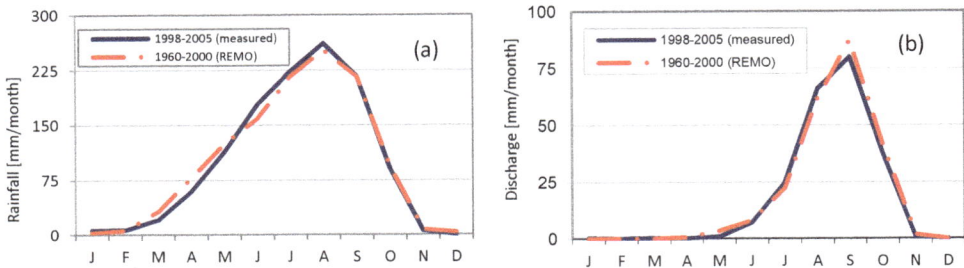

Climate change projections as simulated throughout REMO are very sensitive to a prescribed degradation of land cover. This sensitivity in addition to an increasing greenhouse gases concentrations have resulted in distinctly warmer and drier climates (with frequent droughts) for the investigated period 2000–2050 over West Africa, reductions in annual rainfall amounts of about 20%–25% of the 20th century annual amounts.

For the Ouémé-Bonou catchment REMO projects a decrease of annual rainfall between 9% and 12% for the scenario B1 and for the period 2010–2030. It increases of up to 4% for the scenario A1B over the period 2010–2014, before decreasing of up to 14% between 2015 and 2029. Maximum and minimum temperatures are expected to increase of up to 2.5 °C over the next 40 years (Figure 9).

Many recent research studies attempted to simulate West African future rainfall and climate parameters throughout the 21st century using atmosphere-ocean global climate models and relying on greenhouse gas emissions scenarios as outlined in the Intergovernmental Panel on Climate Change archives for Assessment Reports (AR3 & 4). The recent Coordinated Regional Climate Downscaling Experiment-CORDEX initiative from the World Climate Research Program promotes running multiple RCM simulations at 50 km resolution for multiple regions including West Africa, highly expected to bring clarifications and improve the projections [56]. The CORDEX initiative includes the study of uncertainty due to structural errors of different GCMs and/or RCMs. Beyond CORDEX and apart from Paeth *et al.* [36,54], who nested REMO in the global circulation model ECHAM5/MPI-OM as described above, Patricola and Cook [57] also attempt to overcome the limitations of global models by nesting a higher resolution regional model, the Weather Research and Forecasting (WRF) model on a grid of 90 km resolution, over West Africa and for the second half of the 21st century. They found a very mixed rainfall change signal characterized by June–July drought, followed by copious rainfall towards the end of the summer [58]. Although focused on different time periods, both studies [36,54,57] favor desiccation, albeit with caveats regarding intraseasonal and spatial variability.

Figure 9. Projected changes in annual precipitation and near-surface temperatures until 2050 over tropical and northern Africa due to increasing greenhouse gas concentrations and man-made land cover changes [55]. The scenario A1B describes a globalized world of rapid economic growth and comparatively low population growth. The scenario B1 also characterizes a future globalized world with a low population growth.

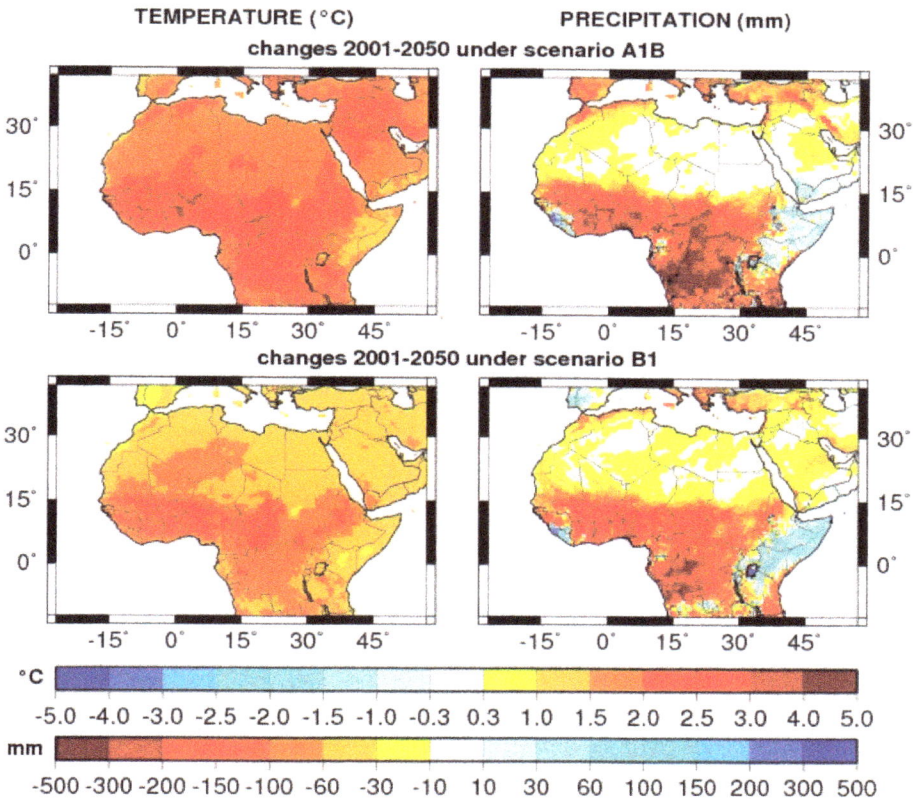

As an alternative to the above-described studies using relatively high-resolution regional climate models, many other studies [59–61] used atmosphere-ocean global climate models to run climate change experiments and have rather concluded a wetter climate for the first half of the 21st century in reference to the 20th century contrary to REMO-based projections presented in Paeth *et al.* [36]. Cook and Vizy [62] have concluded no impact of climate change on projected West Africa rainfall, while Biasutti *et al.* [63] have argued towards uncertain rainfall projections.

The land use/cover classification used, considers 17 land use/cover types (Table 6 and Figure 10). A subsequent accuracy check shows that the overall accuracy is high (87%) [64]. The land use/cover scenarios were computed in the framework of the European Union funded project RIVERTWIN [37]. The major driver for land use change is population growth and subsequent conversion of the natural savannah vegetation into settlements, roads, and a mosaic of fields by slash and burn clearance [65]. Two socio-economic scenarios have been set up: (1) La, stronger economic development, controlled urbanization, 3.2% population growth per year; and (2) Lb, weak national economy, uncontrolled settlement and farmland development, 3.5% population growth per year. For each scenario, the population growth has been translated into a specific demand for settlements and agricultural area according to the development of the national framework. This demand has been satisfied according to the proximity to roads and existing villages, new settlements and agricultural areas have been created leading to the land use distribution. The General Directorate for Water has selected several potential sites for future construction of multi-purpose reservoirs for large scale irrigation. Therefore, large areas of natural vegetation were also accordingly converted to croplands. With respect to the scenarios La and Lb, change in the Ouémé land use/cover is expressed by the conversion of the natural vegetation including savannah into agricultural lands and pastures: 10% to 20% for the scenario La and 20% to 40% for the scenario Lb.

Table 6. Land use/cover categories, their area and percentage of total area for the Ouémé-Bonou catchment (49,256 km²). Values displayed in brackets are related to the Donga-Pont catchment (586 km²) [64]. SWAT model was adapted to consider almost all land use classes mentioned in the table.

Land Use Categories	Land Use Code	Area (km²)	Percentage of Total Area
Galery forest	GF	1759 (8.6)	3.98 (1.47)
Humid and dry dense forest	FD	1220 (0.4)	2.76 (0.06)
Swamp formations	FM	17 (0)	0.04 (0)
Riverine formations	FR	107 (0)	0.24 (0)
Woodland and woodland savannah	FCSB	6716 (8.6)	15.2 (1.47)
Flooding savannah	SM	222 (0)	0.5 (0)
Tree and shrub savannah	SA	17231 (285.8)	38.99 (48.74)
Saxicolous savannah	SS	313 (0)	0.71 (0)
Grassland	PH	14 (0)	0.03 (0)
Mosaic of cropland and bush fallow	CJ	13713 (280.1)	31.03 (47.77)
Mosaic of cultivation with Parkia and Cashew trees	CJNA	32 (0)	0.07 (0)
Mosaic of cultivation with palm trees	CJP	1189 (0)	2.69 (0)
Industrial plantations	PI	127 (0)	0.29 (0)
Village plantations	PV	1209 (0.5)	2.74 (0.08)
Barren lands/area without vegetation	BAR	5 (0)	0.01 (0)
Urban and built-up	AG	277 (2.4)	0.63 (0.4)
Water bodies	PE	47 (0)	0.11 (0)

Figure 10. Land use/cover of the Ouémé-Bonou (49,256 km^2) and the Donga-Pont catchments (586 km^2). The legend is fully explained in Table 3. (**a**) reference map (2003); (**b**) La 2015–2019; (**c**) La 2025–2029; (**d**) Lb 2015–2019; (**e**) Lb 2025–2029 [37].

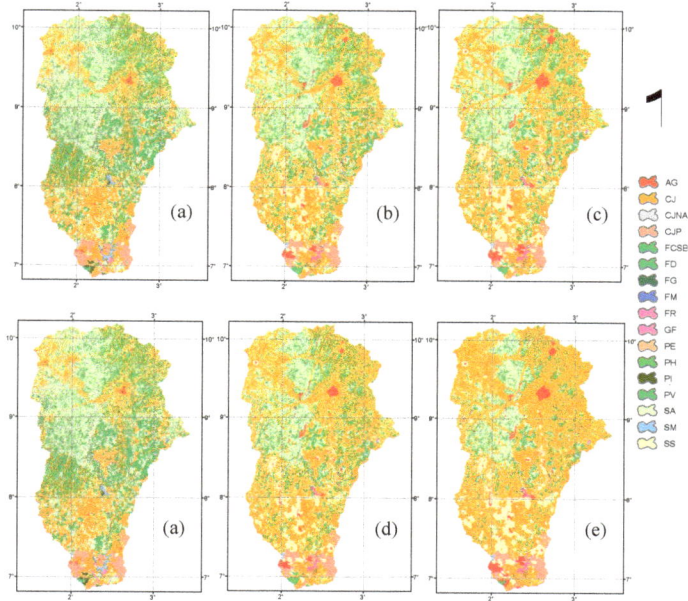

3. Results and Discussion

3.1. Impacts of Climate Change

Climate change impacts on key water balance components and sediment-organic N yield over time windows of 10 to 20 years for the Donga-Pont sub-catchment (Table 7 and Figure 11a) were computed to be consistent with other climate change impact studies. The Table and Figure show a decrease in rainfall all over the 10-year and 20-year time windows, resulting in a reduction of up to 22% of the simulated water-nutrient components for all scenarios except the scenario A1B over the time window 2010–2019. These results suggest significant impacts of year-to-year climate variability. To be consistent with the used land use scenarios 5-year time windows (Tables 8 and 9, Figure 11b,c) will be considered in the discussion as provided in the following paragraphs.

As shown in Tables 8 and 9, the simulated absolute surface runoff ranges from 100 to 140 mm per year for the Donga-Pont catchment (586 km²), while varying from 60 to 80 mm per year for the Ouémé-Bonou catchment (49,256 km²), due to the significantly different catchment sizes with a likely slight effect of the higher rate of agricultural lands in the Donga-Pont catchment compared to the Ouémé-Bonou catchment (cf. Table 6). In fact, due to unsustainable agriculture practices over the region, higher surface runoff may be associated to larger rate of agricultural lands.

Table 7. Simulated SWAT components under climate scenarios (with unchanged land use map derived from 2003 Landsat image). Deviation (in %) from the reference scenario (2000–2009) are shown in brackets.

Donga-Pont (586 km²)	Reference (2000–2009)	A1B (2010–2019)		A1B (2020–2029)		A1B (2010–2029)		B1 (2010–2019)		B1 (2020–2029)		B1 (2010–2029)	
Rainfall (mm a⁻¹)	1233.4	1196.6	(−3)	1107.2	(−10)	1151.9	(−7)	1145.5	(−7)	1126.2	(−9)	1135.8	(−8)
Water yield (mm a⁻¹)	254.7	282.8	(11)	206.6	(−19)	244.7	(−4)	238.2	(−6)	219.4	(−14)	228.8	(−10)
Groundwater flow (mm a⁻¹)	117.6	122.7	(4)	91.4	(−22)	107.0	(−9)	105.3	(−10)	99.5	(−15)	102.4	(−13)
Surface runoff (mm a⁻¹)	137.2	158.8	(16)	113.8	(−17)	136.3	(−1)	131.5	(−4)	118.5	(−14)	125.0	(−9)
Act. Evapotranspiration (mm a⁻¹)	923.7	840.9	(−9)	849.4	(−8)	845.1	(−9)	848.6	(−8)	846.7	(−8)	847.6	(−8)
Sediment yield (ton ha⁻¹ a⁻¹)	0.4	0.5	(17)	0.4	(−5)	0.4	(6)	0.4	(−2)	0.4	(−8)	0.4	(−5)
Organic N load (kg ha⁻¹ a⁻¹)	0.7	0.8	(11)	0.6	(−10)	0.7	(0)	0.7	(−5)	0.6	(−9)	0.7	(−7)

Sediment yield was more important at the Donga-Pont catchment scale, ranging from 0.3 to 0.4 ton·ha⁻¹·a⁻¹ against 0.3 ton·ha⁻¹·a⁻¹ for the Ouémé-Bonou catchment. This can be explained by the higher rate of agricultural lands in the Donga-Pont catchment compared to the Ouémé-Bonou catchment (cf. Table 6). Thus, due to inadequate tillage and unsustainable practices, sediments are very susceptible to loading. Conversely, the simulated organic nitrogen yields for the Ouémé-Bonou catchment (roughly 1.2 ton·ha⁻¹·a⁻¹) were twice the computed amount for the Donga-Pont catchment (0.6 to 0.7 ton·ha⁻¹·a⁻¹). This can be easily understood since due to the higher rate of agricultural lands in the Donga-Pont catchment, the topsoils are more degraded and very poor humus and organic matters are available for loading.

In both investigated scales, annual sediment yield and actual evapotranspiration have significantly decreased (of up to 20%) over the simulated years (2000 to 2029). Groundwater flow decreased significantly from 15% to 22% for the Donga-Pont catchment and from 4% to 17% for the Ouémé-Bonou catchment. Figure 11b,c shows a decreasing trend for all the simulated components (of up to 20%), regardless the different scenarios, but more pronounced for the Donga-Pont catchment.

One may conclude that differences in water balance components as well as nutrient load rate while moving between different catchment scales are functions of rate of agricultural lands, which are more sensitive than natural vegetations. Similar results were found in the upper Ouémé catchment (about 15,000 km²) using the SWAT model and the same IPCC SRES scenarios A1B and B1 [9,15].

Figure 11. Simulated trends under climate scenarios with land use from the year 2003. (**a**) Donga-Pont: trend over 20 years; (**b**) Donga-Pont: trend over a 5-year time window; (**c**) Ouémé-Bonou: trend over a 5-year time window.

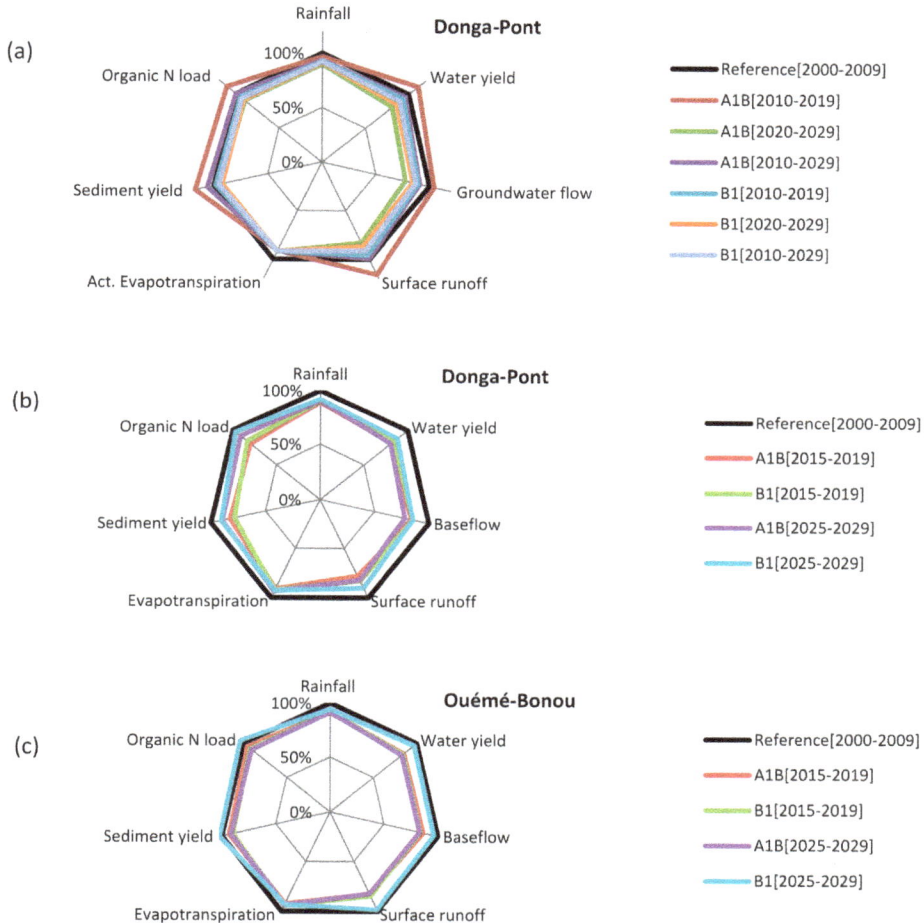

Table 8. Simulated SWAT components under climate scenarios (with unchanged land use map derived from 2003 Landsat image). Deviation (in %) from the reference scenario (2000–2009) are shown in brackets.

Donga-Pont (586 km²)	Reference (2000–2009)	A1B (2015–2019)		B1 (2015–2019)		A1B (2025–2029)		B1 (2025–2029)	
Rainfall (mm a⁻¹)	1233.4	1096.7	(−11)	1113.9	(−10)	1116.0	(−10)	1138.0	(−8)
Water yield (mm a⁻¹)	254.7	206.9	(−19)	215.5	(−15)	205.6	(−19)	225.3	(−12)
Groundwater flow (mm a⁻¹)	117.6	98.9	(−16)	100.3	(−15)	91.6	(−22)	100.5	(−15)
Surface runoff (mm a⁻¹)	137.2	106.7	(−22)	113.7	(−17)	112.6	(−18)	123.4	(−10)
Act. Evapotranspiration (mm a⁻¹)	923.7	839.1	(−9)	849.2	(−8)	863.6	(−7)	854.1	(−8)
Sediment yield (ton ha⁻¹ a⁻¹)	0.4	0.3	(−25)	0.3	(−25)	0.4	(0)	0.4	(0)
Organic N load (kg ha⁻¹ a⁻¹)	0.7	0.6	(−14)	0.6	(−14)	0.6	(−14)	0.7	(0)

Table 9. Simulated SWAT components under climate scenarios (with unchanged land use map derived from 2003 Landsat image). Deviation (in %) from the reference scenario (2000–2009) are shown in brackets.

Ouémé-Bonou (49,256 km²)	Reference (2000–2009)	A1B (2015–2019)		B1 (2015–2019)		A1B (2025–2029)		B1 (2025–2029)	
Rainfall (mm a⁻¹)	1138.9	1035.2	(−9)	1045.2	(−8)	1041.5	(−9)	1074.9	(−6)
Water yield (mm a⁻¹)	224.6	191.4	(−15)	189.0	(−16)	186.2	(−17)	218.0	(−3)
Groundwater flow (mm a⁻¹)	147.6	127.5	(−14)	123.7	(−16)	122.8	(−17)	142.2	(−4)
Surface runoff (mm a⁻¹)	77.0	63.9	(−17)	65.3	(−15)	63.4	(−18)	75.8	(−2)
Act. Evapotranspiration (mm a⁻¹)	794.8	734.3	(−8)	747.5	(−6)	745.8	(−6)	740.2	(−7)
Sediment yield (ton ha⁻¹ a⁻¹)	0.3	0.3	(0)	0.3	(0)	0.3	(0)	0.3	(0)
Organic N load (kg ha⁻¹ a⁻¹)	1.2	1.1	(−8)	1.1	(−8)	1.1	(−8)	1.2	(0)

3.2. Impacts of Land Use Change

As indicated in the Section 2.3, changes in the Ouémé land use according to the scenarios La and Lb are mainly expressed by the conversion of savannah into croplands and pastures in a range of 10% to 20% of the agricultural lands for the scenario La (stronger economic development, controlled urbanization, 3.2% population growth per year) and 20% to 40% for the scenario Lb (weak national economy, uncontrolled settlement and farmland development, 3.5% population growth per year). Accordingly, an increasing surface runoff was simulated (from 2000 to 2029) for both scenarios La and Lb, but in a pronounced way for the scenario Lb (Table 10 and Figure 12). These increases range from 9% to 27% for the Donga-Pont catchment and between 22% and 57% for the Ouémé-Bonou catchment.

In spite of this simulated increase for the surface runoff, water yield has change in a very low rate (roughly ±5%), forcing a decrease of the groundwater flow (between −3% and −33% for the Donga-Pont catchment and between −7% and −18% for the Ouémé-Bonou). This reveals the proof of decrease of the infiltration rate over the study area, and a severe threat to its groundwater systems. Sediment yield has increased from 25% to 75% for the Donga-Pont catchment and from 33% to 66% for the Ouémé-Bonou catchment. Organic nitrogen load has increased from 14% to 43% for the Donga-Pont catchment and from 17% to 58% for the Ouémé-Bonou due to conversion of natural vegetation to new cropland, which are more important at the Ouémé-Bonou scale.

Actual evapotranspiration has decreased slightly from 0.5% to 1.1% for the Donga-Pont catchment and 0.8% to 2% for the Ouémé-Bonou catchment. For this trend the land use scenario Lb differs from the other by showing the highest underlying simulated rates, as already discussed above for the other components.

Table 10. Simulated SWAT components under land use scenarios (with unchanged climate condition of the period 2000 to 2009). Deviation (in %) from the reference scenario (2000–2009) are shown in brackets.

Donga-Pont (586 km²)	Reference (2000–2009)	La (2015–2019)		Lb (2015–2019)		La (2025–2029)		Lb (2025–2029)	
Rainfall (mm a⁻¹)	1233.4	1233.4	(0)	1233.4	(0)	1233.4	(0)	1233.4	(0)
Water yield (mm a⁻¹)	254.7	228.7	(−10)	236.7	(−7)	267.0	(5)	263.4	(3)
Groundwater flow (mm a⁻¹)	117.6	78.8	(−33)	79.0	(−33)	113.6	(−3)	88.8	(−24)
Surface runoff (mm a⁻¹)	137.2	149.1	(9)	157.0	(14)	152.4	(11)	174.3	(27)
Act. Evapotranspiration (mm a⁻¹)	923.7	913.7	(−1)	918.0	(−1)	915.3	(−1)	919.1	(0)
Sediment yield (ton ha⁻¹ a⁻¹)	0.4	0.5	(25)	0.6	(50)	0.5	(25)	0.7	(75)
Organic N load (kg ha⁻¹ a⁻¹)	0.7	0.8	(14)	0.9	(29)	0.8	(14)	1.0	(43)
Ouémé-Bonou (49,256 km²)									
Rainfall (mm a⁻¹)	1138.9	1138.9	(0)	1138.9	(0)	1138.9	(0)	1138.9	(0)
Water yield (mm a⁻¹)	224.6	231.5	(3)	238.8	(6)	233.6	(4)	243.0	(8)
Groundwater flow (mm a⁻¹)	147.6	137.9	(−7)	128.8	(−13)	133.9	(−9)	121.7	(−18)
Surface runoff (mm a⁻¹)	77.0	93.7	(22)	110.0	(43)	99.7	(29)	121.3	(58)
Act. Evapotranspiration (mm a⁻¹)	794.8	788.9	(−1)	782.1	(−2)	787.4	(−1)	778.8	(−2)
Sediment yield (ton ha⁻¹ a⁻¹)	0.3	0.4	(33)	0.4	(33)	0.4	(33)	0.5	(67)
Organic N load (kg ha⁻¹ a⁻¹)	1.2	1.4	(17)	1.7	(42)	1.5	(25)	1.9	(58)

Figure 12. Simulated trends under land use scenarios (with unchanged climate: 2000–2009). (**a**) Donga-Pont; (**b**) Ouémé-Bonou.

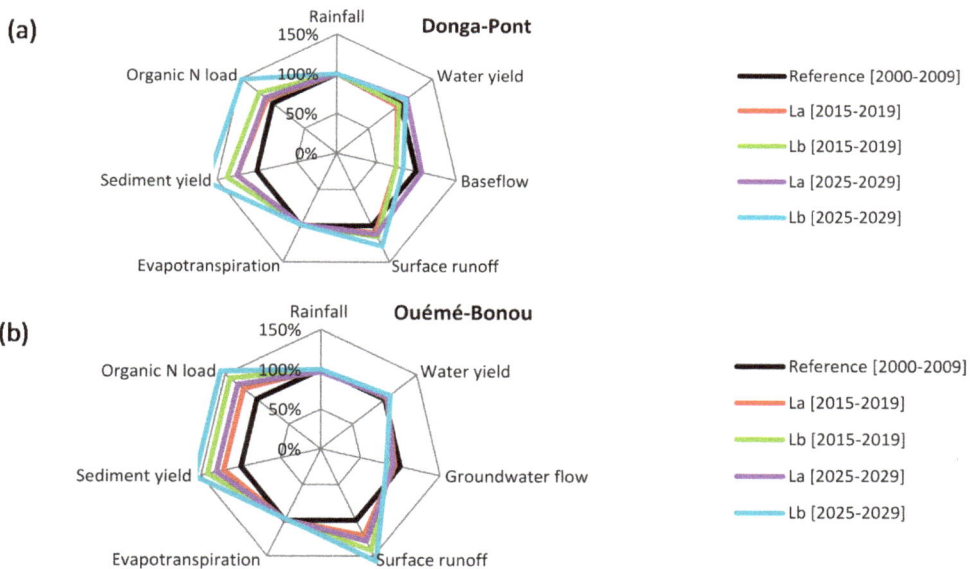

In summary (combining the Sections 3.1 and 3.2), surface runoff was found more sensitive to land use change (+22% to +75% of changes) than climate change (−5% to −20% of changes), evapotranspiration was more sensitive to climate change (−8 to −12% of changes) than land use change (−0.5% to −2% of changes) and groundwater flow was less sensitive to climate change

(−4% to −22%) than land use change (−9% to −57%). Sediment yield was more sensitive to land use (+25% to +75%) than climate change (−25% to 0%). Organic nitrogen load was more sensitive to land use change (+14% to +58%) than climate change (−8% to −14%).

At the large catchment scale (Ouémé-Bonou), sediment yield has decreased (−25% for the climate scenarios and −57% for the land use scenarios) and organic nitrogen load has increased (100% for the climate scenarios and 175% for the land use scenarios) in reference to the meso scale catchment (Donga-Pont).

Li *et al.* [22] modeled the hydrological impact of land use change in the Niger and the Lake Chad basins (West Africa) and found that a deforestation percentage below 50% has no significant impact on the stream flow, but a total deforestation increases the annual stream runoff by 35%–65%. This finding is very contrasting the results obtained here, where a progressive conversion of natural vegetation to cropland to a maximum of 40% over 30 years has increased the surface runoff by 22%–57% and decreased the groundwater flow by 7%–18%.

3.3. Impacts of Combined Climate and Land Use Scenarios

In reference to the results discussed in the Sections 3.1 and 3.2, an overview of combined land use and climate change scenario effects on the degradation trend presented in Table 11 and Figure 13 identified climate change as the major driver of the changes in water yield and actual evapotranspiration. In the meantime, land use changes raises as the major driver affecting surface runoff, groundwater flow, sediment yield and organic nitrogen load. Stronger effects of climate change were computed for the time period 2015–2019 in the Donga-Pont catchment scale, which have been significantly but not completely compensated by the land use change effects over the time period 2025–2029.

At the Donga-Pont scale (Table 12 and Figure 14, time period 2025–2029), annual surface runoff may change from −8% to +17%. The decrease (−8%) (scenarios La + A1B) was mainly driven by climate change, while land use change effects have mainly resulted in an increase (+17%) (scenarios La + B1, Lb + A1B and Lb + B1). Sediment yield increased for all the scenarios from 12.5% to 50%, and organic nitrogen load also has increased from 1.4% to 46% (driven by land use changes). Under climate change effects, water yield has changed from −9% to +1.6%, groundwater flow from −29% to 1% and evapotranspiration from −9% to −7.5%.

At the Ouémé-Bonou scale (Table 12 and Figure 14, time period 2025–2029), annual surface runoff may change from +5.7% to +42% for all the scenarios, driven by land use change. Sediment yield increased for all the scenarios from 15.6% to 41%, and organic nitrogen load also has increased from 15% to 47% (driven by land use changes). Under climate change effects, water yield has changed from −14% to +0.6%, groundwater flow from −30.5% to −16% and evapotranspiration from −8.4% to −7%.

Table 11. Simulated SWAT components under land use and climate scenarios (2015 to 2019). Deviation (in %) from the reference scenario (2000–2009) are shown in brackets.

Donga-Pont (586 km²)	Reference (2000–2009)	La & A1B (2015–2019)		La & B1 (2015–2019)		Lb & A1B (2015–2019)		Lb & B1 (2015–2019)	
Rainfall (mm a⁻¹)	1233.4	1096.7	(−11)	1113.9	(−10)	1096.7	(−11)	1113.9	(−10)
Water yield (mm a⁻¹)	254.7	191.7	(−21)	201.1	(−17)	199.1	(−18)	206.8	(−15)
Groundwater flow (mm a⁻¹)	117.6	74.1	(−30)	76.6	(−28)	74.5	(−30)	75.1	(−29)
Surface runoff (mm a⁻¹)	137.2	116.4	(−14)	123.1	(−9)	123.5	(−9)	130.4	(−4)
Act. Evapotranspiration (mm a⁻¹)	923.7	831.7	(−10)	841.5	(−9)	833.6	(−10)	843.7	(−9)
Sediment yield (ton ha⁻¹ a⁻¹)	0.4	0.5	(20)	0.4	(−8)	0.5	(15)	0.4	(7)
Organic N load (kg ha⁻¹ a⁻¹)	0.7	0.5	(−29)	0.6	(−10)	0.7	(0)	0.7	(3)
Ouémé-Bonou (49,256 km²)									
Rainfall (mm a⁻¹)	1138.9	1035.2	(−9)	1045.2	(−8)	1035.2	(−9)	1045.2	(−8)
Water yield (mm a⁻¹)	224.6	195.9	(−13)	194.1	(−14)	201.1	(−10)	200.1	(−11)
Groundwater flow (mm a⁻¹)	147.6	120.2	(−19)	116.0	(−21)	113.9	(−23)	109.5	(−26)
Surface runoff (mm a⁻¹)	77.0	75.7	(−2)	78.1	(1)	87.3	(13)	90.6	(18)
Act. Evapotranspiration (mm a⁻¹)	794.8	730.8	(−8)	743.4	(−6)	726.2	(−9)	738.1	(−7)
Sediment yield (ton ha⁻¹ a⁻¹)	0.3	0.4	(9)	0.3	(6)	0.4	(25)	0.4	(22)
Organic N load (kg ha⁻¹ a⁻¹)	1.2	1.3	(13)	1.3	(9)	1.5	(29)	1.5	(26)

Figure 13. Simulated trends under land use and climate scenarios (2015 to 2019). (**a**) Donga-Pont; (**b**) Ouémé-Bonou.

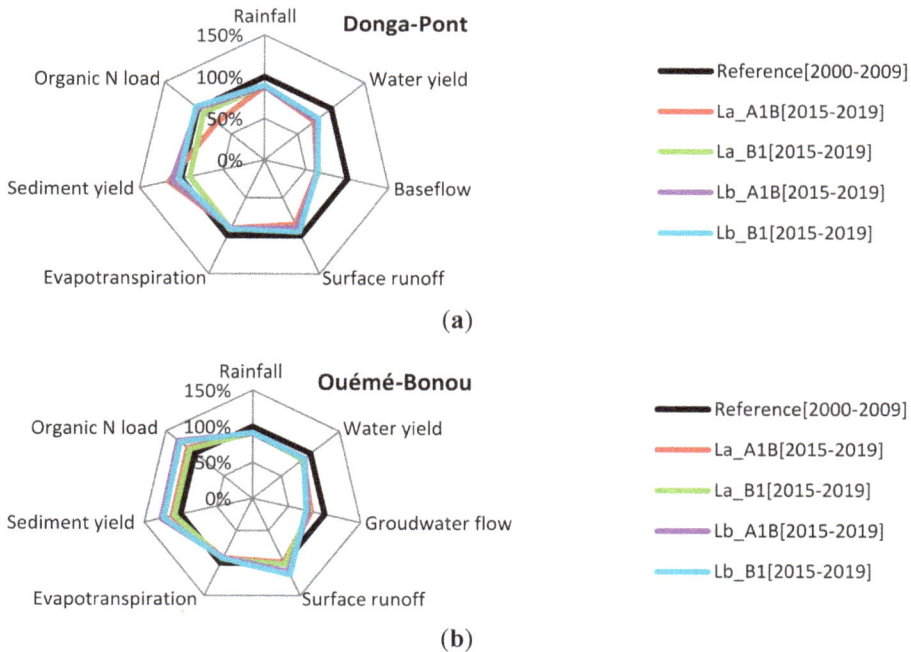

(a)

(b)

Regardless the modeling scale, the simulated impacts for the land use and climate change scenarios, over the simulated period 2000 to 2029 (30 years) may be summarized as follow: (1) surface runoff, groundwater flow, sediment and organic nitrogen load are found mainly sensitive to land use change with roughly −8% to 50% of changes; and (2) water yield and evapotranspiration are found more sensitive to climate change with roughly −31% to +2% of changes as consequence of rainfall reduction (cf. Section 2.3).

Table 12. Simulated SWAT components under land use and climate scenarios (2025 to 2029). Deviation (in %) from the reference scenario (2000–2009) are shown in brackets.

Donga-Pont (586 km²)	Reference (2000–2009)	La & A1B (2025–2029)		La & B1 (2025–2029)		Lb & A1B (2025–2029)		Lb & B1 (2025–2029)	
Rainfall (mm a⁻¹)	1233.4	1116.0	(−10)	1138.0	(−8)	1116.0	(−10)	1138.0	(−8)
Water yield (mm a⁻¹)	254.7	221.9	(−9)	244.9	(1)	221.8	(−9)	246.7	(2)
Groundwater flow (mm a⁻¹)	117.6	95.5	(−10)	107.0	(1)	75.0	(−29)	87.4	(−18)
Surface runoff (mm a⁻¹)	137.2	125.1	(−8)	136.5	(1)	145.9	(7)	158.4	(17)
Act. Evapotranspiration (mm a⁻¹)	923.7	857.4	(−8)	847.3	(−9)	854.7	(−8)	845.2	(−9)
Sediment yield (ton ha⁻¹ a⁻¹)	0.4	0.5	(13)	0.5	(15)	0.6	(45)	0.6	(50)
Organic N load (kg ha⁻¹ a⁻¹)	0.7	0.7	(1)	0.8	(9)	1.0	(36)	1.0	(46)
Ouémé-Bonou (49,256 km²)									
Rainfall (mm a⁻¹)	1138.9	1041.5	(−9)	1060.9	(−7)	1041.5	(−9)	1060.9	(−7)
Water yield (mm a⁻¹)	224.6	193.3	(−14)	215.8	(−4)	201.6	(−10)	223.3	(−1)
Groundwater flow (mm a⁻¹)	147.6	111.9	(−24)	124.1	(−16)	102.6	(−30)	113.8	(−23)
Surface runoff (mm a⁻¹)	77.0	81.4	(6)	91.7	(19)	99.0	(29)	109.6	(42)
Act. Evapotranspiration (mm a⁻¹)	794.8	740.4	(−7)	734.6	(−8)	733.0	(−8)	728.3	(−8)
Sediment yield (ton ha⁻¹ a⁻¹)	0.3	0.4	(16)	0.4	(22)	0.4	(34)	0.5	(41)
Organic N load (kg ha⁻¹ a⁻¹)	1.2	1.4	(15)	1.5	(25)	1.6	(37)	1.7	(47)

Figure 14. Simulated trends under land use and climate scenarios (2025 to 2029). (**a**) Donga-Pont; (**b**) Ouémé-Bonou.

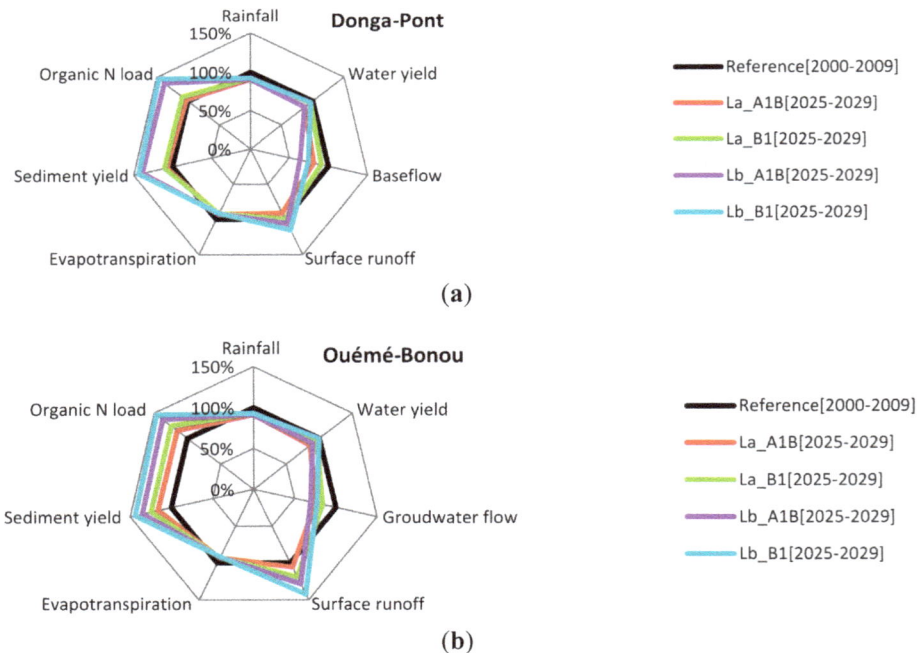

(a)

(b)

In addition, the results suggest that variables such as surface runoff, groundwater flow, sediment and transported nutrients, mainly sensitive to land use change are significantly affected by scale-dependent rate of agricultural lands, while variables such as water yield and evapotranspiration, mainly sensitive to climate change, have changed almost similarly at both scales. An application of the conceptual semi-distributed model UHP-HRU (Universal Hydrological Program-Hydrological Response Unit) in the same study area, using the physical-based model SIMULAT-H as benchmark [9], has led to a similar conclusion that water yield is more influenced by climate than land use, which more affects runoff components (surface runoff *vs.* interflow and base flow). Results of scenario analysis from the same work [9] revealed that the amount of renewable water decreases during the period 2001–2049 in both climate IPCC SRES scenarios A1B and B1, which were also used in the present study.

Sediment yield was more important at the Donga-Pont catchment scale, ranging from 0.4 to 0.6 ton·ha^{-1}·a^{-1} against 0.32 to 0.45 ton·ha^{-1}·a^{-1} at the Ouémé-Bonou catchment scale. This may be explained by decreases of retrained eroded particles at large scale, due to sequestration in the inland valleys, which are important at large scales [9]. According to Vanmaercke *et al.* [66] and de Vente *et al.* [67], high erosion rates are often observed at the local scale, but sediment yield, which is the net transport out of the catchment, decreases with increasing size of the catchment. This is consistent with the results obtained.

The simulated organic nitrogen yield for the Ouémé-Bonou catchment (roughly 1.2 to 1.7 ton·ha^{-1}·a^{-1}) was almost twice the computed amount for the Donga-Pont catchment (roughly 0.5 to 1 ton·ha^{-1}·a^{-1}). This may be the consequence of the higher rate of agricultural lands in the Donga-Pont catchment, resulting in poorer topsoils in terms of humus and organic matters rates available.

Figure 15. Impacts of land use and climate change on Organic Nitrogen patterns in the Ouémé-Bonou catchment (49,256 km^2).

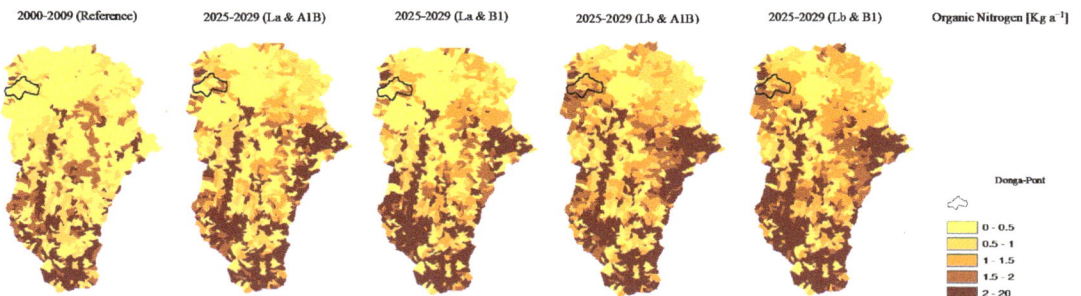

Figure 15 shows organic nitrogen load patterns for the combined scenarios (La & A1B), (La & B1), (Lb & A1B) and (Lb & B1) from 2000 to 2030 within the Ouémé-Bonou catchment. In accordance with the land use dynamic, the simulated pattern reflects higher and increasing degradation in the agricultural land. Surface runoff varies from 0 to 350 mm·a^{-1}, sediment yield varies from 0 to 10 ton·ha^{-1}·a^{-1}, and lost soil organic nitrogen varies from 0 to 20 kg·ha^{-1}·a^{-1}. Lost soil organic nitrogen shows higher dynamic and more threatened areas compared to sediment yield. This is completely consistent with the findings in section 3.3. As land use change was identified as the major driver of the ongoing land degradation, the scenario Lb (pessimistic, weak national economy, uncontrolled settlement and farmland development, 3.5% population growth per year) must be avoided in order to exclude the effects simulated by the combinations (Lb & A1B) and (Lb & B1) for which larger threatened areas are shown in the figures.

In another study, SWAT was applied in the upper Ouémé catchment to evaluate the effects of climate change and land use change on soil erosion [9,15]. In their combined land use and climate scenarios, the soil erosion rates increases with a large variability within the study area, which shows the high impact of land use change. The study has concluded that in areas with a high potential of cropland expansion, future sediment yield will be driven by land use change and may therefore strongly increase. Conversely, in the areas with a low potential for cropland expansion and strong reductions in rainfall, future sediment yield may decrease.

These findings are consistent with the results presented in this study, which furthermore discussed aspects of soil organic nutrients.

4. Conclusions

Soil and water are essential natural resources, available in limited quantities, but are nowadays dangerously exposed to climate and land use change [33]. Balancing the future degradation requires a sufficient understanding of processes behind at different scales [33], a task that may not be possible without an overriding of difficulties related to data availability in a data-poor environment such as Benin.

In this work, a regionalization methodology is used to overcome two difficulties in model setup in the Ouémé catchment: parameter scale-effects and associated uncertainty issues for large scale model application [32] and the lack and non-accurateness of boundary condition data (e.g., stream water-sediment-nutrient measurements). The SWAT model was applied using scale dependent and regression-based parameter models to simulate climate and land use change impacts on water yield, sediment and nutrient loads in Benin at the meso and the regional scale (49,256 km²).

The results revealed significant and increasing impacts over years. Surface runoff, groundwater flow, sediment and organic nitrogen load were affected by land use change (as dominant effects) of -8% to $+50\%$, while water yield and evapotranspiration were affected by climate change (as dominant effects) of -31% to $+2\%$. These rates may be reached gradually over years and according to the scenario data used. It was found that variables such as sediment and soil nutrients, mainly sensitive to land use change were likely functions of scale-dependent cropland rates. Furthermore, higher sediment yields were associated with higher scale-dependent rate of croplands, while higher organic nitrogen loads were associated with higher scale-dependent rate of natural vegetations. Partly the effect of climate change (decrease in surface runoff) and land use change (increase in surface runoff) balance out, resulting in a complex reaction of the system to Global Change. These results are consistent with findings of Hiepe [15] and Speth *et al.* [9]. Nevertheless, it is clear that land use change acts on shorter time scales than climate change causing a higher impact of land use change in the next years.

This work is a significant contribution for supporting sustainable management strategies to drive stronger economic development that considers controlled settlements, controlled farmland extension, less pressures on natural vegetations, and sustainable farming system managements. The study reveals the relevancy and the efficiency of the modeling strategy used for bridging the data gap, since the results are consistent with previous findings within the study area. Future works should focus on the interaction between uncertainties associated with the scenario data and the model structural errors. This will help to clearly quantify the uncertainties in the simulated impacts which are not investigated in this work.

Acknowledgments

This work has been funded by the German Federal Ministry of Education and Research (BMBF) as part of the West African Science Service Center on Climate Change and Adapted Land Use (WASCAL) research program. Many thanks to our partners in Benin and all colleagues of the GLOWA-IMPETUS project-Global Wandel des Wasserkreislaufs-Integratives Management Projekt für einen effizienten und tragfähigen Umgang mit Süßwasser in Westafrika (Global Cange and the Hydrological Cycle—An integrated approach to the efficient management of scarce water resources in West Africa) who provided data and assistance. The first author thanks the German Academic Exchange Service for funding.

Authors Contribution

Aymar Y. Bossa, Euloge K. Agbossou and Bernd Diekkrüger designed the study, developed the methodology, and wrote the manuscript. Aymar Y. Bossa performed the field work, collected the data, and conducted the computer analysis while Euloge K. Agbossou and Bernd Diekkrüger supervised this part of the work.

Conflicts of Interest

The authors declare no conflict of interest.

References

1. United Nations Environment Programme (UNEP). *Global Environment Outlook 4*; Progress Press Ltd: Valleta, Malta, 2007.
2. Bossio, D.; Geheb, K.; Critchley, W. Managing water by managing land: Addressing land degradation to improve water productivity and rural livelihoods. *Agric. Water Manag.* **2010**, *97*, 536–542.
3. Oldeman, L.R. *Guidelines for General Assessment of the Status of Human-induced Soil Degradation*; Working Paper and Reprint 88/4; International Soil Reference and Information Centre: Wageningen, the Netherlands, 1991.

4. Scherr, S.J.; Yadav, S. *Land degradation in the Developing World: Implications for Food, Agriculture, and the Environment to 2020*; Food, Agriculture, and the Evironment Discussion Paper 14; International Food Policy Research Institute (IFPRI): Washington, DC, USA, 1996.

5. Lal, R. Soil Erosion in the Tropics. Principles & Management; McGraw-Hill: New York, NY, USA, 1990.

6. Giertz, S.; Diekkrüger, B.; Steup, G. Physically-based modeling of hydrological processes in a tropical headwater catchment in Benin (West Africa)—Process representation and multi-criteria validation. *Hydrol. Earth Syst. Sci.* **2006**, *10*, 829–847.

7. Laouina, A.; Coelho, C.; Ritsema, C.; Chaker, M.; Nafaa, R.; Fenjiro, I.; Antari, M.; Ferreira, A.; van Dijck, S. Dynamique de l'eau et gestion des terres dans le contexte du changement global, analyse agro-hydrologique dans le bassin du Bouregreg (Maroc). *Sécheresse* **2004**, *15*, 66–77. (In French)

8. Roose, E.; de Noni, G. Recherches sur l'érosion hydrique en Afrique: Revue et perspectives. *Sci. Chang. Planétaires Sécheresse* **2004**, *15*, 121–129. (In French)

9. Speth, P., Christoph, M., Diekkrüger, B., Eds. *Impacts of Global Change on the Hydrological Cycle in West and Northwest Africa*; Springer Publisher: Heidelberg, Germany, 2010.

10. Chen, L.; Zhu, A.X.; Qin, C. Identification of critical source areas of soil erosion on moderate fine spatial scale in Loess Plateau in China. *African J. Agric. Res.* **2012**, *7*, 2962–2970.

11. Ouyang, W.; Skidmore, A.K.; Hao, F.; Wang, T. Soil erosion dynamics response to landscape pattern. *Sci. Total Environ.* **2010**, *408*, 1358–1366.

12. United Nations World Water Assessment Programme. *The Impact of Global Change on Erosion and Sediment Transport by Rivers: Current Processes and Future Challenges*; the United Nations World Water Development Report 3; United Nation Educational: Paris, France, 2009.

13. Yang, D.; Kanae, S.; Oki, T.; Koike, T.; Musiake, K. Global potential soil erosion with reference to land use and climate changes. *Hydrol. Process.* **2003**, *17*, 2913–2928.

14. Lawal, O.; Gaiser, T.; Nuga, B. Estimation of potential soil losses on a regional scale: A case of Abomey-Bohicon Region, Benin Republic. *Agric. J.* **2007**, *2*, 1–8.

15. Hiepe, C. Soil Degradation by Water Erosion in a Sub-Humid West-African Catchment, a Modelling Approach Considering Land Use and Climate Change in Benin. Ph.D. Thesis, University of Bonn, Bonn, Germany, 19 December 2008.

16. Bossa, A.Y.; Diekkrüger, B.; Giertz, S.; Steup, G.; Sintondji, L.O.; Agbossou, E.K.; Hiepe, C. Modeling the effects of crop patterns and management practices on N and P loads to surface water and groundwater in a semi-humid catchment (West Africa). *Agric. Water Manag.* **2012**, *115*, 20–37.

17. Keyzer, M.A.; Sonneveld, B.G.J.S.; Pande, S. The Impact of Climate Change on Crop Production and Health in West Africa: An Assessment for the Oueme River Basin in Benin; Centre for World Food Studies: Amsterdam, the Netherlands, 2007.

18. Cornelissen, T.; Diekkrüger, B.; Giertz, S. A comparison of hydrological models for assessing the impact of land use and climate change on discharge in a tropical catchment. *J. Hydrol.* **2013**, *498*, 221–236.

19. Faramarzi, M.; Abbaspour, K.C.; Ashraf Vaghefi, S.; Farzaneh, M.R.; Zehnder, A.J.B.; Srinivasan, R.; Yang, H. Modeling impacts of climate change on freshwater availability in Africa. *J. Hydrol.* **2013**, *480*, 85–101.

20. Chaplot, V. Water and soil resources response to rising levels of atmospheric CO_2 concentration and to changes in precipitation and air temperature. *J. Hydrol.* **2007**, *337*, 159–171.

21. Legesse, D.; Vallet-Coulomb, C.; Gasse, F. Hydrological response of a catchment to climate and land use changes in Tropical Africa: Case study South Central Ethiopia. *J. Hydrol.* 2003, 275, 67–85.

22. Li, K.Y.; Coe, M.T.; Ramankutty, N.; de Jong, R. Modeling the hydrological impact of land-use change in West Africa. *J. Hydrol.* **2007**, *337*, 258–268.

23. Neumann, R.; Jung, G.; Laux, P.; Kunstmann, H. Climate trends of temperature, precipitation and river discharge in the Volta Basin of West Africa. *Int. J. River Basin Manag.* **2007**, *5*, 37–41.

24. Mahe, G.; Paturel, J.; Servat, E.; Conway, D.; Dezetter, A. The impact of land use change on soil water holding capacity and river flow modelling in the Nakambe River, Burkina-Faso. *J. Hydrol.* **2005**, *300*, 33–43.

25. Ward, P.J.; van Balen, R.T.; Verstraeten, G.; Renssen, H.; Vandenberghe, J. The impact of land use and climate change on late Holocene and future suspended sediment yield of the Meuse catchment. *Geomorphology* **2009**, *103*, 389–400.

26. Oguntunde, P.G.; Abiodun, B.J. The impact of climate change on the Niger River Basin hydroclimatology, West Africa. *Clim. Dyn.* **2013**, *40*, 81–94.

27. Blöschl, G.; Sivapalan, M. Scale issues in hydrological modeling: A review. *Hydrol. Proc.* **1995**, *9*, 251–290.

28. Abbott, M.B.; Bathurst, J.C.; Cunge, J.A.; O'Connell, P.E.; Rasmussen, J. An introduction to the European Hydrological System—Systeme Hydrologique Europeen, "SHE", 2: Structure of a physically-based, distributed modeling system. *J. Hydrol.* **1986**, *87*, 61–77.

29. Sivapalan, M.; Viney, N.R. Large scale catchment modeling to predict the effects of land use and climate. *Water J. Aust. Wat. Wastewat. Assoc.* **1994**, *21*, 33–37.

30. Sivapalan, M.; Viney, N.R. Application of a nested catchment model for predicting the effects of changes in forest cover. In *Forest Hydrology*, Proceedings of International Symposium, Tokyo, Japan, October 1994; pp. 315–322.

31. Arnold, J.G.; Srinivasan, R.; Muttiah, R.S.; Williams, J.R. Large area hydrologic modeling and assessment part I: Model development. *J. Am. Water Resour. Assoc.* **1998**, *34*, 73–89.

32. FitzHugh, T.W.; Mackay, D.S. Impacts of input parameter spatial aggregation on an agricultural nonpoint source pollution model. *J. Hydrol.* **2000**, *236*, 35–53.

33. De Vente, J.; Poesen, J.; Verstraeten, G.; Govers, G.; Vanmaercke, M.; van Rompaey, A.; Arabkhedri, M.; Boix-Fayos, C. Predicting soil erosion and sediment yield at regional scales: Where do we stand? *Earth Sci. Rev.* **2013**, *127*, 16–29.

34. Thamm, H.P.; Judex, M.; Menz, G. Modeling of Land-Use and Land-Cover Change (LUCC) in Western Africa using remote sensing. *Photogramm. Fernerkund. Geoinf.* **2005**, *3*, 191–199.

35. Judex, M. Modellierung der Landnutzungsdynamik in Zentralbenin mit dem XULU-Framework. Ph.D. Thesis, University of Bonn, Bonn, Germany, 11 April 2008.

36. Paeth, H.; Born, K.; Podzun, R.; Jacob, D. Regional dynamic downscaling over West Africa: Model evaluation and comparison of wet and dry years. *Meteorol. Zeit.* **2005**, *14*, 349–367.

37. Regional Model for Integrated Water Management in Twinned River Basins (RIVERTWIN). *Adapted and Integrated Model for the Ouémé River Basin, Final Report*; Institute for Landscape Planning and Ecology: Stuttgart, Germany, 2007.

38. Bossa, A.Y.; Diekkrüger, B. Estimating scale effects of catchment properties on modeling soil and water degradation. In *Managing Resources of a Limited Planet: Pathways and Visions under Uncertainty*, Proceedings of the International Environmental Modelling and Software Society (iEMSs), Sixth Biennial Meeting, Leipzig, Germany, 1–5 July 2012; Seppelt, R., Voinov, A.A., Lange, S., Bankamp, D., Eds.; International Environmental Modelling and Software Society: Leipzig, Germany, 2012; pp. 2974–2981.

39. Wright, E.P.; Burgess, W. The hydrogeology of crystalline basement in Africa. *Geol. Soc. Spec.* **1992**, *66*, 1–27.

40. Faure, P.; Volkoff, B. Some factors affecting regional differentiation of the soils in the Republic of Benin (West Africa). *Catena* **1998**, *32*, 281–306.

41. Junge, B. Die Böden des oberen Ouémé-Einzugsgebietes in Benin/Westafrika—Pedologie, Klassifizierung, Nutzung und Degradierung. Ph.D. Thesis, Institut für Bodenkunde, Bonn, Germany, 23 July 2004.

42. Soil Conservation Service Engineering Division. *Urban Hydrology for Small Watersheds*; Technical Release 55; U.S. Department of Agriculture: Washington, DC, USA, 1986.

43. Sloan, P.G.; Moore, I.D. Modelling subsurface stormflow on steeply sloping forested watersheds. *Water Resour. Res.* **1984**, *20*, 1815–1822.

44. Williams, J.R. Sediment-yield prediction with universal equation using runoff energy factor. In *Present and Prospective Technology for Predicting Sediment Yield and Sources*. Proceedings of the sediment yield workshop, USDA Sedimentation Lab, Oxford, MS, USA, 28–30 November 1972; U.S. Department of Agriculture: Washington, DC, USA, 1975; pp. 244–252.

45. McElroy, A.D.; Chiu, S.Y.; Nebgen, J.W.; Aleti, A.; Bennett, F.W. *Loading Functions for Assessment of Water Pollution from Nonpoint Sources*; EPA document EPA 600/2-76-151; Midwest Research Institute: Kansas City, KS, USA, 1976.

46. Williams, J.R.; Hann, R.W. *Optimal Operation of Large Agricultural Watersheds with Water Quality Constraints*; Tech. Rep. No. 96; Texas Water Resources Institute: Temple, TX, USA, 1978.

47. Ritchie, J.T. A model for predicting evaporation from a row crop with incomplete cover. *Water Resour. Res.* **1972**, *8*, 1204–1213.

48. Christoph, M.; Fink, A.; Diekkrüger, B.; Giertz, S.; Reichert, B.; Speth, P. IMPETUS: Implementing HELP in the Upper Ouémé Basin. *Water SA* **2008**, *34*, 481–490.

49. Igué, A.M. Soil Information System for the Oueme Basin; Institut National de la Recherche Agricole du Bénin (INRAB): Cotonou, Benin, 2005.

50. Bossa, A.Y.; Diekkrüger, B.; Igué, A.M.; Gaiser, T. Analyzing the effects of different soil databases on modeling of hydrological processes and sediment yield in Benin (West Africa). *Geoderma* **2012**, *173–174*, 61–74.

51. Abbaspour, K.C. *SWAT Calibration and Uncertainty Programs—A User Manual*; Department of Systems Analysis, Integrated Assessment and Modeling (SIAM), Eawag, Swiss Federal Institute of Aquatic Science and Technology: Duebendorf, Switzerland, 2008.

52. Lam, Q.D.; Schmalz, B.; Fohrer, N. Modelling point and diffuse source pollution of nitrate in a rural lowland catchment using the SWAT model. *Agric. Water Manag.* **2010**, *97*, 317–325.

53. Food and Agriculture Organization (FAO). *Control of Water Pollution from Agriculture—FAO Irrigation and Drainage Paper 55*; Food and Agriculture Organization of the United Nations Rome: Rome, Italy, 1996.

54. Paeth, H.; Born, K.; Girmes, R.; Podzun, R.; Jacob, D. Regional climate change in tropical Africa under greenhouse forcing and land-use changes. *J. Clim.* **2009**, *22*, 114–132.

55. Patricola, C.M.; Cook, K.H. Northern African climate at the end of the twenty-first century: An integrated application of regional and global climate models. *Clim. Dyn.* **2010**, *37*, 1165–1188.

56. Druyan, L.M. Studies of 21st-century precipitation trends over West Africa. *Int. J. Climatol.* **2011**, *31*, 1415–1424.

57. Held, I.M.; Delworth, T.L.; Lu, J.; Findell, K.L.; Knutson, T.R. Simulation of Sahel drought in the twentieth- and twenty-first centuries. *Proc. Natl. Acad. Sci. USA* **2005**, *102*, 17891–17896.

58. Kamga, A.F.; Jenkins, G.S.; Gaye, A.T.; Garba, A.; Sarr, A.; Adedoyin, A. Evaluating the National Center for Atmospheric Research climate system model over West Africa: Present-day and the 21st century A1 scenario. *J. Geophys. Res.* **2005**, *110*, doi:10.1029/2004JD004689.

59. Hoerling, M.; Hurrell, J.; Eischeid, J.; Phillips, A. Detection and attribution of 20th century Northern and Southern African rainfall change. *J. Clim.* **2006**, *19*, 3989–4008.

60. Cook, K.H.; Vizy, E.K. Coupled model simulations of the West African monsoon system: Twentieth- and twenty-first-century simulations. *J. Clim.* **2006**, *19*, 3681–3703.

61. Biasutti, M.; Held, I.M.; Sobel, A.H.; Giannini, A. SST forcings and Sahel rainfall variability in simulations of the twentieth and twenty- first centuries. *J. Clim.* **2008**, *21*, 3471–3486.

62. Maraun, D.; Wetterhall, F.; Ireson, A.M.; Chandler, R.E.; Kendon, E.J.; Widmann, M.; Brienen, S.; Rust, H.W.; Sauter, T.; Themeßl, M.; *et al.* Precipitation downscaling under climate change: Recent developments to bridge the gap between dynamical models and the end user. *Rev. Geophys.* **2009**, *48*, doi:10.1029/2009RG000314.

63. Paeth, H. Key factors in African climate change evaluated by a regional climate model. *Erdkunde* **2004**, *58*, 290–315.

64. Igue, A.M.; Houndagba, C.J.; Gaiser, T.; Stahr, K. Land use/cover map and its accuracy in the Oueme Basin of Benin (West Africa). *Int. J. Agri. Sci.* **2012**, *2*, 174–184.

65. Götzinger, J. Distributed Conceptual Hydrological Modeling-Simulation of Climate, Land Use Change Impact and Uncertainty Analysis. Ph.D. Thesis, University of Stuttgart, Stuttgart, Germany, 19 July 2007.

66. Vanmaercke, M.; Poesen, J.; Verstraeten, G.; de Vente, J.; Ocakoglu, F. Sediment yield in Europe: Spatial patterns and scale dependency. *Geomorphology* **2011**, *130*, 142–161.

67. De Vente, J.; Poesen, J.; Verstraeten, G. The application of semi-quantitative methods and reservoir sedimentation rates for the prediction of basin sediment yield in Spain. *J. Hydrol.* **2005**, *305*, 63–86.

Vulnerability Assessment of Environmental and Climate Change Impacts on Water Resources in Al Jabal Al Akhdar, Sultanate of Oman

Mohammed Saif Al-Kalbani, Martin F. Price, Asma Abahussain, Mushtaque Ahmed and Timothy O'Higgins

Abstract: Climate change and its consequences present one of the most important threats to water resources systems which are vulnerable to such changes due to their limited adaptive capacity. Water resources in arid mountain regions, such as Al Jabal Al Akhdar; northern Sultanate of Oman, are vulnerable to the potential adverse impacts of environmental and climate change. Besides climatic change, current demographic trends, economic development and related land use changes are exerting pressures and have direct impacts on increasing demands for water resources and their vulnerability. In this study, vulnerability assessment was carried out using guidelines prepared by United Nations Environment Programme (UNEP) and Peking University to evaluate four components of the water resource system: water resources stress, water development pressure, ecological health, and management capacity. The calculated vulnerability index (VI) was high, indicating that the water resources are experiencing levels of stress. Ecosystem deterioration was the dominant parameter and management capacity was the dominant category driving the vulnerability on water resources. The vulnerability assessment will support policy and decision makers in evaluating options to modify existing policies. It will also help in developing long-term strategic plans for climate change mitigation and adaptation measures and implement effective policies for sustainable water resources management, and therefore the sustenance of human wellbeing in the region.

Reprinted from *Water*. Cite as: Al-Kalbani, M.S.; Price, M.F.; Abahussain, A.; Ahmed, M.; O'Higgins, T. Vulnerability Assessment of Environmental and Climate Change Impacts on Water Resources in Al Jabal Al Akhdar, Sultanate of Oman. *Water* **2014**, *6*, 3118-3135.

1. Introduction

Freshwater resources are key ecosystem services which sustain life and all social and economic processes. Their disruption threatens the health of ecological systems, people's livelihoods and general human wellbeing. However, water resources are being degraded as a result of multiple interacting pressures [1], particularly environmental and climate changes. The Fourth and Fifth Assessment Reports of the Intergovernmental Panel on Climate Change (IPCC) played a major role in framing understanding of likely impacts of climate change on human society and natural systems, making it clear that "water is in the eye of the climate management storm" [2–4]. Different possible threats resulting from anthropogenic climate change include temperature increases, shifts of climate zones, sea level rise, droughts, floods, and other extreme weather events [5]. The Earth's surface temperature has increased by about 0.5 °C during the last two decades, and a rise with similar amplitude is expected up to 2025, with direct effects on the global hydrological cycle, impacting water availability and demand [2–4]. Negative impacts on water availability and on the health of freshwater ecosystems will have negative consequences for social and ecological systems and their processes [6]. For example, with an approximately 2 °C global-mean temperature rise, around 59% of the world's population would be exposed to irrigation water shortage [7].

Besides climate change impacts, other drivers of environmental changes such as demographic trends, economic development and urbanization and related land-use changes are exerting pressures and increase demand for water resources [8]. Together, these drivers are stressing water resources far beyond the changes caused by natural global climatic changes in the recent evolutionary past. As a result of rapid population growth and economic development, and mismanagement of water resources, these drivers exert pressures on water resources, changing them both spatially and temporally and causing imbalances between supply and demand in hydrological systems [9]. The net effects can be translated into increases in the vulnerability of water resources systems. These systems are especially vulnerable to such changes because of their limited adaptive capacity, which can create major challenges for future management of water resources for human and ecosystem needs [10]. Therefore, there is a need to assess the vulnerability of water resources in order to enhance management capacity and adapt measures to cope with these changes for sustainable water resources use and management.

Vulnerability is a term commonly used to describe a weakness or flaw in a system; its susceptibility to a specific threat and harmful event. There have been many efforts to use the concept across different fields which are often location or sector specific. A variety of definitions of vulnerability have been proposed in the climate change literature, e.g., [11–18]. Common to most is the concept that vulnerability is a function of the exposure and sensitivity of a system to a climate hazard, and the ability to adapt to the effects of the hazard [15,19]. From a social point of view, vulnerability is defined as the exposure of individuals or collective groups to livelihood stress as a result of the impacts of such environmental change or climate extremes [17,20]. In this context, vulnerability can be explained by a combination of social factors and environmental risk, where risk derives from physical aspects of climate-related hazards exogenous to the social system [21–24]. Vulnerability to climate change is generally understood to be a function of a range of biophysical and socioeconomic factors. It is considered a function of wealth, technology, education, information, skills, infrastructure, access to resources, and stability and management capabilities [14,25]. The IPCC has defined vulnerability to climate change as the degree to which a system is susceptible to, and unable to cope with, adverse effects of climate change, including climate variability and extremes [26,27]. Vulnerability is also a function of the character, magnitude, and rate of climate change and variation to which a system is exposed, its sensitivity, and its adaptive capacity [28,29]. It is widely seen as an integrative concept that can link the social and biophysical dimensions of environmental change [30,31].

Nevertheless, "vulnerability" means different things to different researchers. From a water resources perspective, vulnerability has been defined as "the characteristics of water resources system's weakness and flaws that make the system difficult to be functional in the face of socioeconomic and environmental change" [10] (p. 2). Thus, water resources vulnerability assessment is an investigative and analytical process to evaluate the sensitivity of a water system to potential threats and identify challenges in mitigating the risks associated with negative impacts, in order to support water resources conservation and management under climate and environmental changes [10]. It is a tool to identify potential risks, helping to analyse specific aspects that contribute to overall risk. It therefore provides useful information to the manager about which components should receive more focus, in order to improve water management capacity towards sustainability in adapting to the changing climate and environmental factors.

Most water-stressed arid countries are vulnerable to the potential adverse impacts of climate change; particularly increases in temperatures, less and more erratic precipitation, drought and desertification. This is especially true in arid mountain regions, particularly Al Jabal Al Akhdar where a unique set of water management practices has enabled the development and survival, over centuries, of an agro-pastoral oasis social-ecological system. This study was conducted to assess the environmental and climate change impacts on water resources of Al Jabal Al Akhdar since no vulnerability assessments have been previously conducted in Oman or in this fragile mountain ecosystem.

The overall aim of this study was to estimate the vulnerability index of water resources of Al Jabal Al Akhdar to climate and environmental changes, to establish to what extent these resources are vulnerable and identify the major risks and levels of stress it faces with regard to water stress, development pressure, ecological health and management capacity. These are essential components for computing vulnerability index and assessing water resources in the region. The results should provide decision-makers with options to evaluate the current situation, modify existing policies, and implement adaptation and mitigation measures for sustainable water resources management in the study area.

2. The Study Area

Al Jabal Al Akhdar (Green Mountain) is located in the central part of the northern western Al-Hajar Mountains of the Sultanate of Oman (Figure 1), in the highest portion of 1500 to 3000 m above sea level [32]. It is a long-established agro-pastoral oasis ecosystem which has supported communities for centuries [32]. Until the late 20th century, this social ecological system has been geographically isolated and, compared to many places, relatively closed to the outside world. Historically, water availability has connected the agro-pastoral system and dictated the bounds of agricultural development and the human development. The area is also of particular cultural significance for Omani people for its location, topography, agricultural terraces, biodiversity and climate.

Al Jabal Al Akhdar has a Mediterranean climate. Because of its altitude, temperatures are some 10 to 12 °C lower than in the coastal plains. In general, mean monthly temperatures drop during winter to below 0 °C and rise in summer to around 22 °C. Temperature records from Saiq Meteorology Station (the only one in the area, part of Oman's national climate monitoring network), from 1979 to 2012 show mean monthly air temperatures from February to April from 12.1 to 18.7 °C and around 25 °C during summer (July and August). Minimum temperatures ranged from −0.6 °C in January to 15.9 °C in July, and maximum temperatures from 20.3 °C in January to 33.5 °C in July [33]. Rainfall is highly variable and irregular with an annual mean of about 250–400 mm [33] and is the main source of fresh water. There are two distinct rainfall seasons: the winter season from mid-November through March, and the summer season from

160

mid-June through mid-September. From 1979 to 2012, the average annual rainfall was 295.3 mm, with the highest monthly averages of 45.8 and 42 mm during August and July; and the lowest of 8.8 and 8.2 mm during October and November, respectively [33].

Figure 1. Oman map showing the location of Al Jabal Al Akhdar (in blue rectangle).

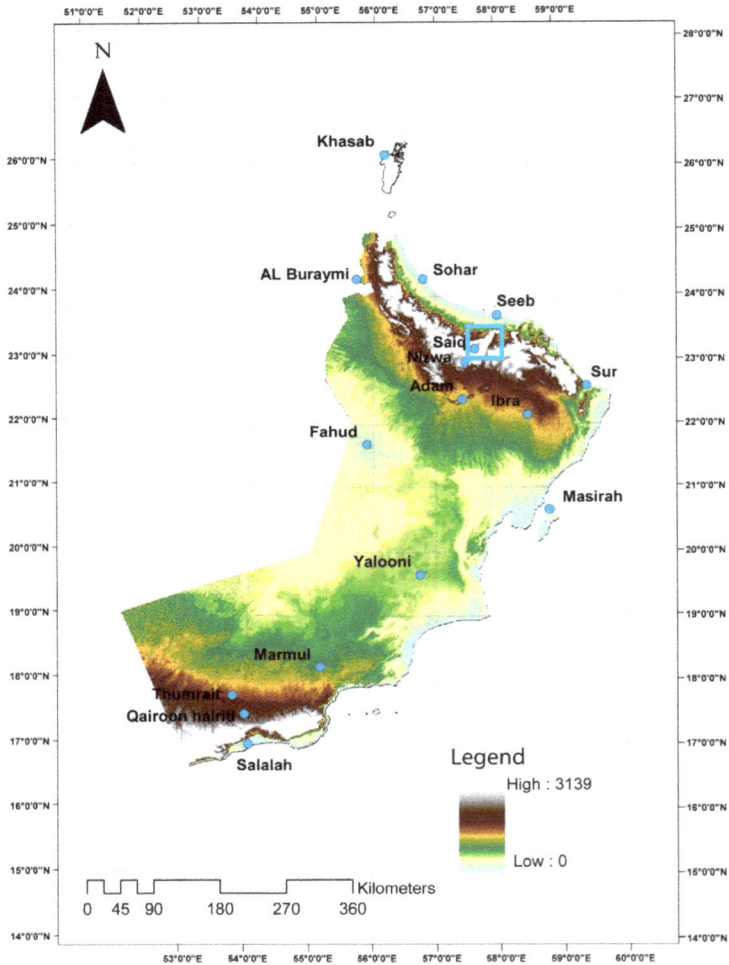

Agriculture is the main economic activity, providing the basis of livelihoods for around 70% of the inhabitants [34]. Although the sector does not contribute much to the national economy (only 3.7% of the total Oman GDP), it is the main dominant water consumer in Oman including the study area (more than 92% of the total available water) [35]. The area produces a variety of fruits, particularly pomegranates and roses (grown for the extraction of rose water), which are sold in the local markets as the major sources of income for farmers. Annual crops such as garlic, onion, maize, barley, oats and alfalfa are sometimes planted in the terraces depending on the availability of water. The area also has several endangered or vulnerable species of flora and fauna that are not found elsewhere in Oman [36].

Livestock husbandry is also an important part of the agriculture in the area, for food and income through the sale of fibre (goat and sheep hair), and provides a source of manure for the cultivation of crops. Goats are the main livestock in local communities, representing more than 80% of the total animal units [37].

Due to its relatively cool weather, especially during summer, and the construction of asphalted access road up to the mountain in 2006, followed by the construction of hotels, many tourists visit the area, mainly to see natural landscapes and agricultural terraces and to camp. There has been an increase in the number of hotels in the study area, from one in 2006 to four in 2014, plus other holiday apartments and rest houses. The number of tourists has increased by 58% from 85,000 in 2006 to 134,000 in 2013 [37].

Natural freshwater resources in Al Jabal Al Akhdar are of three types: groundwater (wells), lotic resources (natural springs and *aflaj*) and lentic resources (man-made dams) [38]. Groundwater is accessed via wells established by a government agency which supplies the water to the communities through networks or water trucks. These wells are the main local source of water for drinking and domestic purposes (municipal, commercial) in the area. *Aflaj* are surface and/or underground channels fed by groundwater or a spring, or streams, built to provide water to the farming communities. The *aflaj* water is managed and distributed to farming areas by local people with no involvement of government in their organizational structure. Dams are artificial structures which are constructed by the government to harvest rainy water. *Aflaj* and dam water are mainly used for agriculture and livestock.

Al Jabal Al Akhdar has limited and highly variable water supplies: the most significant parameters influencing freshwater availability and causing environmental stress are the amount and frequency of rainfall. According to the climate change projection for the country [39], the variability of rainfall is expected to further increase, adding more uncertainty and complication to the planning and management of water resources. Furthermore, the population of the area has grown rapidly, from less than 2000 in 1970 to over 7000 in 2010 [40]. Socioeconomic development and related land-use changes due to the expansion of infrastructure and services, construction and commercial activities and urbanization have direct impacts in terms of increasing demand for water resources [37]. The establishment of settlements has been influenced strongly by the availability of water for drinking and domestic uses; all people have access to safe and good quality drinking water [37]. Together, the anthropogenic activities and climate change have affected the availability of water resources, and if these trends continue, the area's ecosystems and residence's households will be further affected. Vulnerability assessment of the environmental and climate change impacts on water resources is therefore essential to inform sustainable water resources management in the area.

3. Methodology

The methodological guidelines for "Vulnerability Assessment of Freshwater Resources to Environmental Change", developed by United Nations Environment Programme (UNEP) and Peking University [10] were used to assess the vulnerability of water resources of Al Jabal Al Akhdar to environmental change and climate impacts. According to the guidelines, the vulnerability of water resources can be assessed from two perspectives: the main threats to water resources and their development and utilization dynamics; and the region's challenges in coping with these threats. The threats can be assessed in terms of resource stresses (RS), development pressure (DP), ecological health (EH) and management capacity (MC). Thus, the vulnerability index (VI) of the water resources can be expressed as: $VI = f(RS, DP, EH, MC)$ [10].

Each component of VI has several parameters: $RS = f$ [water stress (RS_s) and water variation (RS_v)]; $DP = f$ [water exploitation (DP_s) and safe drinking water inaccessibility (DP_d)]; $EH = f$ [water pollution (EH_p) and ecosystem deterioration (EH_e)]; $MC = f$ [water use inefficiency (MC_e), improved sanitation inaccessibility (MC_s), and conflict management capacity (MC_g)]. In accordance with the vulnerability assessment guidelines, a number of governing equations were applied to estimate these parameters and VI (Table 1).

RS determines the water resources availability to meet the pressure of water demands for the growing population taking into consideration the rainfall variability. Therefore, it is influenced by the renewable water resources stress (RS_s) and water variation parameter resulting from long-term rainfall variability (RS_v). RS_s is expressed as per capita water resources and usually compared to the internationally agreed water poverty index of per capita water resources (1700 m^3/person/year) [10]. As Oman is part of West Asia, characterized by scarce water resources, the more appropriate and realistic value of 1000 m^3/person/year [9] was used. RS_v was estimated by the coefficient of variation (CV) of the rainfall record from 1979 to 2012, obtained from Saiq Meteorology Station. The CV was estimated by the ratio of the standard deviation of the rainfall record to the average rainfall (Table 1).

DP was estimated in terms of the overexploitation of water resources (DP_s) and the provision and accessibility of safe drinking water supply (DP_d). DP_s was estimated by the ratio of the total water demands (domestic, commercial, agriculture) to the total renewable water resources (Table 1). DP_d is defined here as the provision of adequate drinking water supplies to meet the basic needs for the society, in regard to how the water development facilities address the population needs [9]. The lack of safe water accessibility was estimated by the ratio of the percentage of population lacking accessibility to the size of the population (Table 1).

EH was measured in terms of the water quality/water pollution parameter (EH_p) and the ecosystem deterioration parameter (EH_e). EH_p was estimated by the ratio of the total untreated wastewater discharge in water receiving systems to the total available renewable water resources (Table 1). The amount of untreated wastewater is estimated as the difference between the generated wastewater collected by the system and amount of wastewater that received treatment. EH_e is defined in this study as the ratio of land area without vegetation coverage (*i.e.*, total land area except that covered with pastures and cultivated areas) to the total land area of Oman ($309,500 \text{ km}^2$) (Table 1).

Table 1. Equations used for calculation of all categories and parameters of vulnerability index of water resources in the study area.

Category	Parameter	Equation	Description
Resource Stress (RS)	RS$_s$	RS$_s$ = (1000 − R)/1000	R: Total renewable water resources per capita (m^3/person/year)
	RS$_v$	RS$_v$ = CV/0.3 CV = S/μ	CV: Coefficient of variation μ: Mean rainfall (mm) S: Standard deviation
Development Pressures (DP)	DP$_s$	DP$_s$ = WRs/WR	WRs: Total water demands WR: Total renewable water resources
	DP$_d$	DP$_d$ = P$_d$/P	P$_d$: Population without access to improved drinking water sources P: Total population of the area
Ecological Health (EH)	EH$_p$	EH$_p$ = (WW/WR)/0.1	WW: Total untreated wastewater WR: Total renewable water resources
	EH$_e$	EH$_e$ = A$_d$/A	A$_d$: Land area without vegetation coverage A: Total area of the country
Management Capacity (MC)	MCe	MC$_e$ = (WE$_{wm}$ − WE)/WE$_{wm}$	WE: GDP value produced from 1 m^3 of water WE$_{wm}$: Mean WE of West Asia countries
	MC$_s$	MC$_s$ = P$_d$/P	P$_d$: Population without access to improved sanitation P: Total population of the area
	MC$_g$	MC$_g$ = parameter matrix	Matrix scoring criteria (Table 2)

$$VI = \sum_{i=1}^{n} \left[\left(\sum_{j=1}^{m_i} x_{ij} * w_{ij} \right) * w_i \right]$$

n: number of parameter category
m_i: number of parameters in ith category
X_{ij}: value of jth parameter in ith category
W_{ij}: Weight given to jth parameter in ith category
W_i: Weight given to ith category

MC assesses the vulnerability of water resources by evaluating the current management capacity to cope with three critical issues: efficiency of water resources use; human health in relation to accessibility to adequate and safe sanitation services; and overall conflict management capacity. Thus, MC was measured with Water use inefficiency parameter (MC$_e$), Improved Sanitation inaccessibility parameter (MC$_s$), and Conflict Management Capacity Parameter (MC$_g$). MC$_e$ was estimated in terms of the financial contribution to gross domestic product (GDP) of one cubic meter of water in any of the water consuming sectors compared to the world average for a selection of countries [10]. Since the agriculture sector is the major consumer of water in Oman, including the study area, it was used to indicate the financial return from the water use. Therefore, MC$_e$ was calculated using US $40 as the mean GDP value produced from 1 m^3 of water for the countries of West Asia [9] (Table 1). MC$_s$ was used as a typical value to measure the capacity of the management system to deal with livelihood improvement in reducing pollution levels. Improved sanitation was defined here as facilities that hygienically separate human excreta from human, animal and insect contact, including sewers, septic tanks, flush toilets, latrines and simple pits [10]. MC$_s$ was estimated as the ratio of proportion of the population without accessibility to improved sanitation facilities to the total population of the area (Table 1). MC$_g$ demonstrates the capacity of a water resources management system to deal with conflicts. A good management system can be assessed by its effectiveness in institutional arrangements, policy formulation, communication mechanisms, and implementation efficiency [10]. The parameter was defined here as the capacity of the area to manage competition over water utilization among different consuming sectors. MC_g was determined based on water assessment survey [37] and expert consultation [41] using conflict management capacity scoring criteria ranging from 0.0 to 0.25 (Table 2), taking into consideration the interrelation of all variables in this table. These aspects were assigned scoring criteria ranging from 0 to 1 giving weights to each parameter.

Table 2. Conflict management capacity parameter assessment matrix (Source: [10]).

Category of Capacity	Description	Scoring Criteria		
		0.0	0.125	0.25
Institutional capacity	Trans-boundary institutional arrangement for coordinated water resources management	Solid institutional arrangements	Loose institutional arrangements	No existing institutions
Agreement capacity	Writing/signed policy/ agreement for water resources management	Concrete/detailed agreement	General agreement only	No agreement
Communication capacity	Routine communication mechanism for water resources management (annual conferences, *etc.*)	Communications at policy and operational levels	Communications only at policy level or operational level	No communication mechanism
Implementation capacity	Water resources management cooperation actions	Effective implementation of basin-wide river projects/programs	With joint project/ program, but poor management	No joint project program

Because the process of determining relative weights can be biased, making it difficult to compare the final results, equal weights were assigned among the parameters in the same category, and also among different categories. According to the guidelines [10], the weight of 0.25 was assigned across all categories (RS, DP, EH, and MC). For parameters RS_s, RS_v, DP_s, DP_d, EH_p and EH_e, the weight of 0.5 was applied, and for parameters MC_e, MC_s, and MC_g, the weight of 0.33 was assigned. The total weights given to all parameters in each category should be equal to 1, and the total weights given to all categories should be also equal to 1 [10].

The vulnerability index (VI) was finally estimated based on the four categories using the equation in Table 1. VI provides an estimated value ranging from zero (non-vulnerable) to one (most vulnerable) to determine the severity of the stress being experienced by the water resources of the study area. A high VI value shows high resource stresses, development pressures and ecological health, and low management capacities.

4. Results and Discussion

4.1. Resource Stresses

4.1.1. Water Stress Parameter

The calculation of water stress for Oman, including the study area, shows a critical water stress ($RS_s = 0.58$) (Table 3) based on the estimated total renewable water resources per capita of 422.5 m³/person/year [42]. The increase in population and rapid socioeconomic development in Al Jabal Al Akhdar exert pressures on water resources: domestic water consumption increased from 150,000 m³ in 2001 to 580,000 m³ in 2012; an annual increase of 35% per year [37]. Much of this increase may be due to the burgeoning tourist industry. For 1985, 1995, and 2005, the calculated RS_s for Oman were 0.0, 0.30, and 0.36 based on the estimated per capita renewable water resources of 1029.35, 697.76 and 635.84 m³/person/year, respectively [9].

Table 3. Calculated Vulnerability Index with various categories and parameters for the water resources of the study area.

Category	Resource Stress		Development Pressure		Ecological Health		Management Capacity		
Parameter	RSs	RSv	DPs	DPd	EHp	EHe	MCe	MCs	MCg
Calculated	0.580	0.330	0.210	0.000	0.140	0.940	1.000	0.000	0.950
Weight in Category	0.50	0.50	0.50	0.50	0.50	0.50	0.33	0.33	0.33
Weighted	0.290	0.165	0.105	0.000	0.070	0.470	0.330	0.000	0.314
Component Total	0.4550		0.1050		0.5400		0.6435		
Weight for Category	0.25		0.25		0.25		0.25		
Weighted	**0.1138**		**0.0263**		**0.1350**		**0.1609**		
Overall Score	**0.436 (High)**								

Notes: Water Stress (RS_s); Water Variation (RS_v); Water Exploitation (DP_s); Safe Drinking Water Inaccessibility (DP_d); Water Pollution (EH_p); Ecosystem Deterioration (EH_e); Water Use Inefficiency (MC_e); Improved Sanitation Inaccessibility (MC_s); Conflict Management Capacity (MC_g).

4.1.2. Water Variation Parameter

Rainfall amount and availability are the dominant factors in the supply of water resources in the study area. Analysis of rainfall data records from 1979 to 2012 resulted in a water variation parameter (RS_v) of 0.33, based on the estimated CV of 0.10, indicating low rainfall variability. The methodology guidelines [9] designate a set of rainfall variation values for the coefficient of variation as CV = 0.3 or as a CV > 0.3. When CV is > 0.3, RS_v is assigned a highest value of 1, indicating large rainfall variation in time and space; a CV less than 0.3 reflects low variability. However, the study area experienced increasing temperatures over the same period (Figure 2). Minimum, mean and maximum temperatures increased at rates of 0.79, 0.27 and 0.15 °C per decade, respectively. Analysis of rainfall data showed a reduction in water availability, with a general decrease in total rainfall from 1979 to 2012 (Figure 2). Over this period, the average rainfall was 296.7 mm; the highest total was in 1997 (901 mm) and annual rainfall decreased subsequently to 202.8 mm in 2012, with an overall decrease in total rainfall at a rate of −9.42 mm

166

per decade; indicating that the area is vulnerable to climate change as it is an arid mountain region. Projection of future climate in Oman using the IPCC A1B scenario shows an increase in temperature and a decrease in rainfall over the coming decades [39].

Figure 2. Trends in mean air temperature (Tmean) and annual rainfall in Saiq Meteorology Station (World Meteorological Organization (WMO) Index: 41254, Universal Transverse Mercator (UTM) coordinates Latitude: 23°04'28.33" N, Longitude: 57°38'46.63" E, Elevation: 1986 m) from 1979 to 2012 (Data source: [33]).

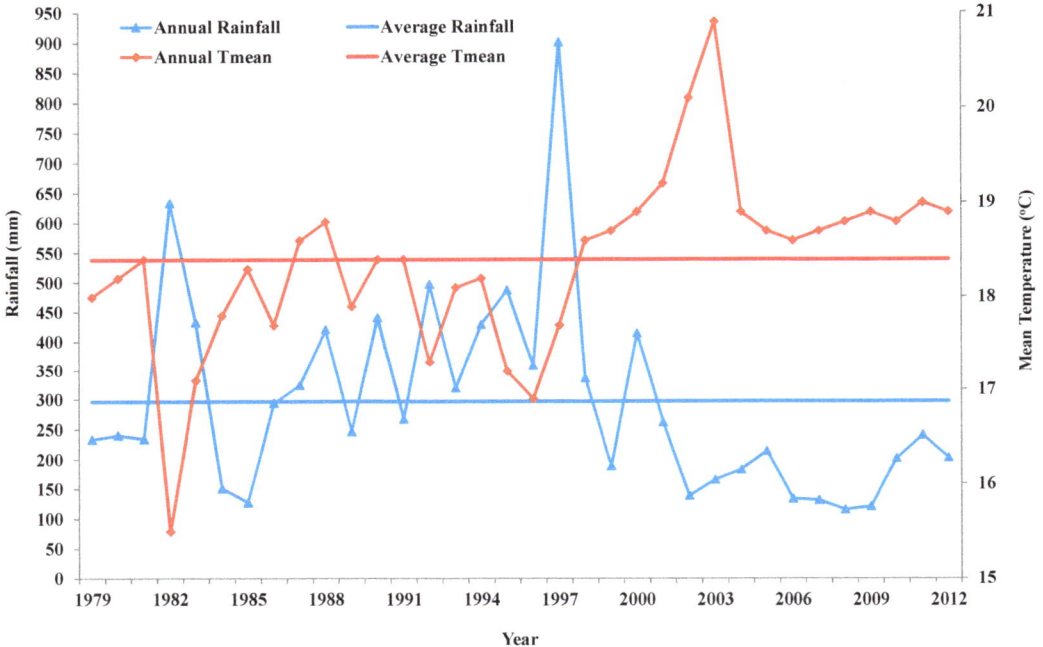

4.2. Water Development Pressures

4.2.1. Water Exploitation Parameter

The assessment of water development pressures indicated that the study area suffers from critical conditions in the development of water resources as determined by the water exploitation parameter ($DP_s = 0.21$) based on total water demands of 14 million m^3/year and the available total water resources of 66 million m^3/year [35] (Table 3), resulting in water shortages for domestic and agricultural purposes. There have been increases in the total population and socioeconomic development as well as increases in construction and commercial activities including hotels, and therefore water consumption by different sectors, causing an imbalance between supply and demand in the absence of the implementation of any conservation and management practices.

4.2.2. Safe Drinking Water Inaccessibility Parameter

The calculated safe drinking water inaccessibility parameter (DP$_d$) was zero since the fundamental needs of the population for water to live are met. There is sufficient infrastructure for providing drinking water throughout the study area; all people have access to safe drinking water. The government supplies drinking water to all households via groundwater wells, and a piped desalinated water project is in progress, to increase the availability of drinking water in the area.

4.3. Ecological Health

4.3.1. Water Pollution Parameter

The estimated water pollution parameter value was (EH$_p$ = 0.14) (Table 3) based on the total untreated wastewater of 945,250 m^3/year [43] and the total available water resources of 66 million m^3/year, given that the urban water usage is 1.1 million m^3/year [35]. The analysis indicates low water pollution risks, which may be attributed to investments in wastewater treatment facilities: the government has established three wastewater treatment plants in the area with tertiary treatment levels and some sewerage systems, and all modern houses and other establishments have septic tanks. However, more investments are needed to increase the proportion of sewer networks connected with the treatment plants. Moreover, some septic tanks in old houses have unlined foundations [43], and need to be reconstructed to avoid pollution to groundwater aquifers.

4.3.2. Ecosystem Deterioration Parameter

Ecosystem deterioration due to the absence of adequate vegetation cover and modified natural landscape is a critical parameter in Oman including the study area, causing severe problems in supporting the functioning of ecosystems. EH$_e$ was calculated as 0.94, based on the evaluation report of the land degradation and desertification in Arab Region [44] including Oman, as there is no available data on ecosystem deterioration for Al Jabal Al Akhdar. There are some indications of ecosystem deterioration in the study area due to decreased rainfall over the last three decades and therefore a decline of groundwater levels and the drying up of most *aflaj* [37]. The population growth, associated with anthropogenic activities and socioeconomic development, and overgrazing, as well as water overconsumption and expansion of land uses through sustained urbanization, have contributed directly or indirectly to the vulnerability of the water resources. The world map of the status of human-induced soil degradation [45] shows that the primary factor contributing to soil degradation in the Al-Hajar Mountains is loss of topsoil through water erosion, with 25%–50% of the area affected by a moderate degree of degradation. According to the study of desertification in the Arab Region by ACSAD (1997) as reported by [46], 89% of Oman was considered as desertified and 7.67% as vulnerable to desertification.

4.4. Management Capacity

4.4.1. Water Use Inefficiency Parameter

Based on the 2013 GDP of Oman (US$80.6 billion) [47] and the total water withdrawal of 1321 million m³/year [42], the calculated water use inefficiency (MC_e) was zero. This is in agreement with [9] which concluded that Oman showed the greatest efficiency gains (decreasing inefficiency) in the West Asia region between 1985 and 2005 (decrease of 25.37%). This was attributed to the uptake of more modern and efficient irrigation infrastructure systems.

However, this parameter was not calculated for the study area since it is based on the country scale and cannot be estimated at a regional scale. In Al Jabal Al Akhdar, farmers still use a traditional method of irrigation by flooding, with no application of modern irrigation technology or investments in improving irrigation infrastructure systems. Based on water assessment survey [37] and personal communication [41] with the author of the UNEP report [9] on this situation, MC_e for the study area was estimated as 1, representing high water use inefficiency. This indicates unsustainable water resources management practices in the absence of a comprehensive water sector plan and strategy, leading to reduced water availability and increased vulnerability.

4.4.2. Improved Sanitation Inaccessibility Parameter

The entire population of the study area has access to sanitation facilities, such as sewer systems, septic tanks and wastewater treatment plants ($MC_s = 0$) (Table 3), indicating adequate management regarding livelihood improvement through government investment in sanitation infrastructure. The availability of this infrastructure reduces pollution levels and preserves water resources, complemented by the implementation of policies and measures which may reduce the vulnerability of water resources to environmental and climate changes. However, more investments to expand the sewerage systems, connecting all households and other establishments to the wastewater treatment plants, are needed.

4.4.3. Conflict Management Capacity Parameter

The study area has no competition over water utilization with the neighboring regions. However, there is competition over water utilization between different sectors (agriculture and domestic). Agriculture is the dominant water consumer, with no application of conservation mechanisms and proper management capacity. There is also an increase in the domestic water consumption from groundwater wells, due to an increase in population and number of hotels and commercial activities, and there is no clear strategy for the development of the area [37]. Therefore, the assessment of MC_g showed a high vulnerability situation in regard to conflict management capacity ($MC_g = 0.95$) since this parameter takes into consideration the interrelation of different categories including institutional, agreement, communication and implementation capacity.

4.5. Vulnerability Index

Based on the available data, the calculated VI is 0.436, in the range of 0.4–0.7 which is classified as high based on the reference sheet for the interpretation of VI [10], indicating that the water resources of Al Jabal Al Akhdar are highly vulnerable and experiencing high stresses. Ecosystem deterioration is the dominant parameter, contributing 27% (Figure 3a). The area has also been experiencing a high degree of water use inefficiency, conflict management capacity and water stress representing 19%, 18% and 17%, respectively (Figure 3a), influencing the overall vulnerability on water resources. Comparison of the share of the different category groups to the final VI showed that the management capacity contributes most to the water resources vulnerability and is the dominant category (37%), followed by ecological health with 31% and water resources stress with 26% (Figure 3b).

Figure 3. (**a**) Percentage of the weighted parameters for Vulnerability Index; (**b**) Share of the percentage of the weighted categories to the final Vulnerability Index for the study area.

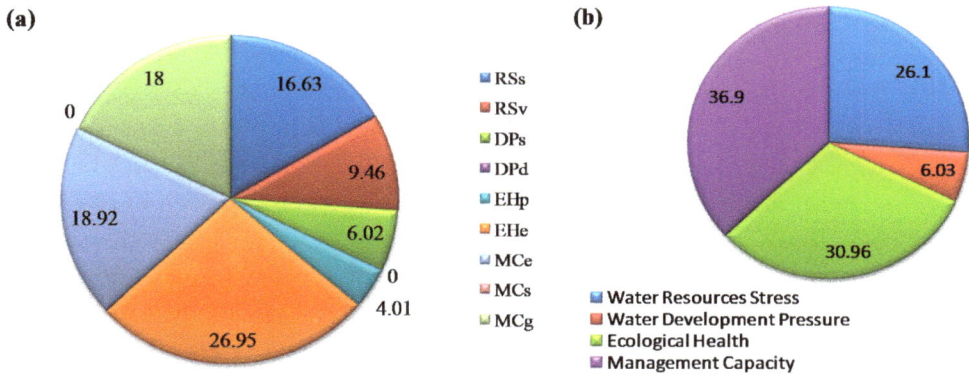

Notes: Water Stress (RSs); Water Variation (RSv); Water Exploitation (DPs); Safe Drinking Water Inaccessibility (DPd); Water Pollution (EHp); Ecosystem Deterioration (EHe); Water Use Inefficiency (MCe); Improved Sanitation Inaccessibility (MCs); Conflict Management Capacity (MCg).

5. Conclusions and Recommendations

This is the first comprehensive vulnerability assessment of water resources in Al Jabal Al Akhdar, or Oman. The results have served to highlight which aspects of water management (resources stress, development pressure, ecological health, and management capacity) contribute most to the vulnerability of water resources and to understand the various risks and thus to suggest potential areas to best focus management efforts. The vulnerability assessment indicated high VI (0.436). Ecosystem deterioration is the dominant parameter contributing 27% to the vulnerability index. The water resources of the area have also been experiencing a high degree of water use inefficiency, conflict management capacity and water stress, influencing the overall vulnerability index by 19%, 18% and 17%, respectively. Management capacity is the dominant category, representing 37% of the category groups, driving the vulnerability of the water resources, which are also highly influenced by the ecological health (31%) and water resources stress (26%). These could be used as indicators for the vulnerability of water resources to environmental and climate changes in the study area. Nevertheless, it must be recognized that due to the lack of availability of local data, some of the inputs to the assessment are at national scale.

There is a clear need for policies and technical solutions to mitigate the pressures (water over consumption and inefficient use, ecosystem deterioration, climate change) which make the water resources more vulnerable. A longer term strategic development plan should be made, with a focus on management capacity to deal with the main threats of conflicts between water consuming sectors, as well as implementation of effective management practices in line with the integrated water resources management approach. Additional effort is needed to improve irrigation water use efficiency, conservation technologies, rainwater harvesting, and reuse of treated wastewater and grey water to relieve some of the agricultural pressures on water resources. There is also an urgent need for mitigation and adaptation to climate change impacts since the region is expected to face further increases in temperatures and decreases in rainfall over the coming decades.

The major contribution of ecosystem deterioration to the overall index suggests that, in order to sustain the ecological health of the area, more efforts are needed to conserve and rehabilitate vegetation cover and implement best practices for land use management and strategic development. More investments are also required to expand sewer networks along with the effective use of wastewater treatment facilities to protect freshwater from pollution. Full coordination, integration and awareness on climate change adaptation should be strongly connected to planning, policies and water management programs at all levels and across all sectors. Further research is needed to provide local, rather than national, input data particularly on the deterioration of ecosystem and vegetation cover, long-term climatic data, and socioeconomic trends, to identify the main driving forces that increase the vulnerability of the water resources, in order to define the optimal approaches for climate change adaptation, to be implemented into operational and sustainable water resources management for the green mountain.

Acknowledgments

The authors would like to thank the Centre for Mountain Studies, Perth College, The University of the Highlands and Islands for covering the cost to publish this article in an open access journal. The authors would also like to thank the Directorate General of Meteorology and Air Navigation, Public Authority of Civil Aviation, Muscat, Oman, for providing rainfall and temperature data for the study area. Furthermore, the authors greatly appreciate the careful revisions of the reviewers.

Author Contributions

Mohammed Al Kalbani collected and analyzed the data, and applied the methodology in collaboration with Martin Price and Asma Abahussain. He also performed all calculations and checked the results in collaboration with Mushtaque Ahmed and Timothy O'Higgins. In addition, he evaluated the data and interpretation the results with Asma Abahussain, and enhanced the writing and editing of the paper with Martin Price.

Conflicts of Interest

The authors declare no conflict of interest.

References

1. Millennium Ecosystem Assessment (MEA). *Ecosystem and Human Wellbeing: Current State and Trends*; Island Press: Washington, DC, USA, 2005.
2. Intergovernmental Panel on Climate Change (IPCC). Summary for Policymakers. In *Climate Change 2007: Impacts, Adaptation and Vulnerability*; Contribution of Working Group II to the Fourth Assessment Report of the Intergovernmental Panel on Climate Change; Parry, M.L., Canziani, O.F., Palutikof, J.P., van der Linden, P.J., Hanson, C.E., Eds.; Cambridge University Press: Cambridge, UK, 2007; pp. 7–22.
3. Intergovernmental Panel on Climate Change (IPCC). Summary for Policymakers. In *Climate Change 2007: The Physical Science Basis*; Contribution of Working Group I to the Fourth Assessment Report of the Intergovernmental Panel on Climate Change; Solomon, S., Qin, D., Manning, M., Chen, Z., Marquis, M., Averyt, K.B., Tignor, M., Miller, H.L., Eds.; Cambridge University Press: Cambridge, UK, 2007; pp. 1–18.
4. Gosling, S.N.; Warren, R.; Arnell, N.W.; Good, P.; Caesar, J.; Bernie, D.; Lowe, J.A.; van der Linden, P.; O'Hanley, J.R.; Smith, S.M. A review of recent developments in climate change science. Part II: The global-scale impacts of climate change. *Prog. Phys. Geogr.* **2011**, *35*, 443–464.
5. Intergovernmental Panel on Climate Change (IPCC). Climate Change 2014. In *Mitigation of Climate Change*; Contribution of Working Group III to the Fifth Assessment Report of the Intergovernmental Panel on Climate Change; Edenhofer, O., Pichs-Madruga, R., Sokona, Y., Farahani, E., Kadner, S., Seyboth, K., Adler, A., Baum, I., Brunner, S., Eickemeier, P., *et al.*, Eds.; Cambridge University Press: Cambridge, UK and New York, NY, USA, 2014.
6. Kundzewicz, Z.W.; Mata, L.J.; Arnell, N.W.; Döll, P.; Kabat, P.; Jiménez, B.; Miller, K.A.; Oki, T.; Sen, Z.; Shiklomanov, I.A. Freshwater resources and their management. In *Climate Change: Impacts, Adaptation and Vulnerability*; Contribution of Working Group II to the Fourth Assessment Report of the Intergovernmental Panel on Climate Change; Parry, M.L., Canziani, O.F., Palutikof, J.P., van der Linden, P.J., Hanson, C.E., Eds.; Cambridge University Press: Cambridge, UK, 2007; pp. 173–210.
7. Rockstrom, J.; Falkenmark, M.; Karlberg, L.; Hoff, H.; Rost, S.; Gerten, D. Future water availability for global food production: The potential of green water for increasing resilience to global change. *Water Resour. Res.* **2009**, *45*, 1–16.
8. Gain, A.K.; Giupponi, C.; Renaud, F.G. Climate Change Adaptation and Vulnerability Assessment of Water Resources Systems in Developing Countries: A Generalized Framework and a Feasibility Study in Bangladesh. *Water* **2008**, *4*, 345–366.
9. United Nations Environment Programme (UNEP). *Assessment of Freshwater Resources Vulnerability to Climate Change: Implication on Shared Water Resources in West Asia Region*; UNEP: Nairobi, Kenya, 2012; pp. 1–164.

10. Huang, Y.; Cai, M. *Methodologies Guidelines: Vulnerability Assessment of Freshwater Resources to Environmental Change*; United Nations Environment Programme (UNEP) and Peking University, China; UNEP, Regional Office for Asia and the Pacific: Bangkok, Thailand, 2009; pp. 1–28.

11. Ribot, J. The causal structure of vulnerability: Its application to climate impact analysis. *GeoJournal* **1995**, *35*, 119–122.

12. Downing, T.E.; Patwardhan, A. *Vulnerability Assessment for Climate Adaptation*; APF Technical Paper 3, Final Draft; United Nations Development Programme: New York, NY, USA, 2003.

13. Bankoff, G.; Frerks, G.; Hilhorst, D. *Mapping Vulnerability: Disasters, Development and People*; Earthscan: London, UK, 2004.

14. O'Brien, K.; Eriksen, S.; Schjolden, A.; Nygaard, L. *What's in a Word? Conflicting Interpretations of Vulnerability in Climate Change Research*; CICERO Working Paper 2004:04; Center for International Climate and Environmental Research: Oslo, Norway, 2004.

15. Brooks, N.; Adger, W.N.; Kelly, P.M. The determinants of vulnerability and adaptive capacity at the national level and the implications for adaptation. *Glob. Environ. Chang.* **2005**, *15*, 151–163.

16. Adger, W.N. Social capital, collective action, and adaptation to climate change. *Econ. Geogr.* **2003**, *79*, 387–404.

17. Adger, W.N.; Huq, S.; Brown, K.; Conway, D.; Hulme, M. Adaptation to climate change in the developing world. *Prog. Dev. Stud.* **2004**, *3*, 179–195.

18. Eakin, H.; Luers, A. Assessing the vulnerability of social-environmental systems. *Annu. Rev. Environ. Resour.* **2006**, *31*, 365–394.

19. Brooks, N. *Vulnerability, Risk and Adaptation: A Conceptual Framework*; Working Paper 38; Tyndall Center for Climate Change Research: Norwich, UK, 2003.

20. Kelly, P.M.; Adger, W.N. Theory and practice in assessing vulnerability to climate change and facilitating adaptation. *Clim. Chang.* **2000**, *47*, 325–352.

21. Wisner, B.; Blaikie, P.; Cannon, T.; Davis, I. *At Risk: Natural Hazards, People's Vulnerability and Disasters*; Routledge: London, UK, 2004.

22. Füssel, H.M.; Klein, R.J.T. Climate change vulnerability assessments: An evolution of conceptual thinking. *Clim. Chang.* **2006**, *75*, 301–329.

23. Eakin, H. The social vulnerability of irrigated vegetable farming households in central Puebla. *J. Environ. Dev.* **2003**, *12*, 414–429.

24. Adger, W.N. Vulnerability. *Glob. Environ. Chang.* **2006**, *16*, 268–281.

25. O'Brien, K.; Eriksen, S.; Nygaard, L.; Schjolden, A. Why different interpretations of vulnerability matter in climate change discourses. *Clim. Policy* **2007**, *7*, 73–88.

26. Schneider, S.H.; Semenov, S.; Patwardhan, A.; Burton, I.; Magadza, C.H.D.; Oppenheimer, M.; Pittock, A.B.; Rahman, A.; Smith, J.B.; Suarez, A.; Yamin, F. Assessing Key Vulnerabilities and the Risk from Climate Change. In *Climate Change 2007: Impacts, Adaptation and Vulnerability*; Contribution of Working Group II to the Fourth Assessment Report of the Intergovernmental Panel on Climate Change; Parry, M.L., Canziani, O.F., Palutikof, J.P., van der Linden, P.J., Hanson, C.E., Eds.; Cambridge University Press: Cambridge, UK, 2007; pp. 779–810.

27. Adger, W.N.; Agrawala, S.; Mirza, M.M.Q.; Conde, C.; O'Brien, K.; Pulhin, J.; Pulwarty, R.; Smit, B.; Takahashi, K. Assessment of Adaptation Practices, Options, Constraints and Capacity. In *Climate Change 2007: Impacts, Adaptation and Vulnerability*; Contribution of Working Group II to the Fourth Assessment Report of the Intergovernmental Panel on Climate Change, Parry, M.L., Canziani, O.F., Palutikof, J.P., van der Linden, P.J., Hanson, C.E., Eds.; Cambridge University Press: Cambridge, UK, 2007; pp. 717–743.

28. McCarthy, J.J.; Canziani, O.F.; Leary, N.A.; Dokken, D.J.; White, K.S. *Climate Change 2001: Impacts, Adaptation, and Vulnerability*; Cambridge University Press: Cambridge, UK, 2001.

29. Lavell, A.; Oppenheimer, M.; Diop, C.; Hess, J.; Lempert, R.; Li, J.; Muir-Wood, R.; Myeong, S. Climate Change: New Dimensions in Disaster Risk, Exposure, Vulnerability, and Resilience. In *Managing the Risks of Extreme Events and Disasters to Advance Climate Change Adaptation*; A Special Report of Working Groups I and II of the Intergovernmental Panel on Climate Change (IPCC); Field, C.B., Barros, V., Stocker, T.F., Qin, D., Dokken, D.G., Ebi, K.L., Mastrandrea, M.D., Mach, K.J., Plattner, G.K., Allen, S.K., *et al.*, Eds.; Cambridge University Press: Cambridge, UK, and New York, NY, USA, 2012; pp. 25–64.

30. Turner, B.L.; Kasperson, R.E.; Matson, P.A.; McCarthy, J.J.; Corell, R.W.; Christensen, L.; Eckley, N.; Kasperson, J.X.; Luers, A.; Martello, M.L.; *et al.* A framework for vulnerability analysis in sustainability science. *Proc. Natl. Acad. Sci. USA* **2003**, *100*, 8074–8079.

31. Ionescu, C.; Klein, R.J.T.; Hinkel, J.; Kavi Kumar, R.S.; Klein, R. Towards a Formal Framework of Vulnerability to Climate Change. Available online: www.usf.uni-osnabrueck.de/projects/newater/downloads/newater_wp02.pdf (accessed on 9 September 2014).

32. Luedeling, E. Sustainability of Mountain Oases in Oman: Effects of Agro-Environmental Changes on Traditional Cropping Systems. Master's Thesis, The University of Kassel, Kassel, Germany, 2007.

33. *Director General of Meteorology and Air Navigation (DGMAN)*; Public Authority of Civil Aviation: Muscat, Oman, 2014.

34. Al-Riyami, Y. *Agriculture Development in Al Jabal Al Akhdar*; Working paper presented in the "Symposium of Economic Development in Al Jabal Al Akhdar"; Oman Chamber of Commerce and Industry: Nizwa Branch, Oman, 2006. (In Arabic)

35. MacDonald, M. *Water Balance Computation for the Sultanate of Oman*; Ministry of Regional Municipality and Water Resources: Muscat, Oman, 2013.

36. Patzelt, A. Syntaxonomy, Phytogeography and Conservation Status of the Montane Flora and Vegetation of Northern Oman - A Centre of Regional Biodiversity. In Proceedings of the International Conference on Mountains of the World: Ecology, Conservation and Sustainable Development, Muscat, Sultanate of Oman, 10–14 February 2008; Victor, R., Robinson, M.D., Eds.; Sultan Qaboos University: Muscat, Sultanate of Oman, 2009.

37. Al Kalbani, M.S. Integrated Environmental Assessment and Management of Water Resources in Al Jabal Al Akhdar Using the DPSIR Framework, Policy Analysis and Future Scenarios for Sustainable Development. Ph.D. Thesis, University of Aberdeen, Aberdeen, UK, in progress.

38. Victor, R.; Ahmed, M.; Al Haddabi, M.; Jashoul, M. Water Quality Assessments and Some Aspects of Water Use Efficiency in Al Jabal Al Akhdar. In Proceedings of the International Conference on Mountains of the World: Ecology, Conservation and Sustainable Development, Muscat, Sultanate of Oman, 10–14 February 2008; Victor, R., Robinson, M.D., Eds.; Sultan Qaboos University: Muscat, Sultanate of Oman, 2009.

39. Al-Charaabi, Y.; Al-Yahyai, S. Projection of Future Changes in Rainfall and Temperature Patterns in Oman. *J. Sci. Clim. Chang.* **2013**, *4*, 154–161.

40. National Centre for Statistics and Information (NCSI). *Census 2010: Final Results, General Census on Population, Housing & Establishments 2010*; NCSI: Muscat, Oman, 2012.

41. Abahussain, A.A. Department of Natural Resources and Environment, College of Graduate Studies, Arabian Gulf University, Kingdom of Bahrain, Personal communication, 2014.

42. Food and Agriculture Organization of the United Nations (FAO). AQUASTAT: FAO's Global Water Information System. Available online: http://www.fao.org/nr/water/aquastat/data/query/results.html (accessed on 14 September 2014).

43. Ministry of Regional Municipalities and Water Resources (MRMWR). *Data on Wastewater Treatment Plants and Sewer Networks*; Unpublished Data, 2014.

44. Arab Center for the Study of Arid Zones and Dry Lands (ACSAD); Council of Arab Ministers Responsible for the Environment (CAMRE); United Nations Environment Programme (UNEP). *State of Desertification in the Arab World (Updated Study)*; Arab Center for the Study of Arid Zones and Dry Lands: Damascus, Syria, 2004. (In Arabic).

45. Oldeman, L.R.; Hakkeling, R.T.A; Sombroek, W.G. *World Map of the Status of Human-induced Soil Degradation*; Global Assessment of Soil Degradation (GLASOD), International Soil Reference and Information Centre, United Nations Environment Programme: Nairobi, Kenya, 1991.

46. Abahussain, A.A.; Abdu, A.Sh.; Al-Zubari, W.K.; El-Deen, N.A; Abdul-Raheem, M. Desertification in the Arab Region: analysis of current status and trends. *J. Arid Environ.* **2002**, *51*, 521–545.

47. The World Bank. Data. Available online: http://data.worldbank.org/indicator/NY.GDP.MKTP.CD (accessed on 14 September 2014).

Water Resources Response to Changes in Temperature, Rainfall and CO₂ Concentration: A First Approach in NW Spain

Ricardo Arias, M. Luz Rodríguez-Blanco, M. Mercedes Taboada-Castro, Joao Pedro Nunes, Jan Jacob Keizer and M. Teresa Taboada-Castro

Abstract: Assessment of the diverse responses of water resources to climate change and high concentrations of CO_2 is crucial for the appropriate management of natural ecosystems. Despite numerous studies on the impact of climate change on different regions, it is still necessary to evaluate the impact of these changes at the local scale. In this study, the Soil and Water Assessment Tool (SWAT) model was used to evaluate the potential impact of changes in temperature, rainfall and CO_2 concentration on water resources in a rural catchment in NW Spain for the periods 2031–2060 and 2069–2098, using 1981–2010 as a reference period. For the simulations we used compiled regional climate models of the ENSEMBLES project for future climate input data and two CO_2 concentration scenarios (550 and 660 ppm). The results showed that changes in the concentration of CO_2 and climate had a significant effect on water resources. Overall, the results suggest a decrease in streamflow of 16% for the period 2031–2060 (intermediate future) and 35% by the end of the 21st century as a consequence of decreasing rainfall (2031–2060: −6%; 2069–2098: −15%) and increasing temperature (2031–2060: 1.1 °C; 2069–2098: 2.2 °C).

Reprinted from *Water*. Cite as: Arias, R.; Luz Rodríguez-Blanco, M.; Taboada-Castro, M.M.; Nunes, J.P.; Keizer, J.J.; Taboada-Castro, M.T. Water Resources Response to Changes in Temperature, Rainfall and CO₂ Concentration: A First Approach in NW Spain. *Water* **2014**, *6*, 3049-3067.

1. Introduction

The study of the effects of climate change on the quantity and quality of water resources has attracted a great deal of attention in recent years, particularly at a regional and global scale [1]. There is a general consensus that the Earth will be subject to warming, leading to changes in global climate patterns [1]. Different responses to global warming are expected from different regions of the world. For Europe, predictions on the evolution of temperature and rainfall, based on models of varying resolution and uncertainty, warn of the possibility of increased aridity in the coming decades. This effect is particularly evident for the southern regions where future climate change is projected to worsen conditions in a region already vulnerable to climate variability as well as lower water availability [1]. In general, the models (e.g., ENSEMBLES) predict a rise in mean temperature and a reduction in rainfall for Spain, with an increase in the number of extreme years in terms of maximum temperature, flood and droughts [2,3], although there are divergences depending on the model and the greenhouse gas emission scenario used. Moreover, projections of the future evolution of rainfall are more speculative than those for temperature, especially for smaller regions, but rainfall patterns are expected to alter in intensity and amount [1,2]. Changes in the temporal and spatial distribution of rainfall can also increase inter-annual rainfall variability as well as the risk of heavy rainfall events and droughts. These changes in temperature and rainfall are expected to impact on the hydrological cycle and alter the different processes occurring at catchment scale, including changes in surface runoff, evapotranspiration rates, nutrient enrichment, erosion and sediment transport [4,5], with concomitant effects on human activities and welfare. Human welfare would be impacted by changes in water supplies and demand; changes in opportunities for non-consumptive uses of the environment for recreation and tourism; changes in loss of property and lives from extreme climate phenomena; and changes in human health [1]. An assessment of the vulnerability of water resources to climate change allows anticipating potential negative impacts, and thus planning and establishing preventive actions with time.

Various approaches to assess the effects of climate change on water resources have been used [5–8]. In general, studies have reported that an increase in CO_2, as long as temperature and rainfall remain constant, will cause increases in water yield due to the marked decrease of the stomatal conductance of plants, thus decreasing evapotranspiration [8,9]. On the contrary, others have shown that higher temperatures lead to increased evaporation rates, reductions in streamflow and more frequent droughts [4,5,10]. All these studies indicate that catchment processes may be very sensitive to changes in temperature, rainfall and higher concentrations of CO_2 in the atmosphere. In Europe, most of the investigations on climate change impact on water resources have predicted a general reduction in the annual streamflow (more intense in the dry season) in southern Europe due to lower rainfall amount and higher temperatures [4].

The Iberian Peninsula is located in an area particularly vulnerable to climate change, where climate projections indicate an increase in arid conditions [1]. Several researches have evaluated the effects of climate change on river regimes in the Iberian Peninsula [10–12], but few studies have been conducted in North-Western Spain [13]. In this area, especially in Galicia, most rivers have little or no flow data available for such analyses. Records of river discharges are relatively recent in the region where water resources have been little considered, since Galicia has been considered to have these in abundance and water availability is not a limiting factor for economic and social development. This is undoubtedly the main reason why the research carried out in the Iberian Peninsula in this field has been focused primarily on Mediterranean, semi-arid environments where the scarcity of water gives rise to serious problems [14]. However, the recent floods in autumn 2006 and winter 2013–14 or the extended drought of summer 2007 occurred in Galicia caused significant ecological, economic and social impacts, highlighting the vulnerability of aquatic ecosystems and the resources that depend on them, e.g., aquaculture. This underlines the need to improve the knowledge of river dynamics as a basis for developing watershed management models in the region in order to prevent the risks associated with these natural phenomena, which seem to have increased in intensity recently [15].

In view of the above, this study has attempted to provide an initial estimate of the effects of potential changes in temperature, rainfall and CO_2 concentration on water resources in the Corbeira catchment, a minimally disturbed area located in Galicia (NW Spain). The analysis was performed using the Soil and Water Assessment Tool (SWAT) hydrological model. The SWAT model is widely used for different purposes (modelling runoff, sediment transport, nutrient, pesticides cycle, *etc.*) around the world and has been applied to many different size catchments and under considerably different conditions, usually with satisfactory results [16]. Among its multiple applications, SWAT has been extensively used to evaluate climate change effects on hydrological processes at catchment scale [9,11,16,17] because of its capability to incorporate future climate predictions from climatic models as inputs to the model, and to account for the effects of increased CO_2 on plant development and evapotranspiration. The interest of the study area is due to its location upstream from the Cecebre reservoir, which is an ecosystem of great ecological interest, being a EU Natura 2000 site, as well as the only source of water supply for the city of A Coruña and the surrounding municipalities (450,000 inhabitants). This fact gives special relevance to the findings of this study, from which a trend can be extracted for policy design purposes for the Cecebre reservoir.

2. Materials and Methods

2.1. Description of the Study Area

The study site for this research is the Corbeira catchment, a headwater catchment of the Mero River Basin, the most important water source for the city of A Coruña, NW Spain (Figure 1a). This catchment has an area of 16 km². The altitude ranges from 60 to 474 m (Figure 1b). The slopes are generally steep, with a mean value of 19%. The most common soils (Figure 1c) are Umbrisol and Cambisols [18] with silt and silt-loam texture settled on basic schists of the "Órdenes Complex" [19].

Figure 1. (**a**) Location of the study area; (**b**) Digital elevation map; (**c**) Soil types map; (**d**) Land use map.

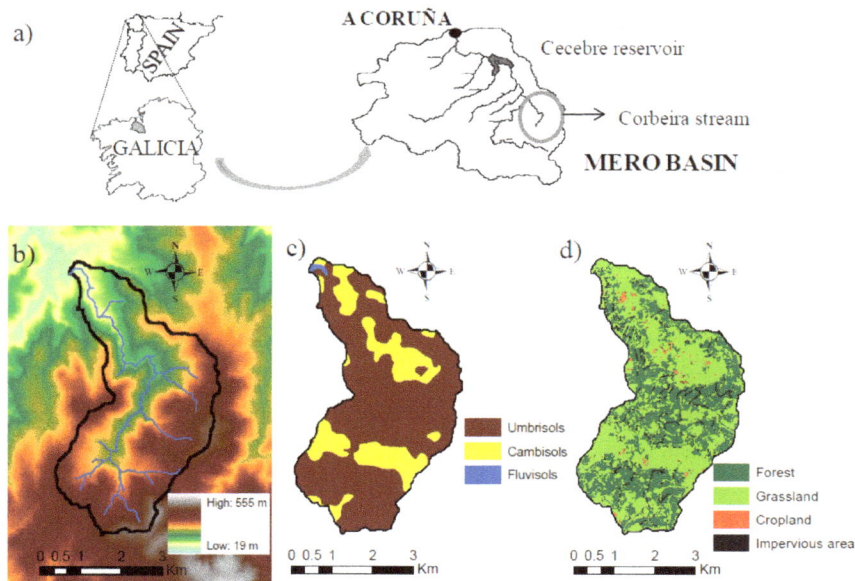

The dominant land uses in this catchment are forestry (65%) and agriculture (30%) (Figure 1d). The forest area mainly consists of commercial eucalyptus plantations, whereas the agriculture area is a patchwork of croplands (4% of total area), growing maize and winter cereal mostly interspersed with grassland (26% of total area). Impervious built-up areas and roads cover about 5% of the whole catchment area and are mainly distributed in the agriculture zone.

The climate is temperate humid, with a mean annual temperature of 13 °C and approximately 1050 mm mean annual rainfall (1983–2009), of which more than 67% falls from October to March.

2.2. Data and Model Setup

2.2.1. The SWAT Hydrological Model

The SWAT model (Soil and Water Assessment Tool) is a process-based and spatially semi-distributed model developed by Agricultural Research Service of the United States Department of Agriculture (USDA) to assess the impact of agricultural management practices on water, sediment and chemical yields in large complex catchments [20], but it is also able to predict water, sediment and nutrient fluxes under different climate change scenarios [8,13,16]. SWAT is a basin-scale, continuous time model operating on a daily step, but it is not designed to simulate detailed, single-event flood routing [21]. The model was selected because it is a dynamic simulation model able to simulate streamflow response to climate change, it is in the public domain, and the generation of input files is eased by GIS-based tools. Although it is frequently applied to medium and large catchments with reasonably good results, it has also been calibrated for small forest catchments with good results [7,22], *i.e.*, with semi-natural land use. So, the SWAT model can be applied for hydrologic simulation in small catchments under different climatic conditions.

In SWAT, the watershed is divided into sub-basins, which are further separated into hydrological response units (HRUs), *i.e.*, areas with a specific combination of soil type, land use and management. Most of the calculations are done at HRU scale and the results integrated at sub-basin scale. SWAT model simulations are divided into two parts: (i) the land phase and (ii) the routing through the river network. The land phase controls the amount of water, sediment and nutrients reaching the main channel, while the second phase defines the movement of water and other elements through the channel to the catchment outlet. The underlying theory detailing transport processes included in the model is available in the SWAT documentation [21].

The simulation of the hydrological cycle is based on the water balance, which is carried out taking into account precipitation, evapotranspiration, surface, lateral and base flow and deep aquifer recharge. The evapotranspiration can be calculated using one of these three methods: Penman-Monteith, Hargreaves and Priestley-Taylor. The Pemman-Monteith method was selected in this case because it uses more physical parameters (daily maximum and minimum temperature, wind speed, humidity and solar radiation). In addition, the Penman-Monteith option in SWAT incorporates the effects of increased CO_2 concentration on plant growth and evapotranspiration. The effect of CO_2 concentration change on plant stomatal conductance is computed by SWAT model using the equation developed by Easterling *et al.* [23], in which increased CO_2 concentrations lead to decreased leaf conductance (doubled CO_2 concentration leads to general decrease of stomatal conductance by 40%) which in turn results in a decrease in the potential evapotranspiration calculation. The change in radiation use efficiency of plants is simulated as a function of CO_2 concentration using the method developed by Stockle *et al.* [24]. Surface flow is estimated using a modification of the Soil Conservation Service Curve Number (SCC CN) method; the lateral flow is calculated based on the kinematic storage model, and the peak runoff rate is estimated by a modified rational method (the peak runoff rate is a function of the fraction of daily rainfall falling in the time of concentration for the sub-basin, the daily surface runoff volume, the sub-basin area and the time of concentration for the sub-basin). The water reaching the river network is then routed to the downstream sub-basin of the catchment. Water is routed through the channel using either the variable storage routing method or the Muskingum river routing method, both of which are variations of the kinematic wave model. In this research, the Muskingum method was used.

2.2.2. Model Inputs

The model required an extensive dataset of meteorological data, topography, soil types and land use and management practices. The meteorological data were acquired from the Galicia Meteorological Service (MeteoGalicia), selecting the closest station to the study area (coordinates: 560019 UTMX-29T ED-50, 4788103 UTMY-29T ED-50). The data included daily rainfall, maximum and minimum temperatures, relative humidity, solar radiation and wind speed. The SWAT model includes the weather generator (WXGEN) model to generate climatic data or to fill in gaps in the measured records. To implement this weather generator, SWAT requires long-term monthly statistical information (e.g., mean and standard deviation) for rainfall, maximum and minimum temperature, dew point temperature, solar radiation, and wind speed. Due to the reduced length of the data series of wind speed, relative humidity and solar radiation, for which consistent 30-year time series are not available, SWAT weather generator used in this study was generated on the basis of the climatic data from a meteorological station (1387E) belonging to the Spanish Meteorological Agency (AEMET) located at about 20 km from the study area whose data keep good correspondence with climatic data acquired from the station of MeteoGalicia (R^2 of 0.82, 0.94, 0.91, 0.72, 0.76 and 0.83 for rainfall, maximum temperature, minimum temperature, wind velocity, solar radiation and relative humidity, respectively). The digital elevation map (DEM) was created from the digital level curves (5 m) provided by the Territorial Information System of Galicia and used to provide elevation details for the SWAT and to delineate catchment boundaries. HRUs were delineated from (i) soil maps (1: 50,000) published by the Environment Department of the Xunta de Galicia, based on the FAO classification [18]; (ii) land use map was obtained from the digital processing (using ER Mapper software) of satellite images Landsat (resolution of 25 m) and aerial photographs (flight from summer 2004, with a resolution of 1 m), provided by the Territorial Information System of Galicia (Xunta of Galicia), and their subsequent field validation, which allowed distinguishing 4 classes of land uses (cropland, grassland, forest and impervious areas); and (iii) three classes of slopes (0%–13%; 13%–25%; >25%). Thresholds of 3% for land use and 20% for soil type were used to limit the number of HRUs in the catchment. This resulted in 12 HRUs. Only one sub-basin was defined in the Corbeira catchment.

The input data of physical soil properties were obtained from experimental works conducted in the catchment, whereas hydrological characterization of soils was built from literature data [25,26] or by estimating the parameters from data of texture and organic matter using pedo-transference functions [27,28].

The characteristics of the grasslands and croplands (maize) were taken from the SWAT database (SWAT plant codes used to represent grasslands and maize land covers were *meadow bromegrass* and *corn*, respectively), while eucalyptus characteristics, not included in the SWAT database, were derived from literature [11,29]. Information on agricultural management, such as dates for planting (maize: 1–10 May) and harvesting (grassland: May, August, November; maize: 20–30 September), was compiled from notes recorded during field research after interviewing farmers. Irrigation is rarely carried out in the catchment; hence, it was not modelled.

2.2.3. Calibration, Validation and Evaluation of the Model

The SWAT model was calibrated using daily stream discharge data measured at the Corbeira catchment outlet. Streamflow was separated into two components (baseflow and direct runoff) using a digital filter [30]. The model was set up from March 2001 to October 2010; the first 3 years were treated as the warm-up period for the model. The period from October 2005 to September 2008 was used for calibration and the period from October 2008 to September 2010 for validation.

The most sensitive model parameters were chosen in the calibration procedure based on preliminary sensitivity analysis using the Latin Hypercube One-factor-At-a-Time approach provided in the SWAT sensitivity analysis interface and SWAT model documentation [21]. The performance of the model presented in Table 1 shows that the simulations generated good results in comparison with observed streamflow data according to the evaluation criteria set by Motovilov *et al.* [31] and Moriasi *et al.* [32], who consider that model performance is satisfactory when the regression coefficient (R^2) is higher than 0.75 and Nash-Sutcliffe efficiency (NSE) is higher than 0.5, and if percentage of bias (PBIAS) is between −25% and +25%. Furthermore, the statistical values for the validation period slightly exceeded those of the calibration period, indicating a low over-parameterization. These differences can be due to the different hydrological conditions of the years used in the study, since hydrological models generally reproduce better normal than extreme hydrological conditions and, in our case, the calibration period comprises more extreme conditions than the validation period.

Table 1. Calibration and validation statistics for daily streamflow. R^2: regression coefficient; PBIAS: percentage of bias; NSE: Nash-Sutcliffe efficiency.

Parameter	Calibration	Validation
R^2	0.80	0.84
PBIAS	−1.8	−3.3
NSE	0.80	0.83
Observed mean and range (m^3 s^{-1})	0.18 (0.02–1.42)	0.24 (0.02–1.20)

The model successfully reproduces the measured streamflow and its trend over time (Figure 2). However, there are some cases in which the model does not agree well with measured streamflow, since it underestimates some peak flows during high-flow periods, e.g., in late March 2006 and in middle December 2006, both coinciding with a flood period. This could be attributed to the inability to simulate the soil moisture conditions during heavy rainfall periods. Curve Number only defines three antecedent moisture conditions: I-dry (wilting point), II-mean moisture, and III-wet (field capacity), and for each of them it assumes a unique relationship between rainfall and runoff, despite one same condition comprises different soil moisture contents. However, in this catchment, the relationship between rainfall and runoff increases as the antecedent soil moisture content rises, and consequently, the hydrological behaviour differs depending on the amount of water stored in the soil [33,34]. It could also be because the method used for simulation of runoff in SWAT (Curve Number) does not account for saturation near-stream zones and is not sensitive to rainfall intensity, so given a same amount of rainfall, SWAT computes the same amount of runoff regardless of intensity and duration of rainfall. The model also overestimates some peak discharge during some rainfall events (e.g., October 2006). This effect has been attributed to different causes, such as the short warm-up period of the model, the underestimation of evapotranspiration and overestimation of soil water content [35].

Figure 2. Rainfall, observed and predicted daily streamflow during the calibration (October 2005–Septmber 2008) and validation period (October 2008–September 2010).

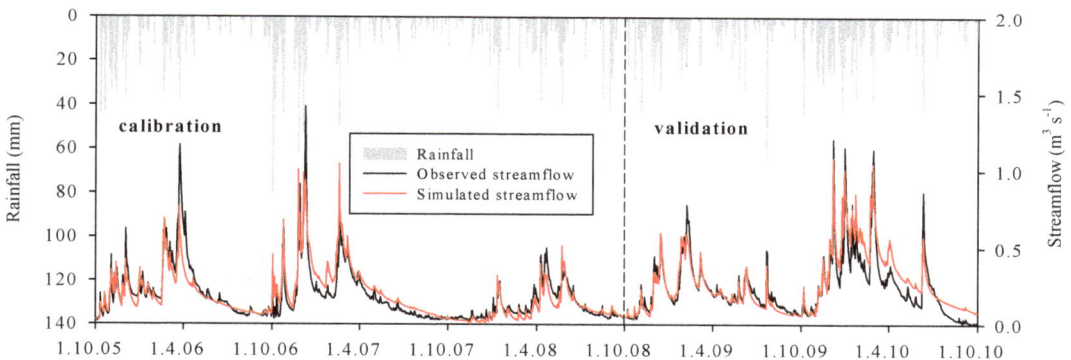

The above results indicate that SWAT is a suitable model to estimate streamflow in the study area; therefore, the calibrated model was used to assess the response of streamflow to future climate change. Several studies have highlighted the degree of uncertainty associated with the evaluation of climate change impacts on hydrology, pointing out that model calibration with present data may result in a bias when applying the model in the future [36,37]. The fact that the model adequately estimates the streamflow either during wet (July 2006) or dry years (August 2007) is a good indicator of its suitability for evaluating the impact of climate change; therefore, it can be used to assess the effects of climate change on streamflow with a reasonable degree of confidence. Anyway, it is appropriate to interpret the effects of climate change in terms of trends, not specific situations.

2.3. Scenarios of Temperature, Rainfall and CO_2 Concentration Changes

In this study, the analysis of climate change impact on water resources has been focused on predicting the potential effects that changes in temperature, rainfall and CO_2 concentration will cause on streamflow. For this purpose, two simulation sets were performed. The first evaluates the effect caused in streamflow by changes in each of the variables, *i.e.*, temperature, rainfall or CO_2 concentration. The second analyses the response of streamflow to simultaneous changes in the three variables.

At present, various methods are used for generating future climate scenarios. These methods include downscaling, change factor method, *etc.* A common procedure to downscale monthly temperature and precipitation of global climate model (GCM) projections to daily time series is stochastic downscaling using a weather generator [38]. Stochastic weather generators can be modified to generate future daily values of climate variables by adjusting historical weather patterns based on predicted future alterations from GCMs or RCMs [39]. In this study, the weather generator included in SWAT [21] was used to create 30-year climate series with changes in input variables. WXGEN uses a first-order Markov chain model to define the day as dry or wet. The daily rainfall amount is estimated based on a skewed or exponential distribution. Daily maximum and minimum temperatures, solar radiation and relative humidity are then generated based on the presence or absence of rain for the day [40]. WXGEN stochastic weather generator is widely used for climate change studies [11,41,42].

To accomplish the analysis of climate change impact on water resources the following steps were performed:

- Once calibrated the SWAT to represent the control conditions (control scenario), the model was run using climate series produced by the weather generator for reference period 1981-2010. Then, the degree of correspondence between simulated streamflow using the stochastic weather generator and simulated streamflow using observed meteorological data were verified, all with the purpose of checking the performance of the SWAT model using a stochastic weather generator to estimate streamflow in the study area. The results of statistical indicators ($r^2 = 0.86$ and NSE $= 0.76$), interpreted according to the criteria proposed by Motovilov *et al.* [31] and Moriasi *et al.* [32], suggest that model performance is satisfactory, indicating that the SWAT weather generator model can be used with a reasonable degree of confidence to analyze climate change scenarios.

- Subsequently, the different climate scenarios were created from the information provided by regional models of the ENSEMBLES project, using the change factors to modify the values of the parameters of the weather generator in a monthly-specific manner.

- Finally, the results for the control scenario were compared with the results of the climate change scenarios in order to quantify the changes on water resources.

The data for the reference period (1981–2001) were obtained from the 1387E meteorological station (AEMET). The different scenarios selected for this study are based on the information provided by the regional models of the ENSEMBLES project [2] for this station (Table 2), for the periods 2031–2060 (intermediate future) and 2069–2098 (distant future). Specifically, the selected scenarios for temperature and rainfall represent the mean and maximum forecasts of the

ENSEMBLES project models (socio-economic A1B scenario) for these variables in the study area. These change factors were used to modify the rainfall and temperature parameters of the weather generator. Other climate variables, such as wind speed, solar radiation, relative humidity and dew point were assumed to be constant throughout future simulation periods. Climate modifications are given as a percentage change in rainfall (rainfall is multiplied by a given factor). Change factors for rainfall were used to alter the frequency and intensity of rainfall. This modification was performed by adjusting the probability of a rainy day followed by another rainy day in the month, and the probability of a rainy day followed by a dry day. These probabilities were obtained by multiplying the baseline probabilities by fifty percent of change factor for rainfall [adjusted probability = baseline probability + (baseline probability × 1/2 of change factor for rainfall)]. Temperature modifications were applied by adding the prescribed change to the weather generator temperature parameters derived from baseline data.

Table 2. Summary of characteristics of the selected RCMs of the ENSEMBLES project.

Name	Institute	GCM	RCM	Time Period
C4IRCA3	C4I [1]	HadCM3Q16	RCA3	1951–2099
CNRM/RM5.1	CNRM [2]	ARPEGE RM5.1	Aladin	1950–2100
DMI/ARPEGE	DMI [3]	ARPEGE	HIRHAM	1951–2100
DMI/BCM DMI	DMI [3]	BCM	DMI-HIRHAM5	1961–2098
DMI/ECHAM5-r3	DMI [3]	ECHAM5-r3	DMI-HIRHAM5	1951–2099
ETHZ/CLM	ETHZ [4]	HadCM3Q0	CLM	1951–2099
ICTP/RegCM3	ICTP [5]	ECHAM5-r3	RegCM3	1951–2100
KNMI/RACMO2	KNMI [6]	ECHAM5-r3	RACMO	1950–2100
MPIM/REMO	MPI [7]	ECHAM5-r3	REMO	1951–2100
SMHI/BCM	SMHI [8]	BCM	RCA	1961–2100
SMHI/ECHAM5-r3	SMHI [8]	ECHAM5-r3	RCA	1951–2100
SMHI/HadCM3Q3	SMHI [8]	HadCM3Q3	RCA	1951–2100

Notes: GCM: global climate models; RCM: regional climate models; [1] Rossby Centre, Swedish Meteorological and Hydrological Institute; [2] National Center of Meteorological. Research, France; [3] Danish Meteorological Institute; [4] Swiss Federal Institute of Technology Zürich; [5] Abdus Salam International Centre for Theoretical Physics, Italy; [6] Royal Netherlands Meteorological Institute; [7] Max Planck Institute for Meteorology, Germany; [8] Swedish Meteorological and Hydrological Institute.

Projected changes in temperature and rainfall in the study area are presented in Figure 3 for the two-time periods (2031–2060 and 2069–2098). All projections show an increase in annual mean temperature and a decrease in rainfall for the two periods, although there is a wide variability among projections, indicating highly uncertain results. The projected temperature changes vary from 0.4–1.8 °C in 2031–2060 to 1.6–3.9 °C in 2069–2098, depending on the climate models. With regard to rainfall, the projections indicate a decrease between 2%–14% in 2031–2060 and 6%–27% in 2069–2098. The mean values of all the projections used predict that temperature will rise by 1.1 °C during 2031–2060 and by 2.2 °C in 2069–2098, while rainfall will decrease by 6% and 27% at the mid and end of the 21st century. CO_2 scenarios change by an increase of 1.5 and 2 times the current CO_2 concentration (330 ppm). It is thought that the selected CO_2 concentrations (550 and

184

660 ppm) give a reasonable representation of future CO_2 conditions for the middle and end of the 21st century under the A1B scenario [3]. Table 3 shows all climate change scenarios used in the SWAT simulations. All climate change scenarios were run for a 30-year period. Land use/land cover was assumed to remain unchanged throughout the simulation.

Figure 3. Climate change scenarios for annual mean temperature and rainfall for: (a) 2031–2060 and (b) 2069–2098 (ENSEMBLES project).

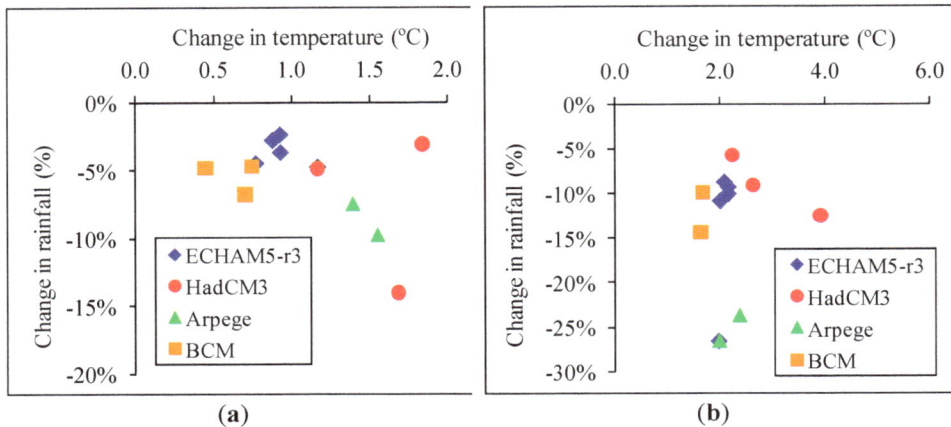

(a) (b)

Table 3. Climate change scenarios used for SWAT simulations.

Scenario	Temperature (°C)	Rainfall (%)	CO_2 Concentration (ppm)
1	1.1 (mean 2031–2060)	0	330
2	1.7 (maximum 2031–2060)	0	330
3	2.2 (mean 2069–2098)	0	330
4	3.9 (maximum 2069–2098)	0	330
5	0	−6 (mean 2031–2060)	330
6	0	−14 (maximum 2031–2060)	330
7	0	−15 (mean 2069–2098)	330
8	0	−27 (maximum 2069–2098)	330
9	0	0	550
10	0	0	660
11	1.1	−6	550
12	1.7	−14	550
13	2.2	−15	660
14	3.9	−27	660
15	1.1	−6	330
16	2.2	−15	330

Note: 0: means no change in the variable.

T-tests were performed to assess if the streamflow estimated from the climate change scenarios and the reference scenario are statistically different from each other. The significance of statistical test was set at $p < 0.01$.

3. Results and Discussion

3.1. Vulnerability of Streamflow to Change in Temperature, Rainfall or CO₂ Concentration

Figure 4 shows the responses of evapotranspiration and streamflow to climate parameter changes. Streamflow will be significantly altered as a result of changes in temperature, rainfall or CO_2 concentration. Streamflow significantly ($p < 0.01$) decreased with the increase in temperature (a larger amount of water is lost through evapotranspiration) and lower rainfall. In both cases, the impact was more pronounced in the period 2069–2098, which showed strong deviation of the climate variables compared to the current conditions (Table 2).

Increasing temperature by 1.1 and 2.2 °C (scenarios 1 and 3) decreased streamflow rates by 13% and 29%, respectively; while 6% and 15% drops in rainfall (scenarios 5 and 7) resulted in a streamflow decrease of 9% and 25%. These results suggest that streamflow in the Corbeira catchment will be more sensitive to the average increase in temperature than to the average decrease in rainfall, highlighting the role of evapotranspiration in the water cycle. However, when compared with the worst case scenarios (scenario 4: T^a + 3.9 °C, scenario 8: P − 27%), streamflow is more sensitive to a reduction in rainfall (Figure 4), which is in accordance with earlier findings in the literature that underline the major role played by rainfall on streamflow changes [5].

Figure 4. Response of evapotranspiration and streamflow to changes in temperature, rainfall and CO_2 concentration based on the scenarios defined in Table 3.

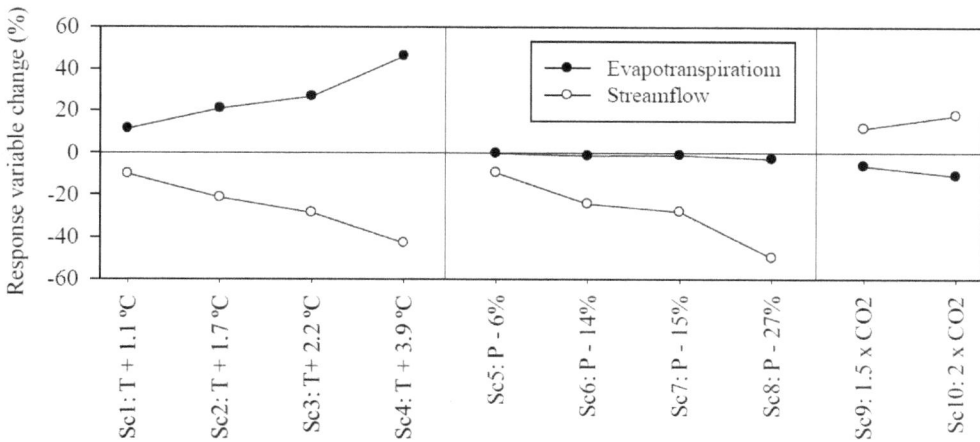

Increase in temperature led to increase in biomass production in some cases and decrease in others, depending on the crop type [6]. This behaviour depends on the temperature reached regarding the optimum, minimum and maximum temperatures associated with plant growth, because temperature is one of the most important factors governing plant growth. For the Corbeira catchment, forest biomass production (forest: 65% of the study area) increased with increasing temperature; however, in grassland and crops, biomass production decreased with temperature increases (Figure 5). This could explain the increase in evapotranspiration (2031–2060: 14%, 21%;

2069–2098: 27%, 46%) and consequently the decrease of streamflow with temperature rise (Figure 4). At present, the main limitations for eucalyptus cultivation in Galicia, the main forest specie in the Corbeira catchment, are low temperatures and frost. The temperature rise and consequent decreased risk of frost increase forest productivity, especially in spring. In summer, the biomass growth rate is lower than in other seasons, which may be associated with water or nutrient limitations. However, the model, under these scenarios, estimated only a slight increase in the number of days with water stress.

Figure 5. Response of vegetation biomass to changes in temperature, rainfall and CO_2 concentration based on the scenarios defined in Table 3.

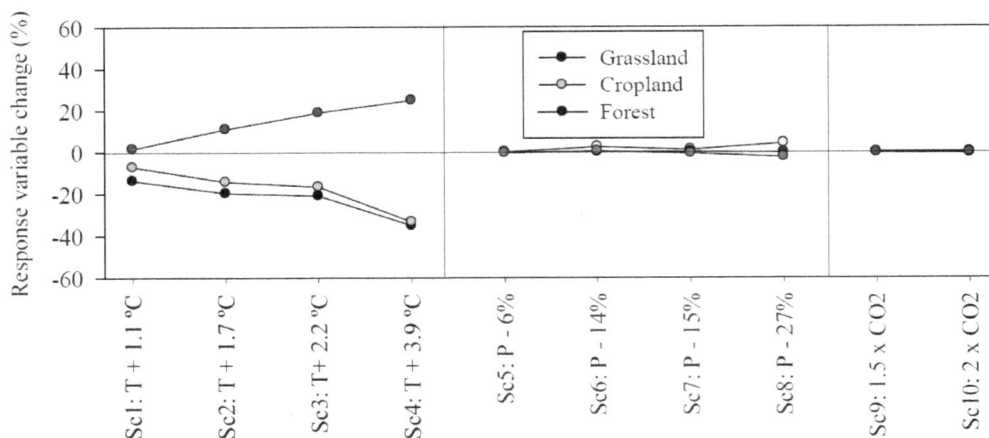

Flow components are affected differently by changes in temperature and rainfall (Figure 6). Thus, a temperature increase had greater impact on subsurface flow (groundwater + lateral flow) because of soil water loss by evapotranspiration, while rainfall decrease had a greater impact on surface runoff. This explains a higher sensitivity of soil water content to temperature changes (Figure 6). These results differ from those of Nunes *et al.* [11] for the Guadiana Basin (southwest Iberian Peninsula) where subsurface flow was mostly affected by reduced rainfall due to the extremely shallow soils in the basin. The soils of the Corbeira catchment, however, are deep and favour the diversion of soil water to evapotranspiration, hence temperature increases mainly affect the subsurface flow.

The streamflow decrease is more significant than that of rainfall (Figures 4 and 6), showing it is not a linear process. It is estimated that for every 1% decrease in rainfall, streamflow falls by approximately 1.5%. These results are close to those obtained by Pruski and Nearing [43], who analysed the effect of rainfall changes on agricultural slopes in different regions of the United States, using the WEPP model. These authors predicted a fall in runoff of 1.97% for every 1% decrease in rainfall. Similarly, Nunes *et al.* [11] reported a fall in runoff of 1.9% and 2.1% for the Guadiana and Tejo basins, respectively.

Figure 6. Response of different flow components and soil water content to changes in temperature, rainfall and CO_2 concentration based on the scenarios defined in Table 3.

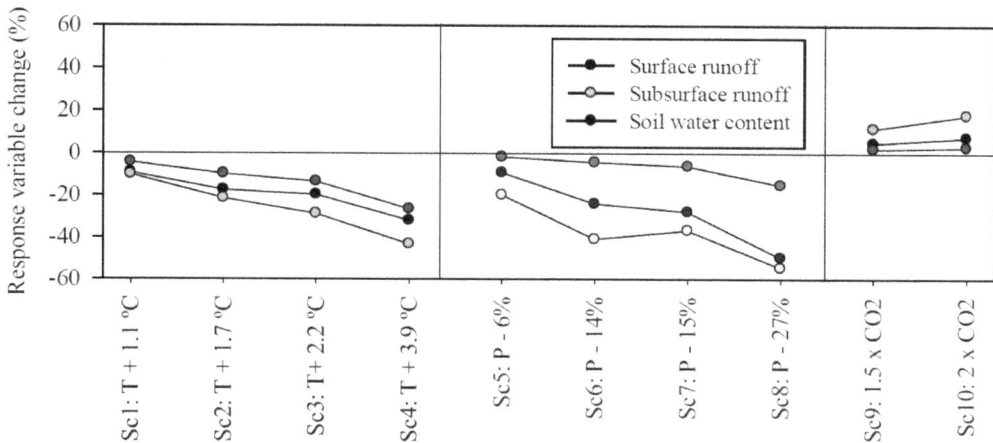

Increasing CO_2 concentration in the atmosphere led to increases in streamflow (10%–15%, Figure 4). The rise in CO_2 concentration could increase vegetation biomass production and evapotranspiration [44]. However, in this study no effect on the vegetation biomass was observed (Figure 5), although a decrease was found in the evapotranspiration (2 × CO_2: −11% evapotranspiration, Figure 4), resulting in higher soil water content and, in turn, higher streamflow (Figure 4). This reduction in the evapotranspiration could be related to stomatal closure of plant leaves in response to increasing CO_2 concentration, as in SWAT a doubled CO_2 concentration leads to a 40% reduction in leaf conductance for all plant species. According to the stomatal conductance optimization hypothesis, the plant stomas are simultaneously maximizing the carbon gain rate while minimizing the rate of water loss [45], *i.e.*, as adaptation mechanism, plants tend to reduce stomatal conductance and suppress transpiration under a high concentration of CO_2. This would lead to greater water-use efficiency by plants (ratios of CO_2 molecules fixed by the plant in relation to the number of water molecules lost by transpiration) allowing a larger amount of water to be available for runoff and recharge.

Moreover, higher CO_2 concentrations can enhance the photosynthesis rate and consequently the vegetation biomass, although it was not observed in this study (Figure 5). This effect, known as the CO_2 fertilization effect, leads to a higher leaf area index (LAI) in the vegetation, which can reduce the radiation reaching the soil surface, thereby reducing soil evaporation, and increasing the streamflow. However, Bunce [46] in a review work concludes that an increase in CO_2 concentrations rarely leads to higher LAI, unless ventilation is artificial, such as it occurs in chambers and greenhouses. In addition, the author indicates that LAI increases above 3–4 m^2 m^{-2} have a minimal effect on evapotranspiration as a result of shade and higher canopy humidity. This conclusion is based on studies of crops in which nutrients are not a limiting factor and, therefore, an even lower response can be expected in natural ecosystems, because nutrients frequently limit plant productivity, thus their responsiveness to CO_2 concentrations.

3.2. Vulnerability of Streamflow to Simultaneous Changes in Climate Parameters

Figure 7 shows the response of evapotranspiration and streamflow to combined simultaneous changes in temperature, rainfall and CO_2 concentration. In comparison with the results obtained when changing climate parameters singly (temperature or rainfall or CO_2 concentration), coupled climate parameter changes had a synergistic effect on streamflow, causing an increase in the vulnerability to change. For scenarios 11 and 13 (mean values), a streamflow decrease of 16% and 35% for the periods 2031–2060 and 2069–2098, respectively is forecasted. For scenarios 12 and 14 (maximum values, unlikely) decreases of 46% and 51% for the same horizons are predicted. Although these results are indicative and should be taken as trend indicators, they show the high vulnerability of the Corbeira stream to climate change, even though climatic variations are relatively low (Table 3). This is consistent with most of the studies on the impact of climate change carried out in the Iberian Peninsula, which have predicted a decline in water resources [7,11,12]. The expected decrease in streamflow in the Corbeira catchment is higher than that estimated for other catchments in NW Spain [12,13], reflecting a greater vulnerability of small catchments to changes in climatic variables, as noted by Beguería *et al.* [47] in other regions of Spain. However, these results should be interpreted with caution. Although the trend seems to be clear, the change percent will vary according to the climate change scenarios considered and the catchment characteristics.

Figure 7. Response of evapotranspiration and streamflow to simultaneous changes in temperature, rainfall and CO_2 concentration based on the scenarios defined in Table 3.

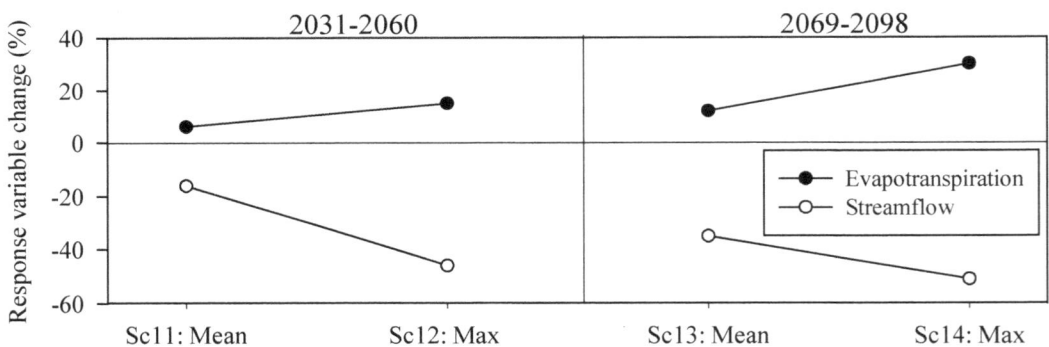

It is generally recognized that the positive effects exerted by an increase in CO_2 concentration on water-plant relationships would be offset by a greater evaporative demand at higher temperatures. Numerous studies indicate that changes in temperature and rainfall alter and, in many cases, limit the effects of CO_2 on plants [11,13]. For example, high temperatures during the flowering period could mitigate the effects of high CO_2 concentration, since they could limit the number, size and grain quality [48]. In order to confirm the effect of a CO_2 increase on streamflow, the model was run using two new scenarios, Sc15, Sc16, (Table 3) with the same rainfall and temperature changes as scenarios 11 and 13, respectively; but with no changes in CO_2 concentration, and both results (with and without increased CO_2) were compared. Not taking into account the effects of CO_2 on plants, there was a decrease in streamflow of 24% (2031–2060, Sc15) and 46% (2069–2098, Sc16) compared to 16% (2031–2060, Sc11) and 35% (2069–2098, Sc13) respectively when enhanced plant photosynthetic water use efficiency (greater stomatal efficiency of plants in response to increased CO_2 concentration) was considered, as for evapotranspiration calculation SWAT takes into account variation of radiation-use efficiency, plant growth, and plant transpiration due to changes in atmospheric CO_2 concentrations. Given the importance exerted by an increase of CO_2 concentration on water resources, this parameter should be considered in any assessment of climate change impact. However, it should be noted that the effects of CO_2 on streamflow might be overestimated, because the SWAT does not assume that leaf area increases with CO_2 concentrations.

4. Conclusions

This work demonstrated the high vulnerability of streamflow to changes in temperature and rainfall in the Corbeira catchment. Furthermore, it was found that an increase in the concentration of CO_2 in the atmosphere could slightly attenuate the effects of climatic variables on water resources. Similarly, both medium and long-term effects of climate change on streamflow can be significant if the forecast temperature and rainfall changes included in this study are met. Overall, the decrease in rainfall was accompanied by a large increase in the evapotranspiration. The combination of these two trends is likely to result in decreased availability of water for crops and natural vegetation. A moderate decrease in streamflow of 16% and 35% is expected for the periods 2031–2060 and 2069–2098, respectively.

In general, it may indicate that this study provides an example of the possible effects of climate change on water resources in the NW Spain, so it can be used as a starting point to improve the understanding of how climate change will impact water resources in this area and provide some data to decision makers. Hydrology and distribution of land uses in the Corbeira catchment are similar to those of the upper and middle Mero River Basin. Therefore, if climate change scenarios adopted in this work occur in the future, significant changes may also occur in the Mero River, affecting the Cecebre reservoir. With increasingly limited water resources and water consumption increasing annually, all the facts point to a situation of greater water unsustainability and therefore to greater environmental unsustainability. This implies that measures, able to solve this situation, should be taken in order to avoid the consequences of a decrease of water resources against an increased demand.

Acknowledgments

This investigation was carried out within the projects REN2003-08143, funded by the Spanish Ministry of Education and Science, and PGIDIT05RAG10303PR and 10MDS103031, financed by Xunta of Galicia. The second author was awarded a Postdoctoral research contract (Xunta of Galicia). We appreciate the detailed review and suggestions of two anonymous reviewers, which improved the manuscript.

Author Contributions

All authors contributed to the design and development of this manuscript under the supervision of M. Teresa Taboada Castro.

Conflicts of Interest

The authors declare no conflict of interest.

References

1. Intergovernmental Panel on Climate Change (IPCC). *Impacts, Adaptation, and Vulnerability.* Part A: Global and Sectoral Aspects. Contribution of Working Group II to the Fifth Assessment Report of the Intergovernmental Panel on Climate Change; Field, C.B., Barros, V.R., Dokken, D.J., Mach, K.J., Mastrandrea, M.D., Bilir, T.E., Chatterjee, M., Ebi, K.L., Estrada, Y.O., *et al*, Eds.; Cambridge University Press: Cambridge, UK; New York, NY, USA, 2014.

2. Van der Linden, P.; Mitchell, J.F.B. *ENSEMBLES: Climate Change and its Impacts: Summary of Research and Results from the ENSEMBLES Project*; Met Office Hadley Centre: Exeter, UK, 2009; p. 160.

3. Houghton, J.T.; Ding, Y.; Griggs, D.J.; Noguer, M.; van der Linden, P.J.; Dai, X.; Maskell, K.; Johnson, C.A. Climate Change: The Scientific Basis. In *Contribution of Working Group I to the Third Assessment Report of the Intergovernmental Panel on Climate Change*; Cambridge University Press: Cambridge, United Kingdom and New York, NY, USA, 2001; p. 881.

4. Lehner, B.; Doll, P.; Alcamo, J.; Henrichs, T.; Kaspar, F. Estimating the impact of global change on flood and drought risks in Europe: A continental integrated analysis. *Clim. Chang.* **2006**, *75*, 273–299.

5. Kundzewics, Z.; Mata, L.; Arnell, N.; Döll, P.; Jimenez, B.; Miller, K.; Oki, T.; Sen, Z.; Shiklomanov, I. The implications of projected climate change for freshwater resources and their management. *Hydrolog. Sci. J.* **2008**, *53*, 3–10.

6. Pruski, F.F.; Nearing, M.A. Climate-induced changes in erosion during the 21st century for eight U.S. locations. *Water Resour. Res.* **2002**, *38*, 34-1–34-12.

7. Zabaleta, A.; Meaurio, M.; Ruiz, E.; Antigüedad, I. Simulation climate change impact on runoff and sediment yield in a small watershed in the Basque Country, northern Spain. *J. Environ. Qual.* **2013**, *43*, 235–245.

8. Chaplot, V. Water and soil resources response to rising levels of atmospheric CO_2 concentration and to changes in precipitation and air temperature. *J. Hydrol.* **2007**, *337*, 159–171.

9. Butcher, J.P.; Johnson, T.E.; Nover, D.; Sarkar, S. Incorporating the effects of increased atmospheric CO_2 in watershed model projections of climate change impacts. *J. Hydrol.* **2014**, *513*, 322–334.

10. López-Moreno, J.I.; Goyette, S.; Beniston, M. Climate change prediction over complex areas: Spatial variability of uncertainties and expected changes over the Pyrenees from a set of regional climate models. *Int. J. Climatol.* **2008**, *28*, 1535–1550.

11. Nunes, J.P.; Seixas, J.; Pacheco, N.R. Vulnerability of water resources, vegetation productivity and soil erosion to climate change in Mediterranean watersheds. *Hydrol. Process.* **2008**, *22*, 3115–3134.

12. Estrela, T.; Pérez-Martin, M.A.; Vargas, E. Impacts of climate change on water resources in Spain. *Hydrolog. Sci. J.* **2012**, *57*, 1154–1167.

13. Raposo, J.R.; Dafonte, J.; Molinero, J. Assessing the impact of future climate change on groundwater recharge in Galicia-Costa, Spain. *Hydrogeol. J.* **2013**, *21*, 459–479.

14. García Ruiz, J.M.; López-Moreno, J.I.; Vicente-Serrano, S.M.; Lasanta-Martínez, T.; Beguería, S. Mediterranean water resources in a global change scenario. *Earth Sci. Rev.* **2011**, *105*, 121–139.

15. Taboada, J.J. Riesgos asociados a fenómenos meteorológicos extremos. In *Riesgos Naturales en Galicia. El Encuentro Entre Naturaleza y Sociedad*; Fra-Paleo, U., Ed.; Universidad de Santiago de Compostela, Servicio de Publicaciones e Intercambio Científico: Santiago de Compostela, Spain, 2010; pp. 25–45.

16. Gassman, P.W.; Reyes, M.R.; Green, C.H.; Arnold, J.G. The Soil and Water Assessment Tool: Historical development, applications, and future research directions. *Trans. ASABE* **2007**, *50*, 1211–1250.

17. Wu, Y.; Liu, S.; Gallant, A.L. Predicting impacts of increased CO_2 and climate change on the water cycle and water quality in the semiarid James River Basin of the Midwestern USA. *Sci. Total Environ.* **2012**, *430*, 150–160.

18. Food and Agriculture Organization (FAO). *World Reference Base for Soil Resources*; World Soil Resources Reports 106; FAO: Rome, Italy, 2014.

19. Instituto Tecnológico Geominero de España (IGME). *Mapa Geológico de España, 1:50000. Hoja 45. Betanzos*; Servicio de Publicaciones del Ministerio de Industria y Energía: Madrid, Spain, 1981.

20. Arnold, J.G.; Srinivasan, R.; Muttiah, R.S.; Williams, J.R. Large area hydrologic modeling and assessment part I: Model development. *J. Am. Water Res. Assoc.* **1998**, *34*, 73–89.

21. Neitsch, S.L.; Arnold, J.G.; Srinivasan, R.; Williams, J.R. *Soil and Water Assessment Tool User's Manual*; Texas Water Resources Institute: Collegue Station, TX, USA, 2002; p. 506.

22. Green, C.; van Griensven, A. Autocalibration in hydrologic modeling: Using SWAT 2005 in small-scale watersheds. *Environ. Model. Softw.* **2008**, *23*, 422–434.

23. Easterling, W.E.; Rosenberg, N.J.; McKenney, M.S.; Jones, C.A.; Dyke, P.T.; Williams, J.R. Preparing the erosion productivity impact calculator (EPIC) model to simulate crop response to climate change and the direct effects of CO_2. *Agric. For. Meteorol.* **1992**, *59*, 17–34.

24. Stockle, C.O.; Williams, J.R.; Rosenberg, N.J.; Jones, C.A. A method for estimating the direct and climatic effects of rising atmospheric carbon dioxide on growth and yield of crops: Part 1—Modification of the EPIC model for climate change analysis. *Agric. Syst.* **1992**, *38*, 225–238.

25. Martínez-Cortizas, A.; Castillo-Rodríguez, F.; Pérez-Alberti, A. Factores que intervienen en la precipitación y el balance de agua en Galicia. *Bol. Asoc. Geógr. Esp.* **1994**, *18*, 79–96.

26. Taboada-Castro, M.M.; Lado-Liñares, M.; Diéguez, A.; Paz, A. Evolución temporal de la infiltración superficial a escala de parcela. In *Avances Sobre el Estudio de la Erosión Hídrica*; Paz, A., Taboada, M.T., Eds.; Universidade da Coruña: A Coruña, Spain, 1999; pp. 101–127.

27. Saxton, K.E.; Rawls, W.J.; Romberger, J.S.; Papendick, R.L. Estimating generalized soil-water characteristics from texture. *Soil Sci. Soc. Am. J.* **1986**, *50*, 1031–1036.

28. Ferrer Julià, M.; Estrela, M.T.; Sánchez, J.A.; García, M. Constructing a saturated hydraulic conductivity map of Spain using pedotransfer functions and spatial prediction. *Geoderma* **2004**, *123*, 257–277.

29. Rodríguez-Suárez, J.A.; Soto, B.; Iglesias, M.L.; Díaz-Fierros, F. Application of the 3PG forest growth model to a Eucalyptus globulus plantation in Northwest Spain. *Eur. J. For. Res.* **2010**, *129*, 573–583.

30. Arnold, J.G.; Allen, P.M.; Muttiah, R.; Bernhardt, G. Automated base flow separation and recession analysis techniques. *Ground Water* **1995**, *33*, 1010–1018.

31. Motovilov, Y.; Gottschalk, G.L.; Engeland, K.; Rodhe, A. Validation of distributed hydrological model against spatial observations. *Agri. For. Meteorol.* **1999**, *98*, 257–277.

32. Moriasi, D.N.; Arnold, J.G.; van Liew, M.W.; Bingner, R.L.; Harmel, R.D.; Veith, T.L. Model evaluation guidelines for systematic quantification of accuracy in watershed simulations. *Transl. ASABE* **2007**, *50*, 885–900.

33. Rodríguez-Blanco, M.L.; Taboada-Castro, M.M.; Taboada-Castro, M.T. Rainfall runoff response and event-based runoff coefficients in a humid area (northwest Spain). *Hydrolog. Sci. J.* **2012**, *57*, 445–459.

34. Palleiro, L.; Rodríguez-Blanco, M.L.; Taboada-Castro, M.M.; Taboada-Castro, M.T. Hydroclimatic response of a humid agroforestry catchment at different times scales. *Hydrol. Process* **2014**, *28*, 1677–1688.

35. Benaman, J.; Shoemaker, C.A.; Haith, D.A. Calibration and validation of soil and water assessment tool on an agricultural watershed in upstate New York. *J. Hydrol.* **2005**, *10*, 363–374.

36. Kirchner, J.W. Getting the right answers for the right reasons: Linking measurements, analyses, and models to advance the science of hydrology. *Water Resour. Res.* **2006**, *42*, doi:10.1029/2005WR004362.

37. Zhang, X.; Xu, Y.P.; Fu, G. Uncertainties in SWAT extreme flow simulation under climate change. *J. Hydrol.* **2014**, *515*, 205–222.

38. Winkler, J.A.; Guentchev, G.S.; Perdinan; Tan, P.N.; Zhong, S.; Liszewska, M.; Abraham, Z.; Niedzwiedz, T.; Ustrnul, Z. Climate scenario development and applications for local/regional climate change impact assessments: An overview for the non-climate scientist. Part I: Scenario development using downscaling methods. *Geogr. Compass* **2011**, *5*, 275–300.

39. Fowler, H.J.S.; Blenkinsop, S.; Tebaldi, C. Linking climate change modelling to impacts studies: Recent advances in downscaling techniques for hydrological modelling. *Int. J. Climatol.* **2007**, *27*, 1547–1578.

40. Sharpley, A.N.; Williams, J.R. *EPIC-Erosion Productivity Impact Calculator, I. Model Documentation*; U.S. Department of Agriculture, Agricultural Research Service: Washington, DC, USA, 1990; p. 235.

41. Ficklin, D.L.; Luo, Y.; Luedeling, E.; Zhang, M. Climate change sensitivity assessment of a highly agricultural watershed using SWAT model. *J. Hydrol.* **2009**, *374*, 16–29.

42. Zhang, H.; Huang, G.H.; Wang, D.; Zhang, X. Uncertainty assessment of climate change impacts on the hydrology of small prairie wetlands. *J. Hydrol.* **2011**, *396*, 94–103.

43. Pruski, F.F.; Nearing, M.A. Runoff and soil-loss responses to changes in precipitation: A computer simulation study. *J. Soil Water Conserv.* **2002**, *57*, 7–16.

44. Rosenzweig, C.; Hillel, D. *Climate Change and the Global Harvest. Potential Impacts of the Greenhouse Effect on Agriculture*; Oxford University Press: New York, NY, USA, 1998; pp. 324.

45. Katul, G.; Manzoni, S.; Palmroth, S.; Oren, R. A stomatal optimization theory to describe the effects of atmospheric CO_2 on leaf photosynthesis and transpiration. *Ann. Bot.* **2009**, *105*, 431–442.

46. Bunce, J.A. Carbon dioxide effects on stomatal responses to the environment and water use by crops under field conditions. *Oecologia* **2004**, *140*, 1–10.

47. Beguería, S.; López-Moreno, J.I.; Lorente, A.; Seeger, M.; García-Ruiz, J.M. Assessing the effect of climate change and land-use changes on streamflow in the central Spanish Pyrenees. *Ambio* **2003**, *32*, 283–286.

48. Hamilton, E.W.; Heckathorn, S.A.; Joshi, P.; Wang, D.; Barua, D. Interactive effects of elevated CO_2 and growth temperature on the tolerance of photosynthesis to acute heat stress in C3 and C4 species. *J. Integr. Plant Biol.* **2008**, *50*, 1375–1387.

Potential Impacts of Climate Change on Precipitation over Lake Victoria, East Africa, in the 21st Century

Mary Akurut, Patrick Willems and Charles B. Niwagaba

Abstract: Precipitation over Lake Victoria in East Africa greatly influences its water balance. Over 30 million people rely on Lake Victoria for food, potable water, hydropower and transport. Projecting precipitation changes over the lake is vital in dealing with climate change impacts. The past and future precipitation over the lake were assessed using 42 model runs obtained from 26 General Circulation Models (GCMs) of the newest generation in the Coupled Model Intercomparison Project (CMIP5). Two CMIP5 scenarios defined by Representative Concentration Pathways (RCP), namely RCP4.5 and RCP8.5, were used to explore climate change impacts. The daily precipitation over Lake Victoria for the period 1962–2002 was compared with future projections for the 2040s and 2075s. The ability of GCMs to project daily, monthly and annual precipitation over the lake was evaluated based on the mean error, root mean square error and the frequency of occurrence of extreme precipitation. Higher resolution models (grid size <1.5°) simulated monthly variations better than low resolution models (grid size >2.5°). The total annual precipitation is expected to increase by less than 10% for the RCP4.5 scenario and less than 20% for the RCP8.5 scenario over the 21st century, despite the higher (up to 40%) increase in extreme daily intensities.

Reprinted from *Water*. Cite as: Akurut, M.; Willems, P.; Niwagaba, C.B. Potential Impacts of Climate Change on Precipitation over Lake Victoria, East Africa, in the 21st Century. *Water* **2014**, *6*, 2634-2659.

1. Introduction

Lake Victoria, Africa's largest fresh water lake covers a surface area of about 68,800 km^2 shared across three East African countries: Uganda (45%), Kenya (6%), and Tanzania (49%). Over 30 million inhabitants depend on Lake Victoria for their livelihoods. Therefore, precipitation changes over the lake are likely to affect the quality of life of many within the East Africa region. Lake Victoria has a complex shoreline structure comprising gulfs and bays that provide potable water abstraction points and also receive municipal and industrial waste from adjacent urban centers.

Due to the vast size of the Lake Victoria basin, it is considered that the average annual lake precipitation almost balances the annual evapotranspiration. Therefore, precipitation variations significantly influence water levels in Lake Victoria. This notion has been applied by several authors to study the water balance of the lake, often translated as changes in the lake levels or outflow regimes—with most variations in the water balance being attributed to the different calculation periods and methods used in estimation of the different balance components *i.e.*, evapotranspiration, inflows, outflows and precipitation [1–5]. About 80% of the Lake Victoria refill is predominantly precipitation compared to the 20% from basin discharge [6]. Satellite remote sensing data was applied in [7] to monitor the water balance of Lake Victoria in comparison to other water bodies in the vicinity—climate forcing explained half of the lake level trends while the outflow patterns were responsible for the other half. Climate forcing is generally affected by the amount of aerosols and greenhouse gases (GHG) in the atmosphere. GHGs absorb and re-emit energy radiated from the Earth's surface, leading to a warming or cooling effect and changes in the Earth's energy balance with time. Increasing greenhouse gas concentrations in the atmosphere leads to warming which in turn causes global atmospheric water vapor and precipitation to increase. Aerosols directly absorb and scatter incoming solar radiation leading to cooling at the surface and a reduction in precipitation. They can also affect precipitation through complex interactions with clouds [8]. At regional scales, changes in precipitation can also be influenced by anthropogenic activities that affect atmospheric transport of water vapor and circulation changes.

The importance of global precipitation changes as addresssed in [8] by the Intergovernmental Panel on climate Change (IPCC) fifth Assessment Report (AR5) suggests a need to understand and project effects of extreme climate conditions. This paper evaluates the newest generation models used in the CMIP5 project with the purpose of studying impact of climate change on the quantity and quality of water in Lake Victoria. Precipitation was aggregated at different temporal scales; daily, monthly and annually. Model evaluation was based on a range of statistical measures and visual graphical comparison for the same aggregation periods in order to postulate possible precipitation changes over Lake Victoria.

2. Data and Methods

2.1. Description of the Study Area

Lake Victoria is located in the upper Nile basin in East Africa within latitudes 00°30′00″ N to 03°00′00″ S and longitudes 31°30′00″ E to 35°00′00″ E. The Lake surface is at an average elevation of about 1135 m.a.s.l (Figure 1). Lake Victoria covers a total catchment area of about 258,000 km^2. The lake itself contributes about 27% of the total catchment area. Generally, the Lake Victoria basin climate is characterized by substantial precipitation occurring throughout the year; however, there are two distinct rainy seasons in which monthly precipitation is generally greater than 10% of the average monthly precipitation. Heavier precipitation occurs in the March-April-May (MAM) season, while the longer rainy season occurs in October-November-December (OND). Climate variability for the lake basin region is influenced by both large-scale and meso-scale circulations resulting from complex interactions of the Inter-Tropical Convergence Zone (ITCZ) and El Nino Southern Oscillation (ENSO), Quasi-biennial Oscillations, large-scale monsoonal winds, and extra-tropical weather systems [9–12].

2.2. Precipitation and Lake Levels

Precipitation over Lake Victoria experienced a predominantly positive trend over the 20th century [12]. A sharp increase in water levels occurred in 1962—it was mainly attributed to the high precipitation in that year and the related high tributary inflows [3,13]. Precipitation occurrence had the largest effect on the lake levels and flow exiting the basin at the Victoria Nile river, while irrigation and hydropower developments had modest effects on these levels and flows [14]. However, commissioning of the Owen Falls Dam, located on the White Nile in 1954 (just prior to the lake rise in 1962) could also have had an impact on the water levels as the lake regained its level as noted by [13,15]. Figure 2 shows the cumulative precipitation and discharge trends from the Lake Victoria catchments compared to the water levels in the lake. It can be deduced that tributary inflows were more significant in increasing lake levels in 1962 and 1998, which years coincided with the El Nino years [9,10] as depicted by the jumps in the cumulative tributary inflows. In conclusion, both human management roles and natural factors affected the lake levels, but precipitation clearly is the major factor. Climate change impact investigations on the Lake Victoria water levels therefore should focus on the future changes in precipitation.

Figure 1. (**a**) Location of Lake Victoria within Africa; (**b**) Coordinates where general circulation model (GCM) precipitation output for Lake Victoria was extracted.

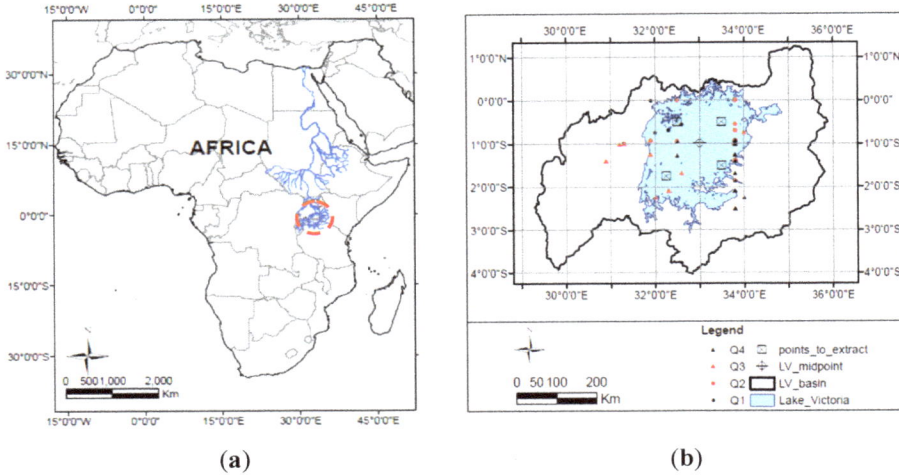

(**a**)   (**b**)

Figure 2. Lake level variations over time compared to cumulative average precipitation over the lake and cumulative total inflow into the lake. The calculated lake level variations are based on the precipitation and inflow [16].

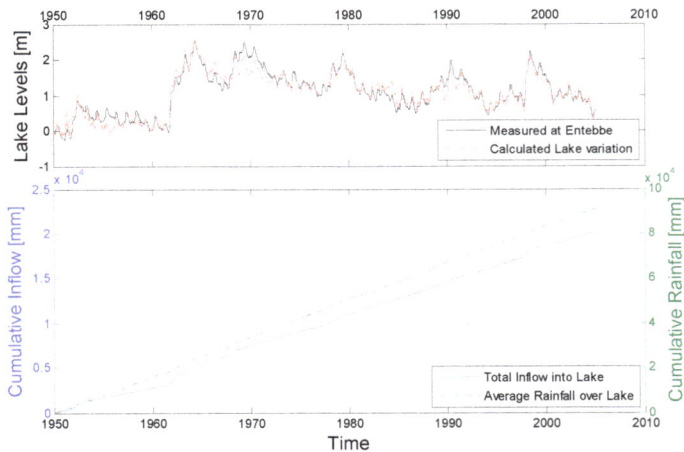

2.3. Climate Model Simulations

General circulation models (GCMs) are numerical models that describe physical processes of the global climate system in the atmosphere, ocean, cryosphere and land surface in response to changing GHG and aerosol concentrations. GCMs provide geographical and physical estimates of regional climate and climate change using three dimensional grids over the globe. The newest generation GCMs are used in the CMIP5 to understand the past and future climate changes. These are the models upon which the recent Fifth Assessment Report (AR5) of the IPCC is based [8]. The CMIPs attempt to address major priorities and incorporate ideas from a wide range of climate modelling communities. The climate change modelling experiments are integrated using atmosphere-ocean global climate models (AOGCMs). These models respond to specified, time-varying concentrations of various atmospheric constituents e.g., GHGs and include interactive representation of the atmosphere, ocean, land and sea ice. CMIP5 also introduces coupling of biogeochemical components to account for closing of carbon fluxes between the oceans, atmosphere and terrestrial biosphere carbon reservoirs for long term simulations in the earth system models. They are capable of using time-evolving emissions of constituents to interactively compute concentrations.

The main difference between these CMIP5 projections and the previous CMIP projections is that their climate change projections include policy intervention and mitigation measures [17]. CMIP5 provides a large set of runs that enable systematic model inter-comparison within each type of experiment and credible multi-model analysis. The core experiments include the historical runs covering much of the industrial period (mid-19th century to the near-present) and future projection simulations forced with specific GHG concentrations and anthropogenic aerosols emissions dubbed "Representative Concentration Pathways" (RCPs) e.g., RCP4.5 and RCP8.5. RCP8.5 is consistent with the high emissions scenario in which the radiative forcing increases throughout the 21st century before reaching 8.5 Wm^{-2} at the end of the century, while RCP4.5 signifies a mid-range mitigations emissions scenario where GHG valuation policies are applied to stabilize atmospheric radiative forcing to 4.5 Wm^{-2} in 2100 (Figure 3). These two CMIP5 scenarios were considered in this study as a basis of exploring climate change impacts and policy issues. RCPs enable investigations of uncertainties related to carbon cycle and atmospheric chemistry. They span a wide range of total forcing values though they do not cover the full range of emissions in the literature, particularly for aerosols [8].

According to [17], a realistic climate model should exhibit internal variability with spatial and temporal structure like the observed. However, in the long-term simulations, timing of individual unforced climate events like El Nino years in the historical runs will rarely (and only by chance) coincide with years of actual occurrence, since historical runs are initiated from an arbitrary point of quasi-equilibrium control run. Hence, the results should be analyzed in probabilistic terms in a similar manner as [18–20].

Figure 3. Representative concentration pathways. (**a**) Changes in radiative forcing relative to pre-industrial conditions; (**b**) Energy and industry CO_2 emissions for the different representative concentration pathway (RCP) candidates. The range of emissions in the recent (post 2001) literature is presented as a thick dashed curve for the maximum and minimum while the shaded area represents the 10th to 90th percentiles [17].

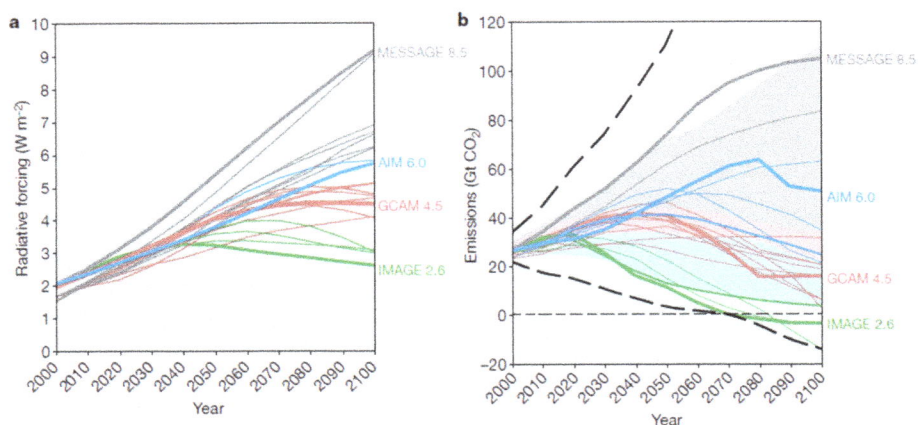

GCMs that were used for both RCP8.5 and RCP4.5 simulations were applied to project the precipitation patterns over Lake Victoria for the 2040s (2020–2060) and 2075s (2055–2095). The historical and future precipitation for the 2040s and 2075s was obtained by simply averaging the simulations from the different GCMs. This method is opposed to applying weighting factors described in [21] and was applied to avoid introducing extra uncertainties as tested by [22]. A total of 42 GCM runs obtained from 26 models simulated by 16 different modeling centers of the CMIP5 archive were used. GCM simulations for the historical period were obtained for the different quarters of Lake Victoria (Figure 1) and averaged to obtain the areal precipitation over the lake.

The performance of GCMs was evaluated based on the historical outputs using the absolute observed precipitation series over the lake for the 41-year period 1962–2002 provided by [12]. Precipitation measurements over the lake are sparse and of low quality. Kizza [16] compared satellite measurements to the lake surrounding observations—TRMM 3B43 product improved the quality of precipitation over the lake by 33% while the PERSIANN product improved the precipitation series by 76%. Kizza *et al.* [12] improved the spatial precipitation input using gridded monthly precipitation with a spatial resolution of 2 km for both ground based and satellite data for the period 1960–2004 providing a plausible lake balance model (Figure 2).

Due to the variation in GCM outputs and for clearer analysis of results, the precipitation simulations by the GCMs were further classified according to the GCM grid sizes. The spatial resolution of the CMIP5 coupled models range from 0.5° to 4° for the atmospheric component and 0.2° to 2° for the ocean component [17]. Table 1 shows an overview of the 26 GCMs used in this study. The model resolutions are classified as follows: Low Resolution (LR) models: grid size >2°,

Medium Resolution (MR) models: 1.5° to 2°, High Resolution (HR) models: <1.5° based on their seasonal performance.

Table 1. Coupled Model Intercomparison Project (CMIP5) general circulation models (GCMs) considered in this study; blue (b), green (g) and red (r).

	Modeling Center	Country	Model	Lat.	Lon.	Res.	Color
i.	Commonwealth Scientific and Industrial Research Organization/ Bureau of Meteorology (CSIRO-BOM)	Australia	ACCESS1.0	1.87	1.25	MR	g
ii.	College of Global Change and Earth System Science, Beijing Normal University	China	BNU-ESM	2.81	2.79	LR	r
iii.	*Centro Euro-Mediterraneo per I Cambiamenti Climatici*	Italy	CMCC-CESM	3.75	3.71	LR	r
		Italy	CMCC-CMS	1.87	1.87	MR	g
iv.	*Centre National de Recherches Meteorologiques / Centre Europeen de Recherche et Formation Avancees en Calcul Scientifique (CNRM/CERFACS)*	France	CNRM-CM5	1.41	1.40	HR	b
v.	Commonwealth Scientific and Industrial Research Organization/ Queensland Climate Change Centre of Excellence (CSIRO-QCCCE)	Australia	CSIRO-Mk3.6	1.87	1.87	MR	g
vi.	Canadian Centre for Climate Modelling and Analysis	Canada	CanESM2	2.81	2.79	LR	r
vii.	Geophysical Fluid Dynamics Laboratory	US-NJ	GFDL-ESM2G	2.5	2.0	LR	r
		US-NJ	GFDL-ESM2M	2.5	2.0	LR	r
viii.	NASA Goddard Institute for Space Studies	US-NY	GISS-E2-H	2.5	2.0	LR	r
		US-NY	GISS-E2-R	2.5	2.0	LR	r
ix.	Met Office Hadley Centre	UK-Exeter	HadCM3	3.75	2.5	LR	r
		UK-Exeter	HadGEM2-CC	1.87	1.25	MR	g
		UK-Exeter	HadGEM2-ES	1.75	1.25	MR	g
x.	*Institut Pierre-Simon Laplace*	France	IPSL-CM5A-LR	3.75	1.89	LR	r
		France	IPSL-CM5A-MR	2.50	1.26	LR	r
		France	IPSL-CM5B-LR	3.75	1.89	LR	r
xi.	Atmosphere and Ocean Research Institute (The University of Tokyo), National Institute for Environmental Studies, and Japan Agency for Marine-Earth Science and Technology	Japan	MIROC-ESM	2.81	2.79	LR	r
		Japan	MIROC5	1.40	1.40	HR	b
xii.	Max Planck Institute for Meteorology (MPI-M)	Germany	MPI-ESM-LR	1.87	1.87	MR	g
		Germany	MPI-ESM-MR	1.87	1.87	MR	g
xiii.	Meteorological Research Institute	Japan	MRI-CGCM3	1.12	1.12	HR	b
xiv.	Norwegian Climate Centre (NCC)	Norway	NorESM1-M	2.50	1.89	LR	r
xv.	Beijing Climate Center, China Meteorological Administration	China	BCC-CSM1.1m	1.12	1.12	HR	b
		China	BCC-CSM1.1	2.81	2.79	LR	r
xvi.	Institute for Numerical Mathematics	Russia	INM-CM4	2.0	1.5	MR	g

2.4. Model Performance Evaluation

Probabilistic analyses were performed to evaluate the effect of climate change on absolute precipitation for different aggregation scales *i.e.*, yearly, monthly, daily; and to investigate reasons for the precipitation changes. The GCM performance was analyzed for the different seasons *i.e.*,

200

January-February (JF), March-May (MAM), June-September (JJAS) and October-December (OND) based on the Normalized Mean Error (NME) and the covariance between the observed and historical GCM output. The NME is defined as the ratio of mean error to sample mean of the observations, while covariance is a measure of how two variables change together—positive covariance implies variables increase or decrease together. Evaluation of GCM performance for annual precipitation was based on the Coefficient of Variation of the Root Mean Square Error (CV(RMSE)) as well. The CV(RMSE) was computed as the ratio of the RMSE to the mean of the observations. The ability of the GCMs to simulate high and extreme precipitation was checked for daily, monthly and annual time scales. For that purpose, precipitation amounts were ranked and plotted against the empirical return period to determine how well the GCMs perform in extreme precipitation distributions. This analysis is useful from a water engineering point of view: If the GCM results would be used for obtaining water engineering design or planning values in terms of precipitation amount for given return periods, the analysis shows the deviations that can be found in these design or planning values.

One important remark should be made about this GCM performance evaluation based on historical precipitation observations: model performance for the historical period, as evaluated here, is not equivalent to future model performance. The latter obviously cannot be validated; that is why the historical analysis is used instead as indicative for future performance.

To determine the influence of future climate change on precipitation, the ratios of potential future simulated precipitation to historical precipitation simulations—hereafter referred to as perturbation factors, were used to project impacts of climate change in the Lake Victoria basin. This approach has been applied to study climate change by several authors e.g., [19,23]. The source of future changes in precipitation reflected in the perturbation factors was further analyzed in the different seasons to understand the influence of individual effects like changes in intensities or number of wet days in each season on the global annual change using Box plots. This analysis aims to provide plausible quantifiable measures of precipitation changes over Lake Victoria in the 21st century.

3. Results and Discussion

3.1. GCM Performance Evaluation

3.1.1. Monthly, Seasonal and Annual Precipitation

The GCM historical and observed series for the period 1962–2002 were aggregated over monthly time scales to evaluate the seasonal variations in the model based precipitation amounts and how much they deviate from the absolute observed values. This was done for the different resolution GCMs (Figure 4a). LR GCMs with grid sizes greater than 2° (>220 km) generally fail to simulate the wetter MAM rainy season depicted by the observed series while the HR GCMs (<165 km) show an acceptable seasonal pattern (Figure 4a). Based on the precipitation results only, the performance of the GCMs improved with the increase in resolution of the GCMs (Figure 4a). The different runs within the same GCMs did not necessarily produce a discrepancy as large as that between the different GCMs (Figure 4a) implying that model parameterization is probably more vital in determining GCM output compared to GCM initializations. For example, CanESM2.1, CanESM2.2, CanESM2.3 have different initializations but not different parameters compared to another model e.g., HadGEM2-CC.1 and HadGEM2-CC.2. From Figure 4a, we can see that differences between HadGEM-CC and CanESM2 models are larger than those arising between different runs within the same model.

Earlier research by [19] reported that there was no strong evidence to suggest that GCM performance improved with higher spatial resolution for the previous generation GCMs (4th Assessment Report of the IPCC based on CMIP3). Of the 18 GCMs of CMIP3 used by [19], only CCSM3.0 can be categorized as HR, based on the definition used in the present manuscript. The three other CMIP3 GCMs (MK3.0, MK3.5, and ECHAM5) similar to CSIRO-Mk3.6 and MPI-ESM-LR in CMIP5 are categorized as MR while the rest fall under LR models. HR and MR GCMs such as CCSM3.0, MK3.0, MK3.5 and ECHAM5 were ranked in the top five performing GCMs while most other GCMs performed poorly for the Katonga and Ruizi catchments, which are located within the Lake Victoria basin [19]. This study conforms to our hypothesis even though the areal extent of these catchments was in the order of 1000–3000 km^2 [19] compared to the 68,800 km^2 expanse of the lake. The improvement in the CMIP5 simulations in which higher spatial resolution coupled models were used to obtain a richer set of outputs cannot be neglected—however, IPCC [8] recognizes an undisputed similarity between CMIP3 and CMIP5 model simulations. This implies that model resolution was vital in determining the GCM performance.

Underestimation of monthly precipitation totals for the LR GCMs can be attributed to the large grid sizes that do not allow simulating different precipitation patterns over the northern and southern parts of the lake since rainfall patterns vary across the Lake Victoria basin. The universal kriging and inverse distance weighting methods used by [16] to obtain the spatial distribution of precipitation over the Lake Victoria basin show influence of the seasonal migration of the ITCZ on the rainy seasons such that the north eastern region generally receives more precipitation compared to the south eastern region. The GCMs underestimate precipitation in the rainy MAM season and the dry JJAS season, but overestimate the variable OND rainy season (Figure 4a,b). With the exception of the HadCM3 model, most GCMs simulate well the variable OND rainy season, which is highly influenced by complex interactions between the Indian and Pacific Oceans, a phenomenon that is well captured by the GCMs. HadCM3, HadGEM2-CC and HadGEM2-ES are developed by the same modeling center using similar radiative forcing. Although these models are essentially different, improved seasonal patterns are noticed in the finer HadGEM2-CC and HadGEM2-ES models compared to the coarser HadCM3 model (Figure 4a).

Figure 4. (**a**) Average monthly precipitation for the different GCMs compared to the observed series, red: Low resolution (LR) GCMs; blue: High resolution (HR) GCMs; green: Medium resolution (MR) GCMs; (**b**) Difference between modeled and observed precipitation for the different classifications of GCMs.

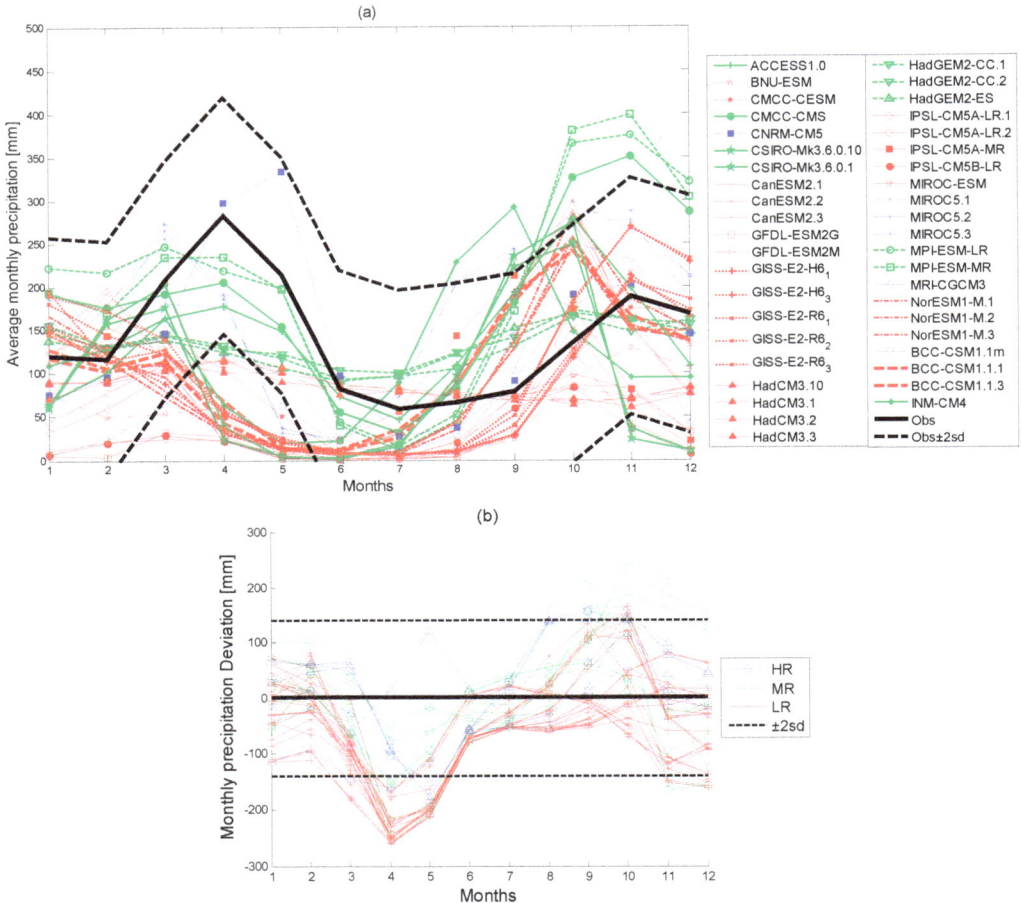

There is a one month lag in the rainy seasons simulated by the GCMs as compared to the observed precipitation (Figure 4a). The monthly precipitation anomalies were calculated to account for the climatology simulated by the different GCMs as a difference between average monthly simulated precipitation, and the average monthly observed precipitation (Figure 4b). Most GCMs simulate the June-February precipitation well (lower monthly anomaly values) while the LR GCMs generally underestimate the MAM season even though there are more LR models compared to HR and MR models (Figure 4b). A more general seasonal check was applied in the JF, MAM, JJAS and OND seasons—to even out effects of the time lag exposed in Figure 4a, as shown in Figure 5a. The LR GCMs generally show lower (often negative) NME and covariance closer to zero or more negative implying that the observed and simulated historical seasonal precipitations did not change together for the LR models. From the covariance results (Figure 5b), it can be concluded that on average the tendency of a linear relationship between the observed and simulated seasonal precipitation decreased with increase in the grid size of the model. Most LR models showed negative covariance, while most HR models showed positive covariance for the average seasonal precipitation.

Figure 5. (**a**) Average seasonal precipitation for the different GCMs compared to the observed series for January-February (JF), March-May (MAM), June-September (JJAS) and October-December (OND) aggregations; (**b**) Covariance *vs.* normalized mean error (NME) for seasonal averages, red: LR GCMs; blue: HR GCMs; green: MR GCMs.

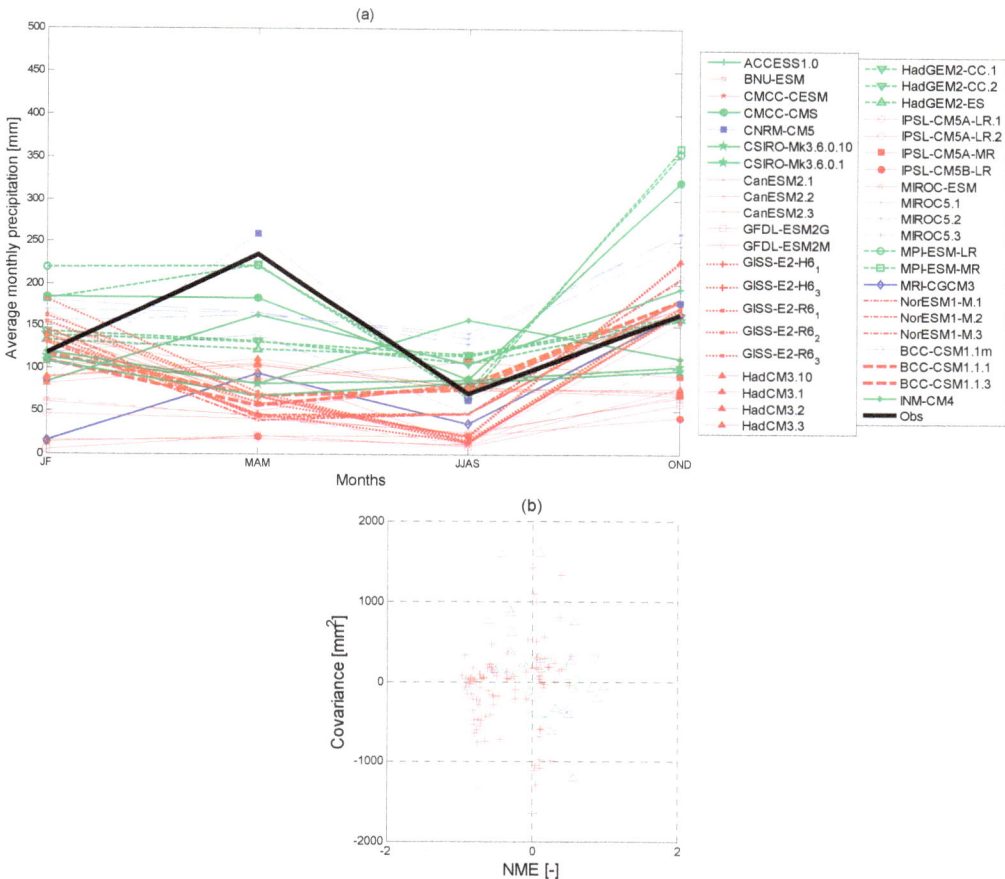

Figure 6 shows the NME and its statistical significance compared to the uncertainty bounds approximated by twice the normalized standard deviation to approximate a 95% confidence interval. The best performing GCMs are again the higher resolution GCMs: CNRM-CM5, ACCESS1.0 and MRI-CGCM3, which lay within plotted uncertainty bounds. Figure 6 also shows the CV(RMSE) on the annual precipitation amounts. The GCMs: GFDL-ESM2G, GFDL-ESM2M, IPSL-CM5A-LR and IPSL-CM5B-LR (all of which are LR models) produce the highest CV(RMSE) and NME, henceforth are considered to be poorly performing. Generally, the GCMs perform better with the annual precipitation simulations compared to seasonal and monthly aggregations based on the NME (−1 to 2.5 for monthly; and −0.7 to 0.5 for annual aggregations in Figure 6).

Figure 6. NME and coefficient of variation of the root mean square error (CV(RMSE)) of average annual precipitation for the GCM simulations compared to the observed series, red: LR GCMs; blue: HR GCMs; green: MR GCMs.

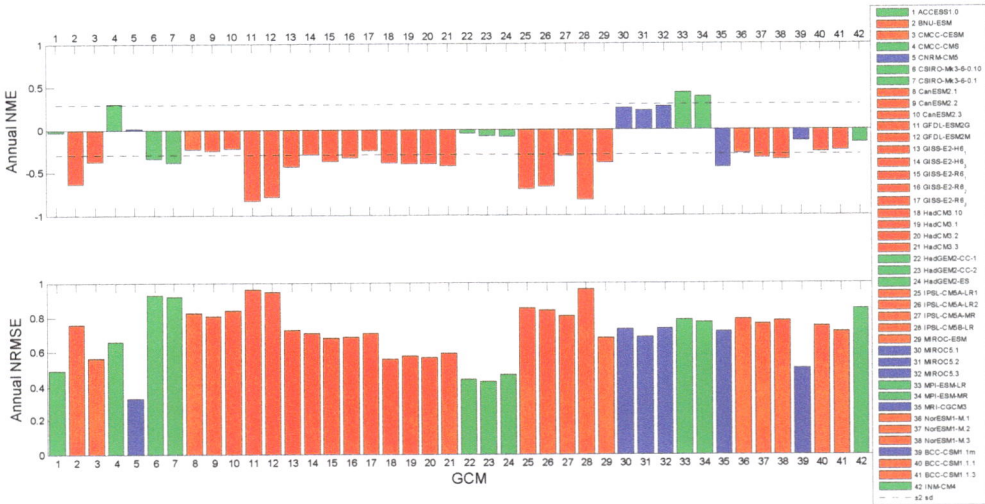

The GCMs show acceptable annual precipitation patterns (observed values located within the interval defined by the standard deviation of the GCM ensemble) but fail to simulate the peak precipitation (Figure 7a,b). The peak precipitation seasons are not well simulated in the GCMs; due to the inability of the GCMs to capture the heavy MAM precipitation even though model representation improves with increased model resolution. Although the peak annual precipitation is not well captured, the general annual variability trend is typically reproduced as it depends on the well simulated OND rainy season rather than the heavy MAM season. The OND and MAM seasons account for more than 65% of the total annual precipitation over the lake, however the variability of precipitation in the OND period has a greater influence on the annual precipitation compared to that in the MAM period [16]. The correlation coefficients between seasonal and annual precipitation totals for the OND and MAM periods were 0.71 and 0.5 respectively, *i.e.*, peaks in annual precipitation totals tended to coincide with peaks in OND rather than MAM seasonal precipitation [24].

Figure 7. Annual variation of observed, and GCM output for (**a**) RCP4.5; (**b**) RCP8.5, GCM output bounds are based on twice the standard deviation.

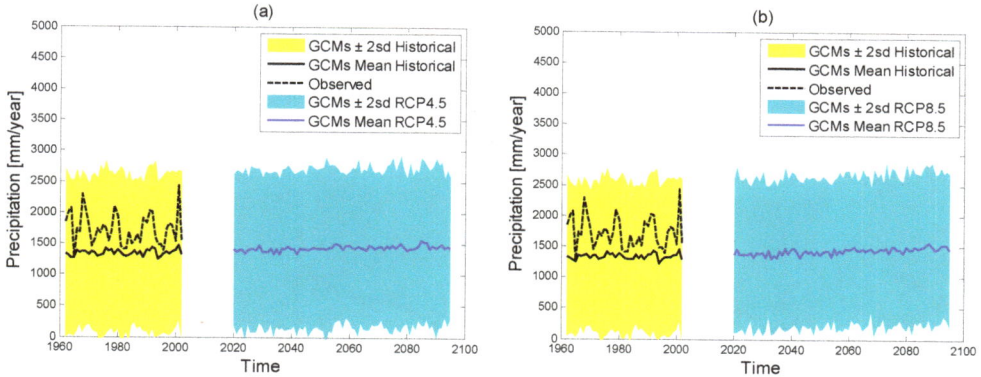

3.1.2. Precipitation Extremes

The LR GCMs underestimate the monthly and annual precipitation amount for given return periods (Figure 8). The underestimation of the precipitation amounts for the higher return periods is probably due to the large variations in topographical and areal properties that are evened out over wider areas while HR GCMs generally provide better simulations for monthly extremes. Some LR GCMs provide satisfactory monthly precipitation extremes, notably BCC-CSM1.1, IPSL-CM5A-MR, GISS-E2-H GISS-E2-R and NorESM-1 (Figure 8a)—this is misleading as monthly precipitation extremes are selected throughout the year yet the observed precipitation peaks in the MAM season may coincide with those in the OND season (Figure 4a). It is not surprising that the LR GCMs consistently simulated lower annual precipitation (Figure 8b). For this reason, monthly precipitation extremes are further classified in the different seasons in Section 3.1.2. CNRM-CM5 and MIROC5 gave the best estimations for both extreme monthly and annual precipitation as shown in Figure 8b.

The GCM performance was evaluated in the wet and dry months *i.e.*, November and July respectively (Figure 9). For the rainy November, the uncertainty in simulating daily precipitation extremes with return periods higher than 4 years is very large, irrespective of the model resolution. The HR models overestimate the extreme precipitation amounts in the wet month of November. Many of these extreme events in the observed series are related to occurrence of El-Niño years as precipitation in the region is strongly quasi-periodic with a dominant ENSO timescale of variability of 5–6 years [9]. The monthly shift in the seasonal variations for the BCC-CSM1.1m, MRI-CGCM3 and BCC-CSM1.1 models was depicted in the daily extreme plots (Figure 9). These HR models overestimate extreme daily precipitation; yet in reality it is due to the one month time lag (Figure 4a). However, in July (the driest month), the GCM performance is very erratic with most LR GCMs underestimating the daily extremes. The uncertainty is higher in the driest month of July which experiences largely varying precipitation (Figure 9b).

Figure 8. (a) Return period of average monthly precipitation; **(b)** Average annual precipitation for the different GCM simulations as compared to the observed series; red: LR GCMs; blue: HR GCMs; green: MR GCMs.

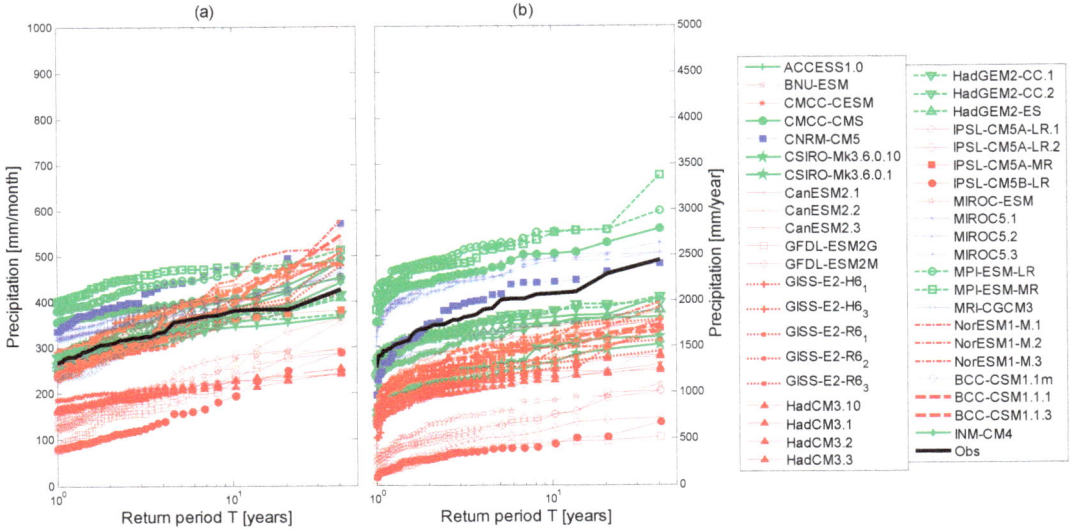

Figure 9. Return period of average daily precipitation for the different GCM simulations for November and July, red: LR GCMs; blue: HR GCMs; green: MR GCMs.

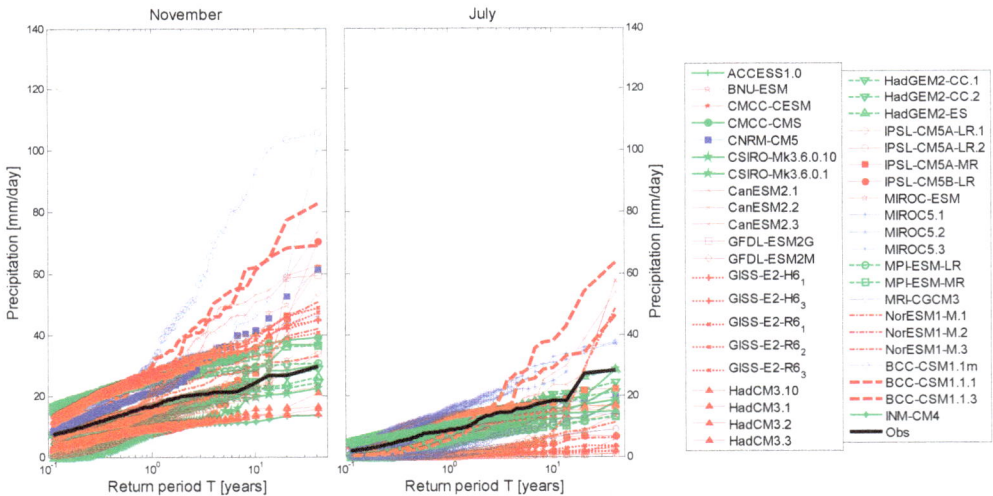

The GCM evaluation showed that GCM outputs provide better results in terms of annual and seasonal precipitation compared to the daily scale analyses (Figures 8,9). Even if model performance cannot be evaluated based on a single index, an array of measures, such as those described in this section provide a good indication of the model overall performance. Higher resolution models provide better estimates of annual, seasonal and monthly precipitation; and precipitation variation (Figures 4,5,8). Table 2 provides a ranking of the good and poor performing GCMs based on annual, seasonal and monthly performance. Differences in latitudes are more significant in GCM precipitation performance than the differences in longitudes. CNRM-CM5 provides the best estimate for annual precipitation and seasonal variation with a minimum time shift in monthly simulations while HadCM3 fails to describe even the basic seasonal variation. As shown by Shaffrey *et al.* [25], reduction of the horizontal resolution e.g., in the HadGEM1 model, may result in reduced SST errors and more realistic approximations of small scale processes, especially the ENSO phenomenon leading to improvement of results simulated by the finer HiGEM model.

Table 2. Ranking of CMIP5 GCMs based on simulation of precipitation over Lake Victoria.

Good Performing GCMs	Long.	Lat.	Resolution	Poor Performing GCMs	Long.	Lat.	Resolution
CNRM-CM5	1.41	1.40	High	HadCM3	3.75	2.5	Low
MIROC5	1.40	1.40	High	IPSL-CM5A-LR	3.75	1.89	Low
BCC-CSM1.1m	1.12	1.12	High	IPSL-CM5B-LR	3.75	1.89	Low
ACCESS1.0	1.87	1.25	Medium	GFDL-ESM2G	2.5	2.0	Low
HadGEM2-CC	1.87	1.25	Medium	GFDL-ESM2M	2.5	2.0	Low
HadGEM2-ES	1.75	1.25	Medium				

3.2. Analysis of Projected Future Precipitation by GCMs

3.2.1. Monthly, Seasonal and Annual Precipitation

The analysis of projected future changes in rainfall shows no significant change for the average monthly precipitation in the 2040s, but a slight increase for the 2075s especially towards the end of the shorter OND rainy season (Figure 10). The historical analysis described earlier in Section 3.1 showed that the precipitation in the OND season is well captured by the GCMs. The magnitude of change is slightly higher for RCP8.5 under which the temperature increase is higher, leading to higher evapotranspiration and precipitation. The effect of GCM uncertainty is found to be far greater than that due to precipitation simulations between the RCP8.5 and RCP4.5 scenarios (Figure 10). Uncertainties in the future simulations are higher for the 2075s than for the 2040s as scenario uncertainty attributed to the uncertainty in emissions of greenhouse gases—hence radiative forcing increases exponentially especially after the 2060s [26].

Perturbation factors for annual precipitation due to climate change are shown in Figure 11. Generally annual precipitation changes converge to the same level for precipitation of return periods greater than two years. For that reason and to simplify the presentation of results, the mean change is computed for the precipitation extremes and plotted for all GCMs in box plots (Figure 11a). Precipitation extremes are defined as events that are larger or equal to those that occur at least once a year. The lower resolution GCMs like IPSL-CM5A-LR, IPSL-CM5B-LR, BNU-ESM, GFDL-ESM2G and GFDL-ESM2M show higher precipitation changes and mostly positive, while the higher resolution GCMs like CNRM-CM5, BCC-CSM1.1m and MIROC5 show precipitation crowding around the unchanged mean climate.

Figure 10. Seasonal variation of average monthly precipitation for observed and GCM historical, RCP4.5 and RCP8.5 series for (**a**) 2040s; (**b**) 2075s. The colored dotted lines indicate the extent of twice the historical standard deviation.

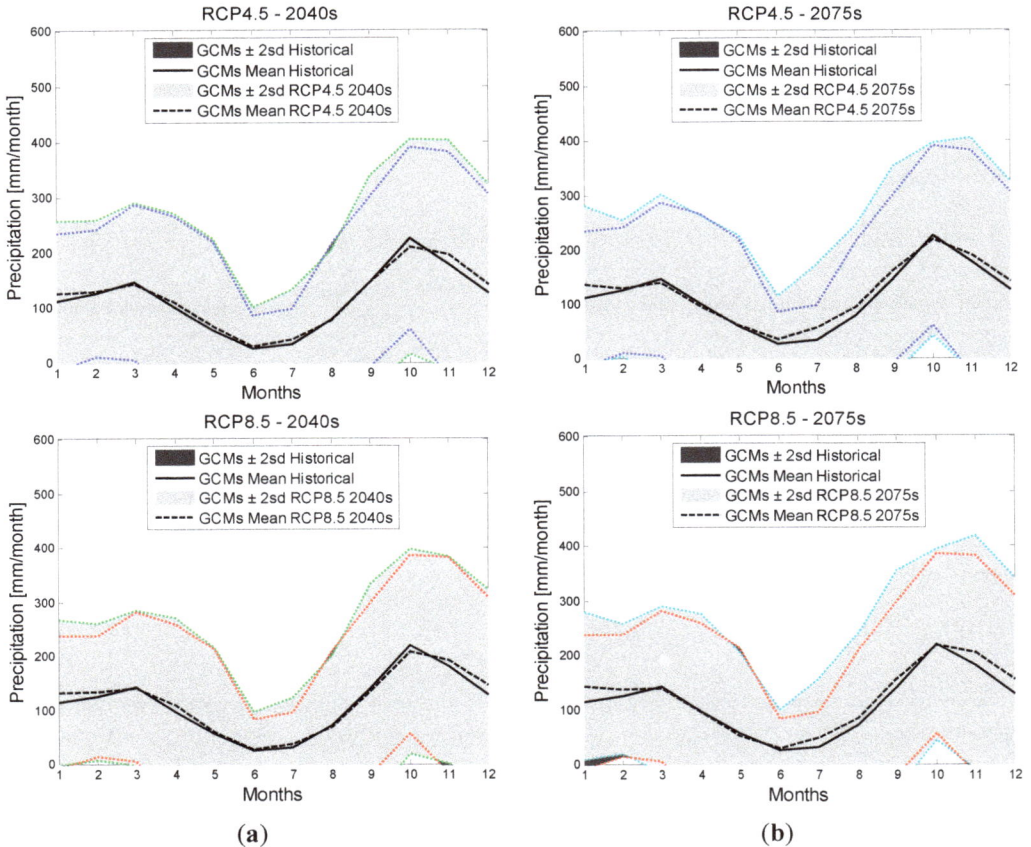

(**a**)   (**b**)

Figure 11 shows an increase in annual precipitation over Lake Victoria in the 21st century for both RCP4.5 and RCP8.5 scenarios (only RCP8.5 shown). For the 2040s, annual precipitation is projected to increase by about 7% for both scenarios, while for the 2075s it is expected to increase by about 10% for the RCP4.5 scenario, and more than 15% for the RCP8.5 scenario. Next to this analysis of annual precipitation, summations of precipitation in the MAM, OND and JJAS seasons were analyzed to determine the perturbation factors for the 2040s and 2075s in order to understand the effect of seasonal precipitation on annual precipitation over the Lake Victoria basin. Figure 12 shows the seasonal change factors for RCP8.5 scenario. Most GCMs generally agree well in the OND rainy season as depicted by the lower divergence and narrower box limits for all resolutions (Figure 12a). Precipitation amounts generally increase in all the seasons for the mitigation RCP4.5 scenario. However, for the RCP8.5 scenario, the seasonal amounts increase only in the rainy seasons (Figure12a). For the dry JJAS season in RCP8.5 scenario, the total seasonal precipitation amount is expected to decrease by about 10% in the 2040s, and increase by about 20% in the 2075s.

Figure 11. (**a**) Perturbation factors for annual precipitation using events with return periods greater than two years for the different GCMs, for the 2040s (notched) and 2075s for RCP8.5; (**b**) Perturbation factors *vs.* return period for annual precipitation for the different GCM simulations for the 2075s under RCP8.5 scenario: Red = LR GCMs, Blue = HR GCMs and Green = MR GCMs.

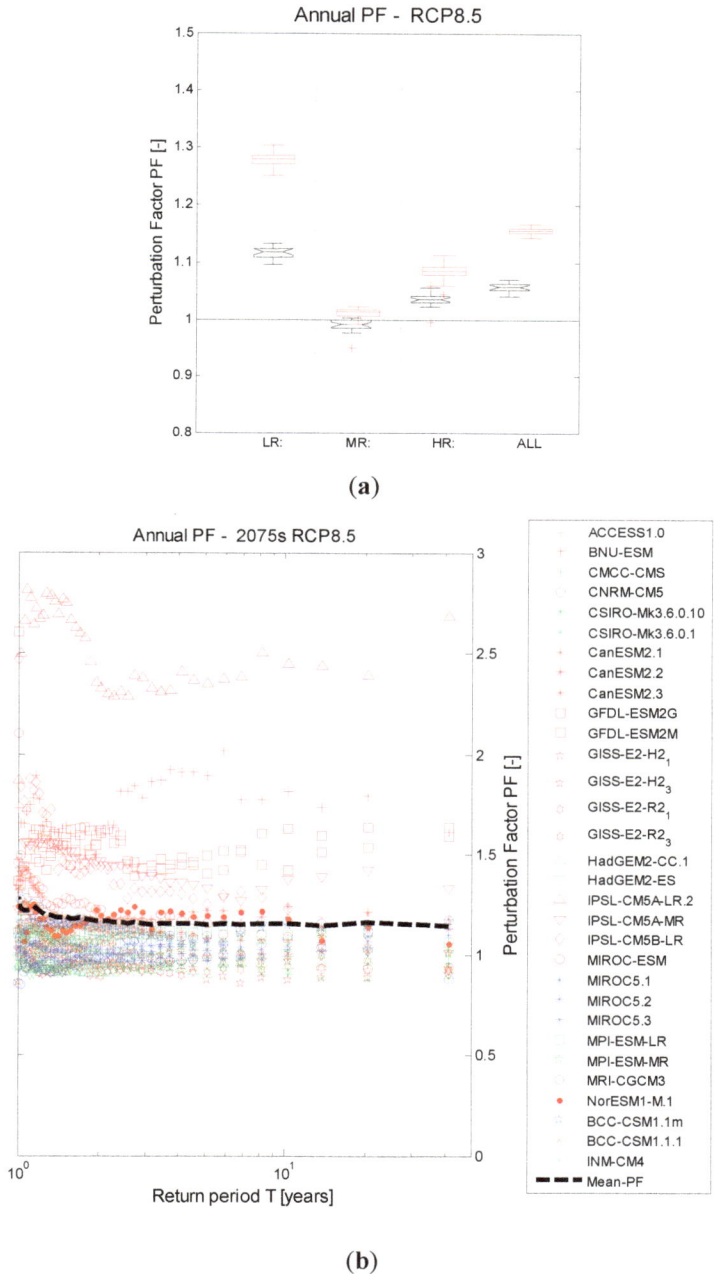

(a)

(b)

Figure 12. (**a**) Perturbation factors for total seasonal precipitation simulated by the different GCMs for the 2040s (notched) and 2075s, for RCP8.5; (**b**) Perturbation factors *vs.* return period for total seasonal precipitation simulated by the different GCMs in the 2075s for RCP8.5, Red = LR GCMs, Blue = HR GCMs and Green = MR GCMs.

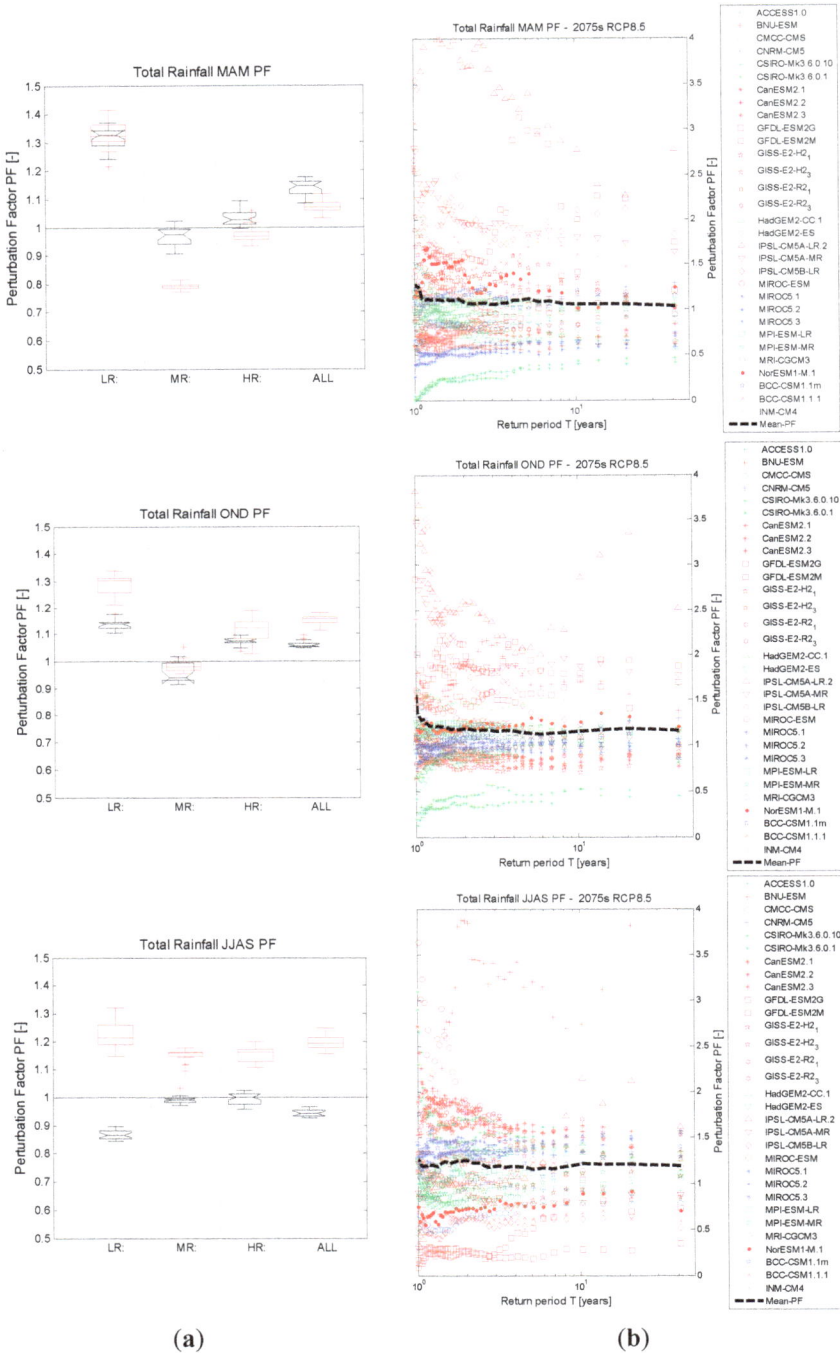

(**a**)   (**b**)

3.2.2. Precipitation Extremes

Changes in Number of Wet Days

A wet day is defined as that having intensity greater than 0.1 mm/day. Precipitation volumes are affected by both the number of wet days and the intensity of precipitation. The number of wet days in the historical and future scenarios was obtained for the different seasons to determine the relative changes in the wet day frequency (Figure 13).

Figure 13. (a) Perturbation factors for number of wet days simulated by GCMs in the different seasons for the 2040s (notched) and 2075s, for RCP8.5; **(b)** Perturbation factors *vs.* return period for number of wet days simulated by the GCMs in the 2075s for RCP8.5 in different seasons, Red = LR GCMs, Blue = HR GCMs and Green = MR GCMs.

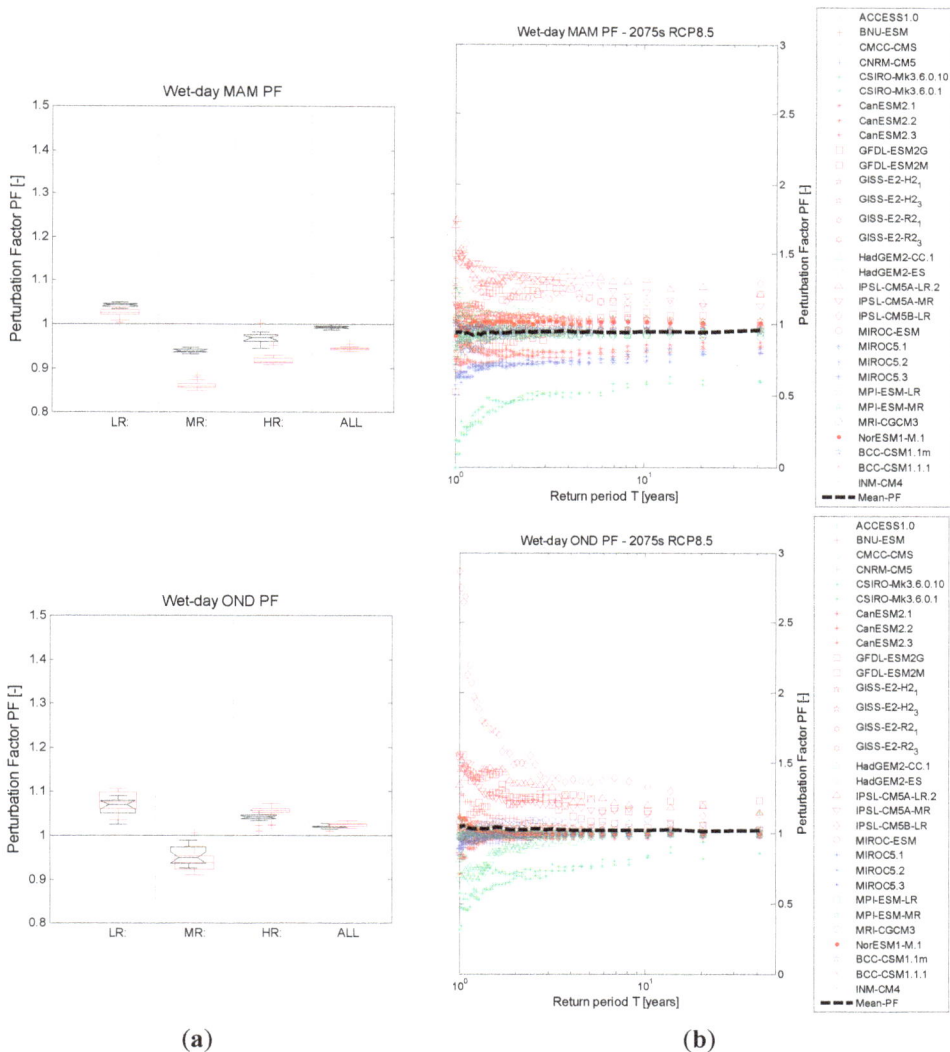

(a)   (b)

Figure 13. *Cont.*

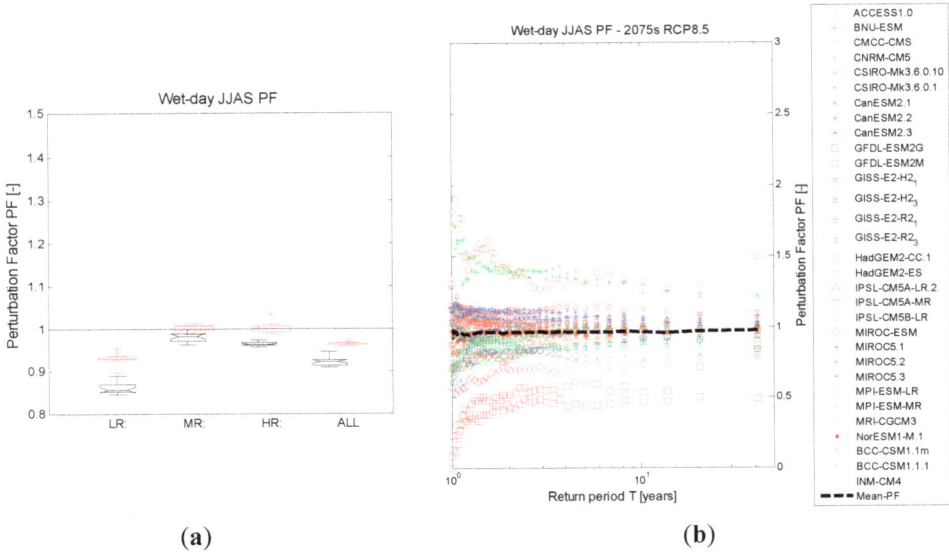

(a)   (b)

The number of wet days decreases in the MAM and JJAS seasons but slightly increases in the OND rainy season for both RCP4.5 and RCP8.5 scenarios (Figure 13). The LR GCMs project greater changes in the number of wet days for the rainy MAM and OND seasons but simulate lower number of wet days in the dry JJAS season probably because larger areas even out localized low precipitation intensities. Annual precipitation over Lake Victoria is estimated to be about 26% higher than over land. This is expected to be associated with 20%–30% more occurrences of cold cloud tops over the lake [27]. It implies that averaging over large areas including land is bound to reduce the precipitation over the lake and sometimes the number of wet days especially in the dry seasons. This again confirms that GCM parameterization and resolution have an important effect on GCM outputs.

Changes in Wet Day Intensities

The daily precipitation intensities in the MAM, OND and JJAS seasons were analyzed to determine the change factors in the 2040s and 2075s under the different scenarios. Figure 14a shows the changes in daily precipitation intensities for the different seasons for events with return periods greater than two years. When changes in wet day intensities *vs.* return periods are analyzed, the intensities are generally seen to increase in the rainy seasons for both RCP4.5 and RCP8.5 scenarios. The daily precipitation extremes increase more towards the 2075s compared to the 2040s especially for RCP8.5 (Figure 14). However, for the dry JJAS season in the RCP8.5 scenario, daily precipitation is generally expected to remain constant in the 2040s and increase by more than 20% in the 2075s explaining the reason for the decrease in the JJAS seasonal precipitation in the 2040s, and increase in the 2075s (Figure 12). The number of wet days slightly decreases in the dry seasons for both decades yet daily precipitation intensities seem to increase more in the late century than the mid-century. In the 2040s, the RCP8.5 and RCP4.5 projections anticipate similar changes in daily extreme precipitation. The difference in relative forcing for the two scenarios in the 2040s is 1 Wm^{-2} compared to 3 Wm^{-2} in the 2075s (Figure 3). This relative difference is consistent with higher temperatures in the 2075s that encourage formation of heavier intense convective storms in the dry season as more moisture is stored in the atmosphere. Therefore, RCP8.5 suggests fewer but heavier intense storms in the 2075s if carbon emissions are not controlled.

Figure 14. (**a**) Perturbation factors for daily precipitation simulated by GCMs in different seasons for the 2040s (notched) and 2075s, for RCP8.5; (**b**) Perturbation factors *vs.* return period for daily precipitation simulated by the GCMs in the 2075s for RCP8.5 in the different seasons, Red = LR GCMs, Blue = HR GCMs and Green = MR GCMs.

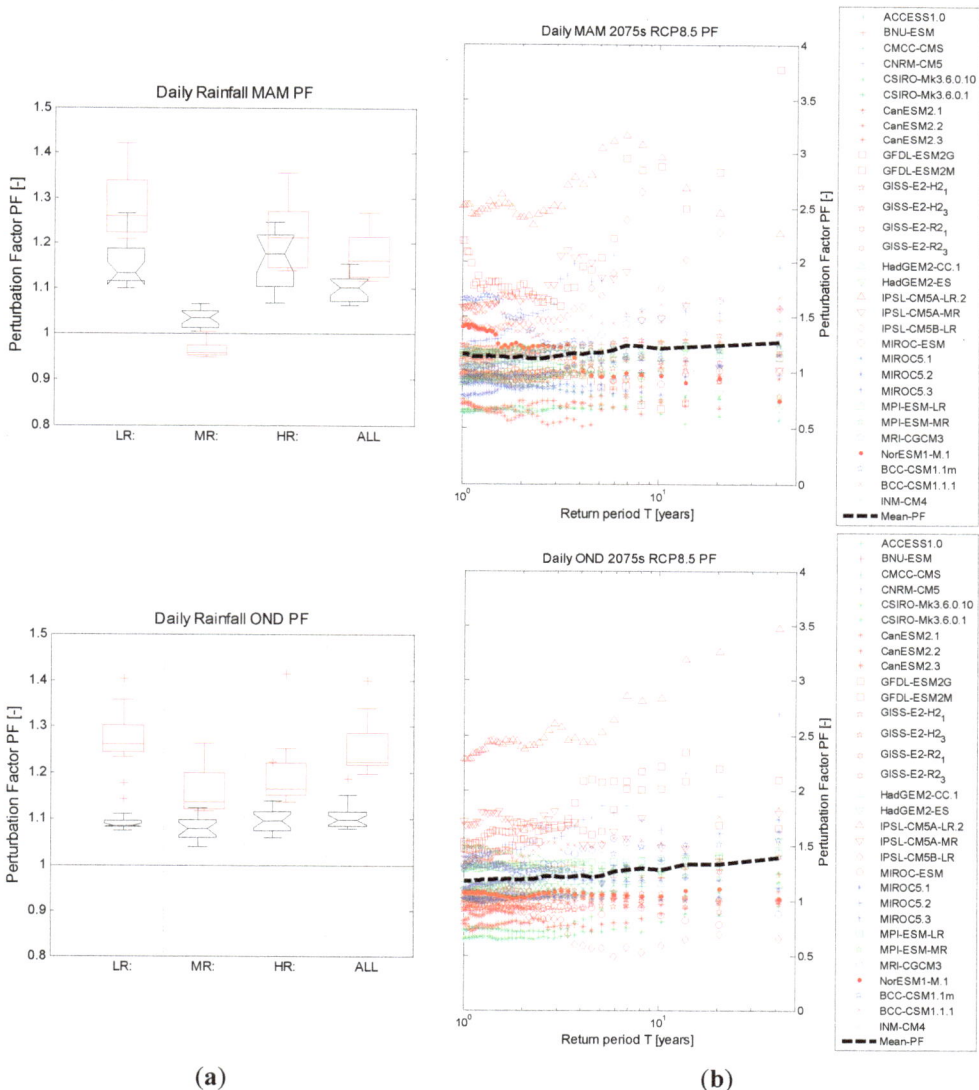

(a)　　　　　(b)

Figure 14. *Cont.*

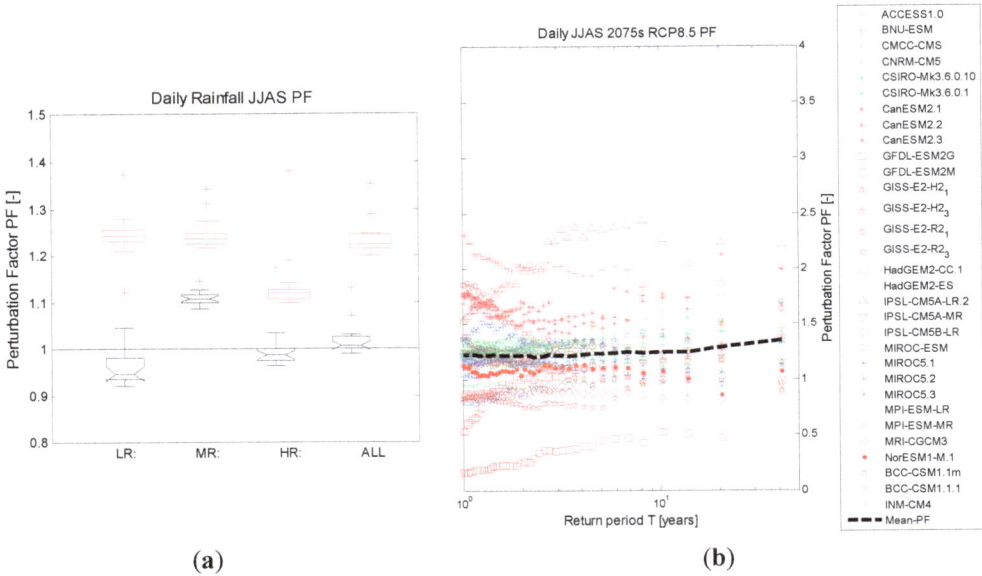

(a)   (b)

Most models show an increase in daily precipitation intensities for the dry JJAS season with some LR GCMs like GFDL-ESM2G, GFDL-ESM2M and IPSL-CM5A-LR strongly deviating from the mean change (Figure 14b). The difference in GCM performance are larger in the JJAS and MAM seasons—which were not well captured by the GCMs. Daily precipitation intensities are expected to increase by about 10%–25% in the OND season, which is consistent with the 10%–20% increase in total precipitation for that season (Figure 12). Despite the 10%–15% increase in MAM daily extremes, seasonal volumes increase by less than 10% in the same season since the number of wet days generally decreases in this season (Figure 13). Notwithstanding, in reality this increment is not expected to have any significant influence on the annual precipitation volumes especially since precipitation in the MAM season is generally underestimated by the GCMs as explained in Section 3.1.

Daily precipitation intensities were also checked in the wettest months (April and November) and dry July month (Figure 15). The large spread of the perturbation factor quartiles in the dry month of July is attributed to division by very low historical rainfall amounts especially for LR GCMs that provide precipitation results averaged over larger areas (Figure 15). The resolutions of the GCMs affect the GCM output so care ought to be taken when choosing GCMs for climate change impact projections. LR GCMs show very large variations from the mean change while the HR GCMs values are crowded around the mean. The large variations are even more pronounced for the RCP8.5 scenario. The large uncertainty in the GCM output for LR models is carried into the computed value for the mean change especially visible in the dry month of July (Figure 15), even when mean change often coincides with the results from the HR GCMs suggesting that finer resolution GCMs are favorable in predicting climate change scenarios.

Figure 15. Perturbation factors for daily precipitation simulated by GCMs in the months of April, November and July for the 2040s (notched) and 2075s, including outliers represented by (+) (**a**) RCP4.5; (**b**) RCP8.5.

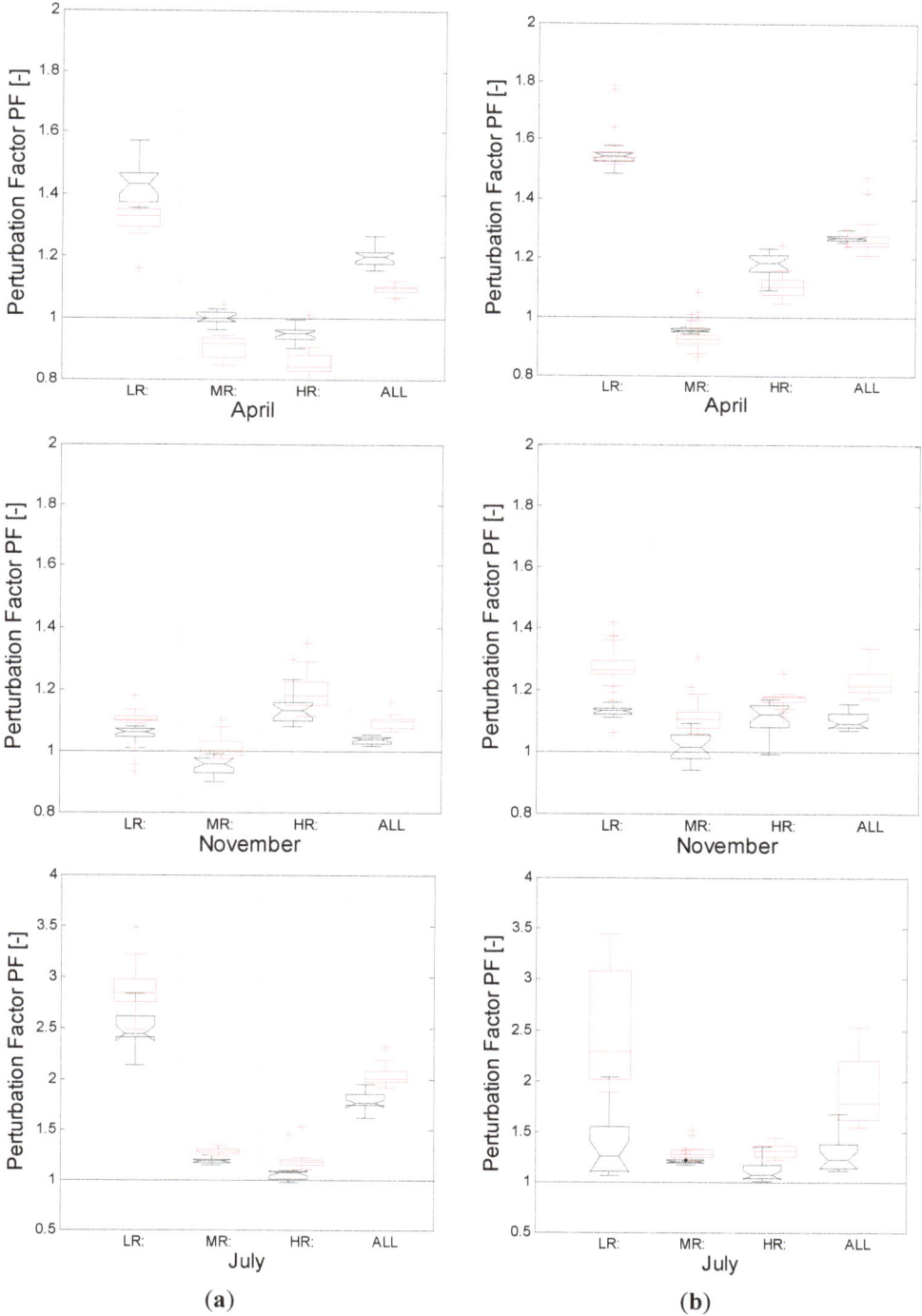

(a)

(b)

3.3. RCP4.5 vs. RCP8.5 and 2040s vs. 2075s Comparison

Figures 16 and 17 summarize the differences in GCM results for the different scenarios and periods based on their resolutions. Precipitation will generally increase in the 21st century for both RCP4.5 and RCP8.5 scenarios—higher increase is generally anticipated for RCP8.5 compared to RCP4.5 (Figure 16). For RCP4.5, annual precipitation is expected to increase by about 7% for both 2040s and 2075s, while RCP8.5 projects about 10% increase in the 2040s and about 18% in the 2075s. This increase is generally attributed to increased precipitation intensities rather than the total number of wet days, as heavier intense storms are expected in the late 21st century according to Section 3.2. The results are consistent with the positive shift in precipitation distribution expected in other parts of East Africa under global warming for the CMIP3 climate models [28]. Generally, the increase in precipitation is more for the 2075s than for 2040s; and this effect is even greater than that arising from differences between RCP8.5 and RCP4.5 scenarios (Figure 17). A high level of uncertainty is presented by the LR GCMs (grid size >2°) compared to the HR and MR GCMs (Figure 17a,b).

Figure 16. Comparison of mean perturbation factors for annual precipitation based on GCM resolutions for the 2040s (notched) and 2075s, (**a**) RCP4.5; (**b**) RCP8.5.

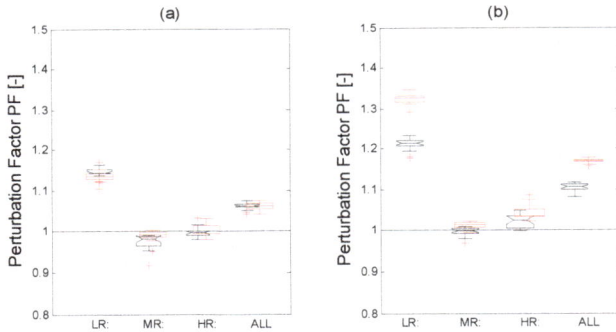

Figure 17. Annual precipitation perturbation factors, (**a**) RCP8.5 *vs.* RCP4.5, for the 2040s (o) and 2075s (+); (**b**) 2075s *vs.* 2040s for RCP4.5 (Δ) and RCP8.5(x), Red = LR GCMs, Blue = HR GCMs and Green = MR GCMs.

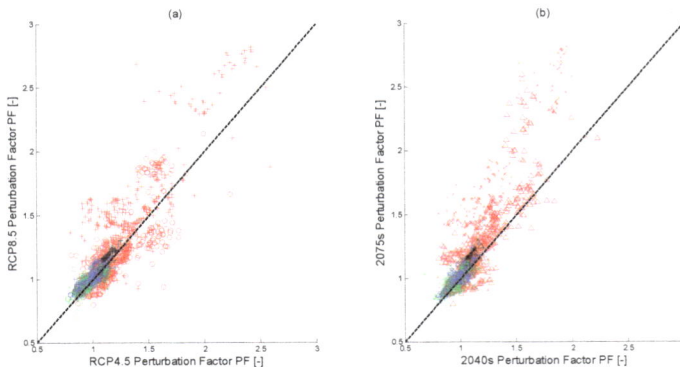

4. Conclusions

The GCM performance over Lake Victoria is highly dependent on the resolution of the GCM, especially the latitudinal scale. High resolution GCMs, namely CNRM-CM5, MIROC5, HadGEM2-CC, HadGEM2-ES and BCC-CSM1.1m gave the best performance in modeling past absolute precipitation over Lake Victoria. Lower resolution GCMs (grid size >2.5°) e.g., GFDL-ESM2G, GFDL-ESM2M, IPSL-CM5A-LR, IPSL-CM5B-LR and HadCM3 produced larger uncertainties in precipitation simulations. Therefore, for future projections of precipitation, high resolution GCMs are favored to provide reliable seasonal results. However, there is need to use a wide range of GCMs, irrespective of their resolution in order to sufficiently capture the uncertainty in climate modeling physics. This uncertainty may be large as also shown in this paper; hence needs to be taken into account in hydrological impact investigations of climate change.

The total annual precipitation is expected to increase by about 6%–8% for the RCP4.5 scenario and about 10%–18% for the RCP8.5 scenario over the 21st century, despite the higher (up to 40%) increase in extreme daily intensities since the number of wet days does not significantly change. This increase is expected to be higher in the late 21st century (2075s) than in the 2040s.

To study future lake level changes, next to precipitation over the lake, also discharges of main inflowing rivers need to be studied. This requires future projections of precipitation and potential evapotranspiration over the Lake Victoria Basin river subcatchments, and impact modeling by means of catchment runoff models. This research provided the baseline for such study by conducting GCM evaluations in precipitation simulation and analysis of future projections.

Acknowledgments

The authors would like to thank the *Vlaamse Interuniversitaire Raad*—Institutional University Cooperation (VLIR-IUC) of Belgium for funding this research under the VLIR-ICP PhD program. We also would like to thank the Ministry of Water and Environment, Uganda, and CMIP5 project for availing the data for this research. The performance evaluations in this study are based on the daily CMIP5 model output, obtained through the Program for Climate Model Diagnosis Intercomparison (PCMDI) portal of the Earth System Grid Federation [29]. We also acknowledge use of the precipitation series dataset provided by Michael Kizza. We thank the anonymous reviewers for their constructive comments that tremendously improved the quality of this paper.

Author Contributions

Patrick Willems and Charles B. Niwagaba supervised the PhD research and reviewed this paper contributing to the write-up.

Conflicts of Interest

The authors declare no conflict of interest.

References

1. Kite, G.W. Recent changes in level of Lake Victoria/Récents changements enregistrés dans le niveau du Lac. *Hydrol. Sci. J.* **1981**, *26*, 233–243.

2. Kite, G.W. Analysis of Lake Victoria levels. *Hydrol. Sci. J.* **1982**, *27*, 99–110.

3. Piper, B.S.; Plinston, D.T.; Sutcliffe, J.V. The water balance of Lake Victoria. *Hydrol. Sci. J.* **1986**, *31*, 25–37.

4. Yin, X.; Nicholson, S.E. The water balance of Lake Victoria. *Hydrol. Sci. J.* **1998**, *43*, 789–811.

5. Tate, E.; Sutcliffe, J.; Conway, D.; Farquharson, F. Water balance of Lake Victoria: Update to 2000 and climate change modelling to 2100/Bilan hydrologique du Lac Victoria: Mise à jour jusqu'en 2000 et modélisation des impacts du changement climatique jusqu'en 2100. *Hydrol. Sci. J.* **2004**, *49*, doi:10.1623/hysj.49.4.563.54422.

6. Awange, J.L.; Ogalo, L.; Bae, K.-H.; Were, P.; Omondi, P.; Omute, P.; Omullo, M. Falling Lake Victoria water levels: Is climate a contributing factor? *Clim. Change* **2008**, *89*, 281–297.

7. Swenson, S.; Wahr, J. Monitoring the water balance of Lake Victoria, East Africa, from space. *J. Hydrol.* **2009**, *370*, 163–176.

8. Intergovernmental Panel on Climate Change (IPCC). *CLIMATE CHANGE 2013: The Physical Science Basis. Contribution of Working Group I to the Fifth Assessment Report of the Intergovernmental Panel on Climate Change*; Cambridge University Press: Cambridge, United Kingdom and New York, NY, USA, 2013; p. 1535.

9. Nicholson, S.E. A review of climate dynamics and climate variability in Eastern Africa. In *Limnology, Climatology and Paleoclimatology of the East African Lakes*; Johnson, T.C., Odada, E.O., Eds.; Gordon and Breach Publishers: Amsterdam, The Netherlands, 1996; pp. 25–56.

10. Indeje, M.; Semazzi, F.H.M.; Ogallo, L.J. ENSO signals in East African rainfall seasons. *Int. J. Climatol.* **2000**, *20*, 19–46.

11. Nicholson, S.E.; Yin, X. Rainfall conditions in Equatorial East Africa during the Nineteeenth century as inferred from the record of Lake Victoria. *Clim. Chang.* **2001**, *48*, 387–398.

12. Kizza, M.; Rodhe, A.; Xu, C.-Y.; Ntale, H.K.; Halldin, S. Temporal rainfall variability in the Lake Victoria Basin in East Africa during the twentieth century. *Theor. Appl. Climatol.* **2009**, *98*, 119–135.

13. Sene, K.J.; Plinston, D.T. A review and update of the hydrology of Lake Victoria in East Africa. *Hydrol. Sci. J.* **1994**, *39*, 47–63.

14. Mutenyo, I.B. Impacts of Irrigation and Hydroelectric Power Developments on the Victoria Nile in Uganda School. Ph.D. Thesis, Cranfield University, Cranfield, UK, 2009; p. 258.

15. Sutcliffe, J.; Parks, Y. *The Hydrology of the Nile*; International Association of Hydrological Sciences IAHS: Wallingford, UK, 1999; Volume 5.

16. Kizza, M. Uncertainty Assessment in Water Balance Modelling for Lake Victoria. Ph.D. Thesis, Makerere University Kampala, Kampala, Uganda, 2012.

17. Taylor, K.E.; Stouffer, R.J.; Meehl, G.A. An overview of CMIP5 and the experiment design. *Bull. Am. Meteorol. Soc.* **2012**, *93*, 485–498.

18. Taye, M.T.; Ntegeka, V.; Ogiramoi, N.P.; Willems, P. Assessment of climate change impact on hydrological extremes in two source regions of the Nile River Basin. *Hydrol. Earth Syst. Sci.* **2011**, *15*, 209–222.

19. Nyeko-Ogiramoi, P.; Ngirane-Katashaya, G.; Willems, P.; Ntegeka, V. Evaluation and inter-comparison of Global Climate Models' performance over Katonga and Ruizi catchments in Lake Victoria basin. *Phys. Chem. Earth* **2010**, *35*, 618–633.

20. Liu, T.; Willems, P.; Pan, X.L.; Bao, A.M.; Chen, X.; Veroustraete, F.; Dong, Q.H. Climate change impact on water resource extremes in a headwater region of the Tarim basin in China. *Hydrol. Earth Syst. Sci.* **2011**, *15*, 3511–3527.

21. Christensen, J.; Kjellström, E.; Giorgi, F.; Lenderink, G.; Rummukainen, M. Weight assignment in regional climate models. *Clim. Res.* **2010**, *44*, 179–194.

22. Shongwe, M.E.; van Oldenborgh, G.J.; van den Hurk, B.; van Aalst, M. Projected changes in mean and extreme precipitation in Africa under global warming. Part II: East Africa. *J. Clim.* **2011**, *24*, 3718–3733.

23. Ntegeka, V.; Baguis, P.; Roulin, E.; Willems, P. Developing tailored climate change scenarios for hydrological impact assessments. *J. Hydrol.* **2014**, *508*, 307–321.

24. Kizza, M.; Westerberg, I.; Rodhe, A.; Ntale, H.K. Estimating areal rainfall over Lake Victoria and its basin using ground-based and satellite data. *J. Hydrol.* **2012**, *464–465*, 401–411.

25. Shaffrey, L.C.; Stevens, I.; Norton, W.A.; Roberts, M.J.; Vidale, P.L.; Harle, J.D.; Jrrar, A.; Stevens, D.P.; Woodage, M.J.; Demory, M.E.; *et al.* HiGEM: The New U.K. High-Resolution global environment model—Model description and basic evaluation. *J. Clim.* **2009**, *22*, 1861–1896.

26. Hawkins, E.; Sutton, R. The potential to narrow uncertainty in regional climate predictions. *Bull. Am. Meteorol. Soc.* **2009**, *90*, 1095–1107.

27. Ba, M.B.; Nicholson, S.E. Analysis of convective activity and its relationship to the rainfall over the rift valley lakes of East Africa during 1983–90 using the meteosat infrared channel. *J. Appl. Meteorol.* **1998**, *37*, 1250–1264.

28. Shongwe, M.E.; van Oldenborgh, G.J.; Hurk, B. Van Den Projected changes in mean and extreme precipitation in Africa under global warming, Part II: East Africa. *J. Clim.* **2011**, *24*, 3718–3733.

29. ESFG ESFG PCMDI. Available online: http://pcmdi9.llnl.gov/esgf-web-fe/live# (accessed on 31 March 2014).

Attribution of Decadal-Scale Lake-Level Trends in the Michigan-Huron System

Janel Hanrahan, Paul Roebber and Sergey Kravtsov

Abstract: This study disentangles causes of the Michigan-Huron system lake-level variability. Regional precipitation is identified as the primary driver of lake levels with sub-monthly time lag, implying that the lake-level time series can be used as a proxy for regional precipitation throughout most of the 1865–present instrumental record. Aside from secular variations associated with the Atlantic Multidecadal Oscillation, the lake-level time series is dominated by two near-decadal cycles with periods of 8 and 12 years. A combination of correlation analysis and compositing suggests that the 8-y cycle stems from changes in daily wintertime precipitation amounts associated with individual storms, possibly due to large-scale atmospheric flow anomalies that affect moisture availability. In contrast, the 12-y cycle is caused by changes in the number of instances, or frequency, of summertime convective precipitation due to a preferred upper-air trough pattern situated over the Great Lakes. In recent decades, the lake-level budget exhibited an abnormal—relative to the remainder of the instrumental record—evaporation-driven trend, likely connected to regional signatures of anthropogenic climate change. The latter effect must be accounted for, along with the effects of precipitation, when assessing possible scenarios of future lake-level variability.

Reprinted from *Water*. Cite as: Hanrahan, J.; Roebber, P.; Kravtsov, S. Attribution of Decadal-Scale Lake-Level Trends in the Michigan-Huron System. *Water* **2014**, *6*, 2278-2299.

1. Introduction

The Laurentian Great Lakes have over 10,000 miles of shoreline which is subject to submersion or drought depending on the water level of the lakes at any given time. Extremes in these levels affect shoreline erosion, cargo ship capacities, hydroelectric power supplies, and recreation for the basins' inhabitants. While commonly considered two separate bodies of water, the second and third largest of the Great Lakes, Lake Michigan and Lake Huron, are connected by the deep Straits of Mackinac which ensures that the water levels of the separate reservoirs remain at equal elevations. Hence, they behave hydraulically like a single lake, and together, Lake Michigan-Huron has over 117,000 km^2 of surface water making it the world's largest freshwater lake [1]. Given this vast surface area, it can reasonably be assumed that climate fluctuations should be well represented in the Michigan-Huron lake-level time series, and indeed several studies have indicated that precipitation is most likely the primary interannual lake-level driver [2–6]. Hence, in addition to addressing immediate socioeconomic impacts that the water levels have on surrounding communities, an understanding of the 1865–present lake-level time series also contributes to a better understanding of historic climate modes and regional precipitation behavior whose impacts extend to other area lakes and groundwater [7], river flows, and agriculture. While this extensive time series likely contains a great deal of information pertaining to past climate behavior, clear illustrations linking interannual lake-level fluctuations to regional precipitation variability have been lacking.

To better isolate climatic connections implicit in the lake-level time series, Hanrahan *et al.* [8] subtracted the outflow-related damping effects from the full lake-level record and determined that the resulting time series exhibits variability over a range of time scales. The longest time scale—secular variations are anti-correlated with North Atlantic sea-surface temperatures (SSTs). The climate mode associated with this SST signal has been dubbed the Atlantic Multidecadal Oscillation (AMO) and is associated with dynamics of the oceanic thermohaline circulation [9–11]. Previous studies have linked AMO phases to precipitation anomalies over large portions of the U.S. [12–15], and this signal is likely transmitted to the lake levels through precipitation changes [8]. In our previous work using Singular Spectrum Analysis (SSA: see Ghil *et al.* [16,17]), we also identified two near-decadal cycles in the Michigan-Huron water levels, namely the 8-y and 12-y cycles [6]. The latter signal is in agreement with Watras *et al.* [7] who identified a ~13-y cycle in lakes and aquifers across the upper Great Lakes' region. The full lake-level time series can thus be decomposed as the sum of SSA reconstructed components associated with each decadal signal, and the residual variability is dominated by the AMO signal.

In this study, we aim to further our understanding of lake-level drivers responsible for decadal-scale lake-level changes. Using regional precipitation datasets, Section 2 establishes a solid connection between precipitation anomalies and lake-level variability, and addresses the issue of a possible time lag between precipitation forcing and lake-level response. In Sections 3 and 4, we explore seasonality of near-decadal lake-level cycles and examine the large-scale atmospheric patterns associated with the corresponding lake-level changes. Section 5 summarizes our results on the connection between regional precipitation and historic lake-level changes, and addresses a recent evaporative trend which is becoming increasingly apparent in the previously precipitation-driven lake-level time series.

2. Michigan-Huron Lake-Level Drivers

The interannual Michigan-Huron lake levels fluctuate in response to the sum of five primary drivers (see Table 1 for a summary of variables used throughout the text): over-lake precipitation P_y, runoff from tributary rivers and streams R_y, evaporative losses $-E_y$, inflow through the St. Marys River from Lake Superior I_y, and outflow to the Mississippi River through the Chicago Diversion and the lower Great Lakes through the St. Clair River $-O_y$; the subscript y denotes the year under consideration. The estimated values for all these quantities were obtained online from the Great Lakes Environmental Research Laboratory (GLERL [18])).

Table 1. Descriptions of variables used throughout the text.

Variable	Description
P_y	Annual total over-lake precipitation [1]
R_y	Annual total runoff [1]
E_y	Annual total evaporative losses [1]
I_y	Annual total inflow (from Lake Superior for Lake Michigan-Huron) [1]
O_y	Annual total outflow (through the St. Marys River and the Chicago Diversion for Lake Michigan-Huron) [1]
$L_{m,y}^G, L_{m,y}^U$	Observed beginning-of-month lake-level as reported by GLERL (G) and monthly-average level by USACE (U)
$dL_{m,y}$	Observed monthly lake-level change as estimated by $L_{m+1,y} - L_{m,y}$
dL_y^s, dL_y^w	Observed seasonal (3-month) lake-level change as estimated by $dL_{m-1,y} + dL_{m,y} + dL_{m+1,y}$, where $m = 7$ for summer (s) and $m = 12$ for winter (w)
dL_y^p	Computed annual lake-level change associated with the precipitation-driven components [1] P_y, R_y, and I_y; see Equation (1)
P_d^r, P_y^r	Daily (d) and annual (y) total regional precipitation depth [2]
dL_y^{pr}	Computed annual lake-level change associated with P_y^r; see Equation (2)
L_y^{rc}	Reconstructed lake-level components associated with the 8-y and 12-y cycles [3]
dL_y^{rc}	Annual change of each reconstructed component as estimated by $L_{y+1}^{rc} - L_{y-1}^{rc}$
P^T	Average total precipitation [2] defined as $\dfrac{\text{accumulated precipitation depth}}{\text{total number of days}}$
P^f	Average precipitation frequency [2], defined as $\dfrac{\text{number of days with precipitation}}{\text{total number of days}}$
P^a	Average precipitation amount [2], defined as $\dfrac{\text{accumulated precipitation depth}}{\text{number of precipitation days}}$
$\hat{P}^f, \hat{P}^a$	Precipitation frequency index $\hat{P}^f$, and amount $\hat{P}^a$, with multidecadal variability removed and averaged over 3 years

Notes: [1] Lake fluxes are provided in linear units (depth) over the lakes surface from GLERL; [2] Computed from the NOAA NCEP CPC gridded precipitation data; [3] Generated from statistically significant spectral peaks identified by Multi-taper method (MTM) analysis in Hanrahan *et al.* [6].

2.1. Drivers of Lake-Level Changes

Three of the lake-level drivers—precipitation, runoff, and inflow—can all be tied to changes in regional precipitation. The over-lake precipitation, along with the river and stream flows that make up the runoff component, all stem from precipitation that occurs within the Michigan-Huron catchment basin. The inflow for Michigan-Huron has been regulated since 1887 by way of navigation locks on the St. Marys River to keep Lake Superior within a specified range of historic water levels [19]. If precipitation is the primary lake-level driver of Lake Superior, long-term precipitation variability over the Superior basin will thus be correlated with the regulated Michigan-Huron inflow.

The fourth lake-level driver, evaporation, is a major contributor to seasonal lake-level variability, but it has historically played a minor role in the behavior of the year-to-year lake level fluctuations. Between 1948 and 2005, annual evaporation totals exhibited a standard deviation of about 0.08 m, while the sum of the precipitation-driven components (precipitation, runoff, and inflow) had a standard deviation of 0.24 m over the same time period. Thus, the expected year-to-year lake-level changes stemming from evaporation variability are considerably smaller than those which occur from precipitation changes. The final lake-level driver, outflow, is a function of the lake-level itself and does not directly respond to atmospheric processes. It instead tends to damp the climate-related lake-level fluctuations arising from the four remaining components [8].

It follows that the historic annual lake-level changes can be largely described by the precipitation-driven components alone. The annual precipitation-driven lake-level changes dL_y^p (Figure 1, black line) can thus be estimated as

$$dL_y^p = P_y + R_y + I_y - \bar{E} - \bar{O} \tag{1}$$

where P_y, R_y, and I_y values are all available for the period 1916–2005, while the historical averages of annual evaporative losses $\bar{E}$, and the outflow average $\bar{O}$ are computed from 1948–2005 values.

Figure 1. Annual lake-level changes estimated from observed levels (dL_y^U; blue x's) and as computed by Equation (1) (dL_y^p; black circles).

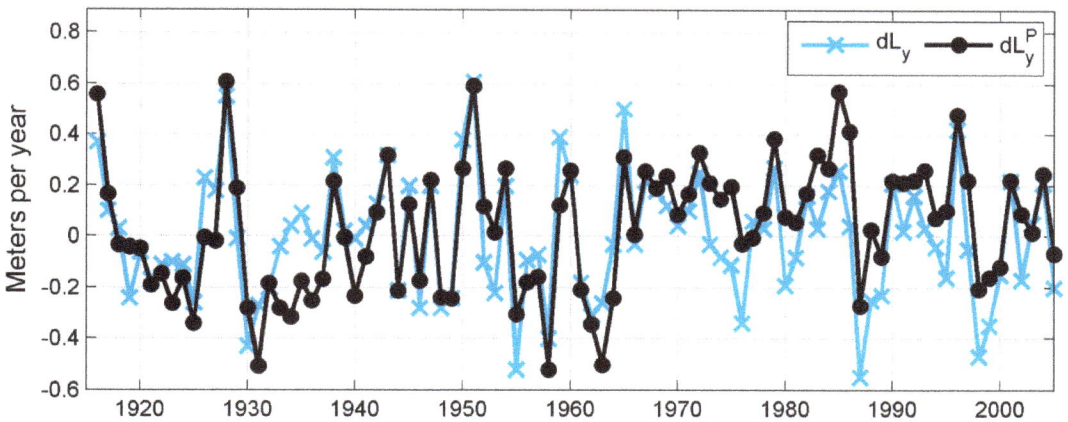

The year-to-year lake-level variability associated with the precipitation-driven components alone (Figure 1; black line), as computed from Equation (1), accounts for about 87% ($r = 0.93$) of the 1916–2005 lake-level behavior as estimated from the monthly-averaged levels obtained from the U.S. Army Corps of Engineers (USACE) dL_y^U (Figure 1; blue line). This indicates that precipitation has indeed had the greatest impact on interannual lake-level changes as it accounts for most of the historic lake-level variability. This precipitation–lake-level relationship is the primary focus of the present study.

2.2. Connecting Regional Precipitation to Lake-Level Fluctuations

While the precipitation-driven components (precipitation, runoff, and inflow) alone can be used to describe much of the historic lake-level behavior, the determination of these individual values is still quite complex. For example, as discussed above, the inflow rates ultimately stem from annual precipitation over the Lake Superior basin, but there are additional factors that affect the amount of water that actually runs into Lake Michigan-Huron. Furthermore, because of ground absorption, evaporation, and other factors, the amount of precipitation that falls onto the land which surrounds Lake Michigan-Huron, is not equal to the amount of water that runs into the lake as runoff. Both the inflow and runoff as reported by GLERL must therefore be directly measured as they flow into the lake, or estimated by nearby flows. Here, we simplify the precipitation–lake-level connection by estimating these fluxes from precipitation indices alone.

We obtained NOAA NCEP Climate Prediction Center (CPC) daily precipitation data from the IRI/LDEO Climate Data Library [20]. This gridded data precipitation data product was based on raw data from NCDC daily co-op stations, the CPC dataset, and hourly precipitation datasets. The precipitation data for overlake gridpoints (Figure 2) were estimated from surrounding land-based stations via Cressman Scheme gridding onto a 0.25 degree grid. For the present analysis, seven of the resulting gridded locations over the Great Lakes' basin were chosen as indicated by the red stars in Figure 2, for 1948–1997. Here, the over-Superior locations were selected to represent the Michigan-Huron inflow and the remaining locations represent the Michigan-Huron over-lake precipitation and runoff. Annual precipitation totals were computed from spatially averaged precipitation over the seven grid locations, resulting in a single 50-y time series of precipitation depths P_y^r. In order to estimate lake-level changes from these regional precipitation values, we must consider that the resulting Michigan-Huron lake-level variations are a multiple of P_y^r, because the surface area of Lakes Michigan and Huron makes up only a small fraction—about 1/5—of the total Michigan-Huron and Superior basin area A. For example, suppose that during a particular period there was an average rainfall of 3 mm in the region, and for simplicity, assume that all of it flowed directly into Lake Michigan-Huron. The total volume of water produced would be $3A$ mm^3, which would equate to a lake-level increase of $3A \times (1/5A)^{-1} = 15$mm, therefore amplifying the lake-level response by a factor of 5.

To compute lake-level changes given P_y^r, we used the historic instrumental record of individual precipitation-related components as reported by GLERL. Over the 1948–1997 time period, the average annual over-lake precipitation was $\bar{P} = 0.840$ m, runoff $\bar{R} = 0.617$ m, and inflow $\bar{I} = 0.604$ m, while the average annual regional precipitation depth from the gridded dataset was $\overline{Pr} = 0.697$ m. The total annual Michigan-Huron over-lake precipitation, runoff, and inflow values, thus equate to about 2.96 times the regional precipitation depth. If this ratio remains relatively constant through our period of interest, lake-level change dL_y^{pr} can be computed given only P_y^r, the historic average annual evaporative losses $\bar{E} = 0.64$ m, and outflow $\bar{O} = 1.48$ m (resulting in the subtraction of a constant value of 2.12 m). This results in the following simplified lake-level equation:

$$dL_y^{pr} = 2.96P_y^r - 2.12 \qquad (2)$$

The lake-level changes estimated from Equation (2) are illustrated in Figure 3 (black line) in conjunction with the observed lake-level change dL_y^y (blue line). To account for possible delays related to snowpack storage, the annual rainfall totals illustrated here are defined with August as the beginning month.

Figure 2. Grid locations considered for regional precipitation estimate using NCEP CPC daily gridded precipitation data. Original image obtained from NOAA/GLERL.

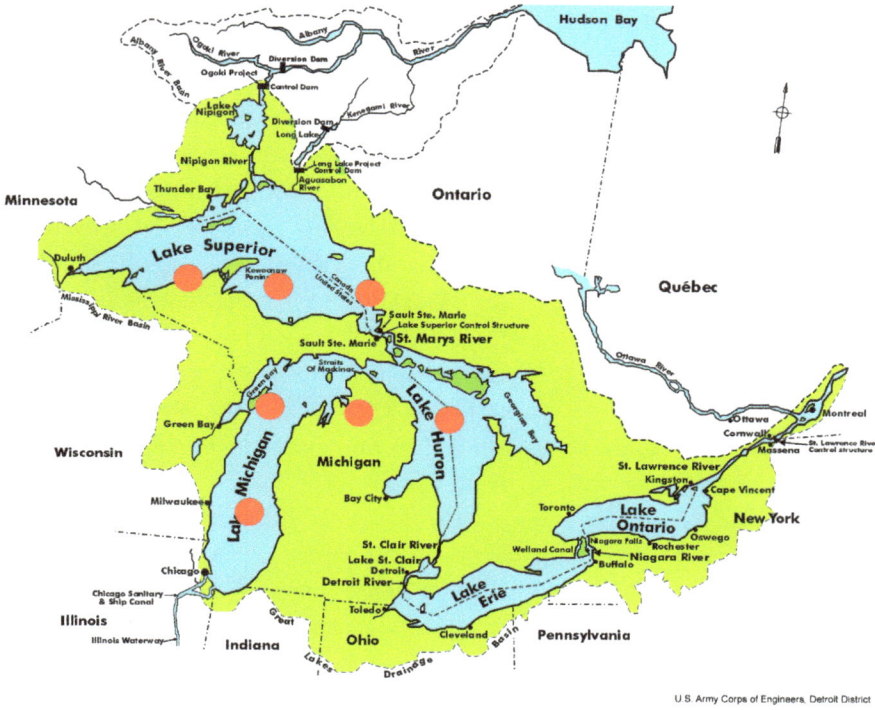

Figure 3. Annual lake-level changes (July–August) estimated from observed levels (dL_y^U; blue x's) and as computed by Equation (2) (dL_y^{pr}; black circles).

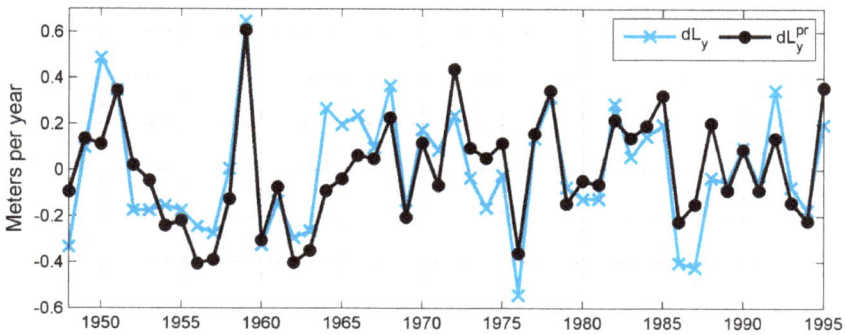

While numerous difficult-to-measure processes are occurring which affect the lake-level water supply (*i.e.*, over-lake and river evaporation, over-land evaporation, channel adjustments, vegetative consumption, groundwater intake, among several others), this simplified regional precipitation index dL_y^{pr} still accounts for the majority of lake-level variability ($r = 0.81$). This ability to connect the lake-level changes directly to a single atmospheric variable is ideal for furthering our understanding of underlying atmospheric behavior. In particular, one can use the lake-level time series, which begins at

1865, and Equation (2) to estimate regional precipitation prior to 1948, where the CPC gridded precipitation dataset is not available, thus providing a much extended period during which to study rainfall variability.

2.3. Lag Time of Precipitation Effects on Lake-Level Changes

To investigate the possibility of a sub-annual time lag between precipitation and lake-level changes, we compared monthly lake-level changes estimated from beginning-of-month lake level data provided by GLERL $dL^G_{m,y}$ to 30-day precipitation totals P^r_d. Correlation coefficients were computed between the 50-year-long monthly time series of $dL^G_{m,y}$ and daily time series of P^r_d. To account for potential end-of-year water-level lag times, the years were overlapped, so that correlations were computed for all months and days July 1948–December 1949 to July 1996–December 1997. The contour plot in Figure 4 illustrates the magnitude of these correlations where months along the horizontal axis refer to the first day of 30-day precipitation totals P^r_d, and those along the vertical axis indicate the month of lake-level change $dL^G_{m,y}$. For example, the lake-level changes in June (from left axis), correlate most strongly with the 30-day precipitation totals that begin on about June 1 of the same year (bottom axis), where $r \approx 0.8$; the points that correspond to this example are indicated by the white arrows. On average, statistically significant correlations ($p < 0.5$) occur where $r > 0.45$. Because the strongest correlations occur during concurrent times, we conclude that the lake levels primarily respond to regional precipitation with a sub-monthly time lag. Therefore, any season-specific variability in the water levels of Michigan-Huron should agree with the timing of regional precipitation drivers.

Figure 4. Correlations between regional precipitation P^r_d that begin on the dates along the horizontal axis, and monthly lake-level changes $dL^G_{m,y}$ that occur during the months along the vertical axis. The shading indicates the strength of correlation (r), as indicate by the bar on the right. The average r-values which significantly exceed zero are 0.45 ($p < 0.05$) and 0.55 ($p < 0.01$). See text for further explanation.

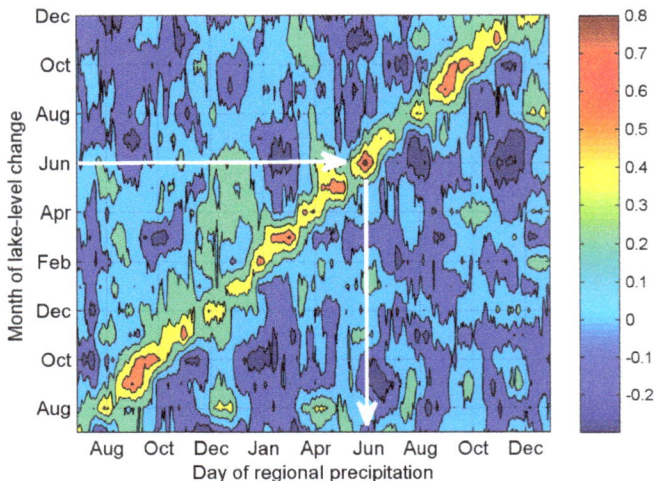

3. Timing of Lake-Level Changes and Precipitation Behavior

3.1. Seasonality of Lake-Level Periodicities

We now investigate potential seasonality in lake-level quasi-periodicities previously identified in Hanrahan et al. [6]. For this analysis, seasonal (3-month) lake-level changes $dL^G_{m,y}$ were compared to the derivatives dL^{rc8}_y and dL^{rc12}_y of the reconstructed lake-level components corresponding to the 8-y and 12-y signals identified by the SSA analysis of the 1865–1999 lake-level time series; see Hanrahan et al. [6], and Ghil et al. [16,17] for details pertaining to the computation of reconstructed components. Correlations between 24 consecutive months of lake-level changes $dL^G_{m,y}$ and the reconstructions dL^{rc}_y are illustrated in Figure 5 for the lake-level change periods of October 1900–1998 to November 1901–1999. For example, the first November correlations (left) are between the November 1900–1998 lake-level changes and the 1901–1999 reconstructions, dL^{rc8}_y (thick green) and dL^{rc12}_y (thin red), respectively. The second November correlations (right) are between the same reconstruction years, and the November 1901–1999 lake-level changes. Because artificially-high correlations may be generated among time series exhibiting red-noise variability, a Monte Carlo test was used to establish significance of the co-variance between the reconstructions and seasonal lake-level behavior by generating 1000 surrogate lake-level time series using the same length and lag-1 autocorrelation as the actual lake-levels. Correlations were computed between the surrogate time series and the reconstructions, which were sorted and then compared to the observed values. The observed values exceeding the 99th percentile of the synthetic values were determined to be significant at the 1% level (black stars in Figure 5).

Figure 5. Correlation between the 8-y (dL^{rc8}_y; thick green) and 12-y (dL^{rc12}_y; thin red) reconstructed lake-level changes, and the observed seasonal (3-month) lake-level changes ($dL^G_{m,y}$). Black stars indicate correlations significant at the 1% level.

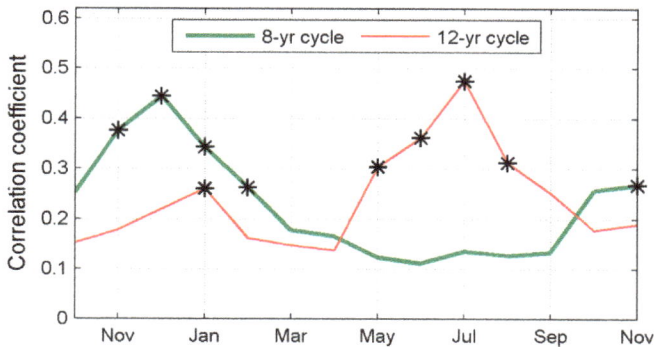

The 8-y reconstruction best correlates with lake-level changes during the preceding winter months of November to January. Conversely, the 12-y cycle is dominant during the summer months, as the peak correlations occur during June to August. The smoothed (3-y average) lake-level change time series that correspond to these seasons (blue) are illustrated in Figure 6a for the winter months dL_y^w (November–January) and Figure 6b for the summer months dL_y^s (June–August), along with the derivatives dL_y^{rc} of the 8-y and 12-y cycles (Figure 6a,b, green and red dashed lines, respectively). We have thus established that the two near-decadal lake-level periodicities occur during different times of the year; the 8-y cycle during winter months, and the 12-y cycle during summer months.

Figure 6. (a) Detrended wintertime (November–January) lake-level changes dL_y^w (blue) and the derivative of the 8-y lake-level reconstruction dL_y^{rc8} (green dashed); **(b)** Detrended summertime (June–August) lake-level changes dL_y^s (blue) and the derivative of the 12-y lake-level reconstruction dL_y^{rc12} (red dashed).

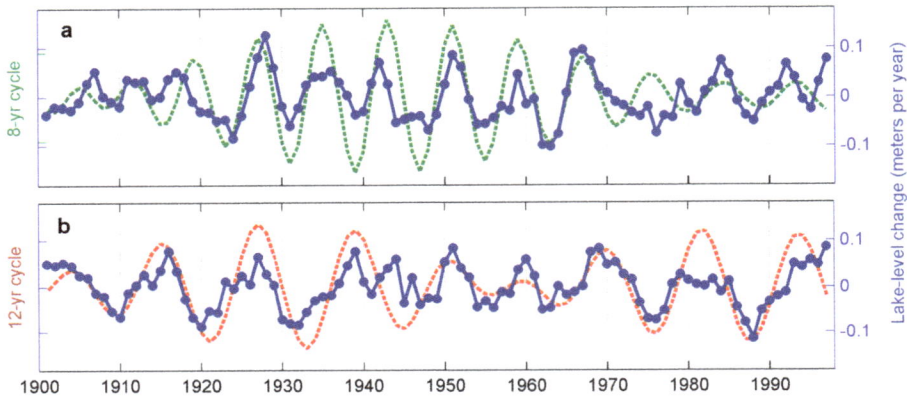

3.2. Seasonality of Precipitation Characteristics

Because regional precipitation drives lake-level changes with sub-monthly time lag (Section 2.3), it follows that the lake-level cycle seasonalities identified above (Section 3.1) should coincide temporally with those of precipitation. In this section, we examine potential seasonal precipitation periodicities by comparing the reconstructed lake-level cycles to two precipitation indices: frequency and amount. Both indices were computed from the daily gridded CPC precipitation data and are defined as follows:

$$\text{Precipitation frequency } (P^f) = \frac{\text{number of precipitation days}}{\text{total number of days}} \tag{3}$$

$$\text{Precipitation amount } (P^a) = \frac{\text{accumulated precipitation depth}}{\text{number of precipitation days}} \tag{4}$$

where a "precipitation day" is defined as having measurable precipitation (at least 1 mm) in one or more of the seven grid locations, and the sum of precipitation is computed as the average precipitation depth over all grid locations summed up over a specified number of days. Hence, P^f indicates the probability of precipitation occurring on any given day, irrespective of the intensity of the precipitation-producing system. Conversely, P^a is related to the intensity of precipitation during individual storm events, as it accounts for the average amount of precipitation that fell during a given precipitation day and location, not for the number of instances or spatial extent.

These metrics are similar to those used by other authors where daily "frequency" and "amount" are sometimes referred to as "relative number of wet days" and "intensity," respectively [21].

To encompass full seasons, P^f and P^a were computed over 120-day moving segments (total number of days = 120) over 1948–1997. As identified in Hanrahan [22], both of these indices exhibit significant multidecadal variability possibly associated with the AMO. To filter out this low-frequency variability, we subtracted 20-y moving averages from both indices. Next, to minimize the highest-frequency variability for the purpose of concentrating on decadal time scales, we computed 3-y moving averages of the final time series. The resulting filtered indices $\hat{P}^f$ and $\hat{P}^a$ are illustrated in Figure 7a,b, respectively.

To evaluate the existence of near-decadal cycles in precipitation frequency and amount, correlation coefficients were computed between the two indices and the reconstructed derivatives. That is, 365 50-year-long time series of $\hat{P}^f$ and $\hat{P}^a$ (vertical axes in Figure 7a,b) which correspond to each day of the year (defined in months on the horizontal axes), were compared to each 50-year-long time series of the 8-y and 12-y reconstructed derivatives, dL_y^{rc8} and dL_y^{rc12}. Precipitation frequency $\hat{P}^f$ best correlates with the 12-y cycle during the summer months (Figure 7c, thin red; black stars indicate correlations significant at the 1% level), and the precipitation amount index $\hat{P}^a$ correlates best with the 8-y cycle during the winter months (Figure 7d, thick green). These results are consistent with the previously-discussed timing of periodicities, where the 12-y and 8-y cycles were linked to the summertime and wintertime lake-level changes, respectively (Section 3.1).

Figure 7. Identification of precipitation characteristics associated with the 8-y and 12-y lake-level cycles. Filtered 120-day moving averages of regional precipitation (**a**) frequency $\hat{P}^f$ and (**b**) amount $\hat{P}^a$; (**c,d**) Correlations between the precipitation indices (from **a** and **b**) and the 8-y cycle dL_y^{rc8} (thick green lines) and 12-y cycle dL_y^{rc12} (thin red lines). Black stars indicate significant correlations at 1% level.

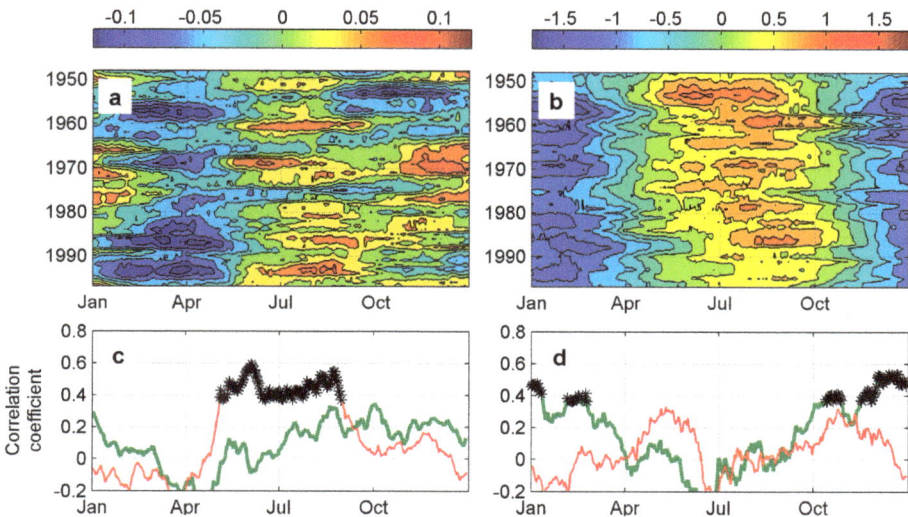

3.3. Verification of Seasonal Characteristics

To further establish significance of these findings, we compared monthly precipitation indices from Equations (3) and (4) during opposing phases of the 8-y and 12-y cycles. The 1948–1997 reconstruction derivatives dL_y^{rc} were divided into two categories: increasing and decreasing lake-level years, as illustrated in Figure 8 (blue circles and red squares, respectively). For the 12-y periodicity (Figure 8b, black line), summertime (June–August) monthly precipitation indices were categorized during 1948–1997, resulting in 25 years, or 75 individual months, per phase. For the 8-y periodicity (Figure 8a, black line), wintertime (September–January) indices were categorized during 1949–1997 (the 1948 winter was not included because November–December data were not available for 1947), resulting in 24 years, or 72 months, per phase.

Figure 8. (a) Derivatives of 8-y lake-level reconstruction dL_y^{rc8} and (b) the 12-y reconstruction dL_y^{rc12} (black lines), and categorization of phases: increasing lake-level years (blue circles) and decreasing lake-level years (red squares).

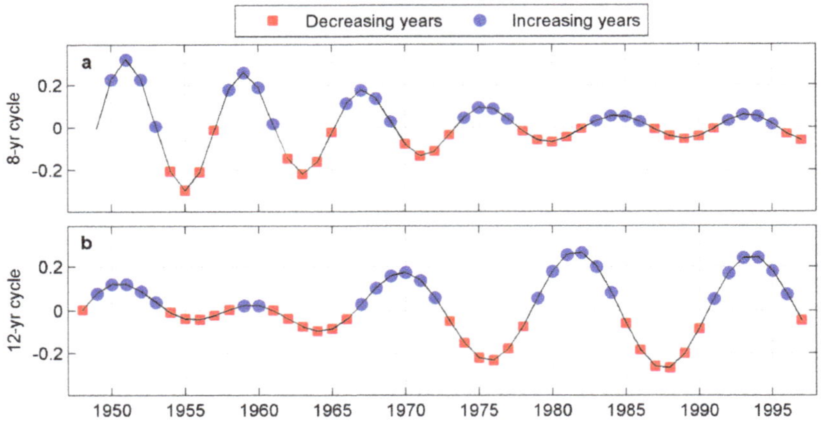

In addition to the precipitation frequency P^f and amount P^a indices, we have also included a total precipitation index P^T, for comparison. This index was defined as the total precipitation amount over the total number of days under consideration:

$$\text{Total precipitation } (P^T) = \frac{\text{accumulated precipitation depth}}{\text{total number of days}} \tag{5}$$

where the denominator includes all days, both precipitating and non-precipitating. When summed up over a month, P^T amounts to the expected total area-averaged precipitation, which can be translated to monthly lake-level changes with Equation (2). On the other hand, the precipitation frequency index P^f, which represents the probability of precipitation occurring during any given day through the month, results in the number of days with some amount of precipitation within the Superior/Michigan-Huron region. The precipitation amount index P^a represents the average daily precipitation amount only during precipitation days.

We computed each index value during years of increasing and decreasing lake levels, over winter (P_w^T, P_w^f, P_w^a) and summer (P_s^T, P_s^f, P_s^a) months. The means were tested for significant differences with a one-sided t-test, between the phases of the 8-y cycle (Table 2) and the 12-y cycle (Table 3). We find that P^T is significantly different between the phases of both periodicities—P_w^T during the 8-y cycle, and P_s^T during the 12-y cycle—indicating that more precipitation fell during the increasing phases of these periodicities than during the decreasing phases. This is in agreement with our results from Section 3.1 that compared seasonal lake-level changes to the lake-level reconstructions. Furthermore, the wintertime precipitation amount index P_w^a is significantly different between the phases of the 8-y cycle, and the summertime precipitation frequency index P_s^f is significantly different between phases of the 12-y cycle. This is in agreement with our findings from Section 3.2 that examined correlations between rainfall indices and lake-level reconstructions.

Table 2. Precipitation indices during winter months associated with the increasing and decreasing phases of the 8-y cycle*.

Index	Average	Increasing	Decreasing	Significance
Total precipitation P_w^T (daily average in mm)	1.58	1.67	1.49	**$p = 0.03$**
Frequency P_w^f (daily probability)	0.72	0.73	0.72	$p = 0.44$
Amount P_w^a (daily average in mm)	2.18	2.29	2.06	**$p = 0.01$**

Note: * p-values in bold indicate significant differences between the increasing and decreasing phases of the 8-y cycle as determined by a one-sided t-test ($p < 0.05$).

Table 3. Precipitation indices during summer months associated with the increasing and decreasing phases of the 12-y cycle*.

Index	Average	Increasing	Decreasing	Significance
Total precipitation P_s^T (daily average in mm)	2.38	2.50	2.26	**$p = 0.02$**
Frequency P_s^f (daily probability)	0.72	0.75	0.70	**$p < 0.01$**
Amount P_s^a (daily average in mm)	3.26	3.32	3.20	$p = 0.20$

Note: * p-values in bold indicate significant differences between the increasing and decreasing phases of the 12-y cycle as determined by a one-sided t-test ($p < 0.05$).

In summary, it is changes in seasonal precipitation totals P^T that are driving the near-decadal lake-level cycles, and these totals vary due to changes both in precipitation frequency P^f and precipitation amount P^a. Consistent with our findings discussed in the previous section, we conclude that precipitation fluctuations associated with the 8-y wintertime cycle are largely stemming from changes in precipitation amounts during the winter months P_w^a. Conversely, the summertime 12-y cycle is primarily being driven by changes in precipitation fluctuations through variations in summertime precipitation frequency P_s^f.

4. Climate Connections

To identify large-scale climate variability associated with the cyclic Michigan-Huron lake-level changes, we compared the winter and summer time series of lake-level tendencies dL_y^w and dL_y^s (Figure 6, blue lines) to the evolution of various atmospheric fields and SSTs. We hypothesize that the actual lake-level changes serve as a more accurate proxy for seasonal precipitation totals due to the spatial restrictions and coarse temporal resolution of the actual precipitation data.

We used monthly averaged 1949–1998 NOAA NCEP-NCAR reanalysis data for 850-mb air temperatures, sea-level pressures (SLPs), and 500-mb geopotential heights, and Kaplan *et al.* [23] gridded SSTs analyzed over the same time period. The climatological monthly means were removed from all indices, and 20-y moving averages were subtracted from each location's time series to concentrate on decadal scales. The anomalous lake-level changes, dL_y^w and dL_y^s, were sorted into years from the largest negative to the largest positive lake-level change, and the first and last 12 years—approximately the upper and lower 25% of all available years—were selected. Next, we generated composite plots to connect the behavior of seasonal lake-level fluctuations to atmospheric and SST variability. The differences between the composites over the years with the greatest positive and negative lake-level changes during winter dL_y^w and summer dL_y^s, are illustrated in Figures 9 and 12, respectively, where dL_y^w was compared to November–January, and dL_y^s was compared to June–August anomalies of SSTs (°C), SLPs (mb), 850-mb air temperatures (°C), and 500-mb heights (m). The white lines encompass areas where the upper and lower means were determined to be significantly different at the 5% confidence level, as determined by a two-sided *t*-test.

4.1. The 8-y Wintertime Cycle

We concluded in Section 3 that the 8-y lake-level cycle is active during November–January, and found variations in wintertime precipitation amounts P_w^a to be the primary driver. Hence, atmospheric connections to Michigan-Huron lake-level changes during these months can help to identify climate modes associated with this periodicity.

4.1.1. Wintertime Anomalies

Composite North Atlantic and North Pacific SST anomalies (Figure 9a) associated with the wintertime lake-level changes dL_y^w have patterns similar to those of the 850-mb temperatures (Figure 9b) and upper-level heights (Figure 9d). The years of greatest wintertime lake-level increase correspond to a warm North Atlantic SST anomaly off the New England coast, which underlies a positive 850-mb temperature and 500-mb height anomaly, both extending back over part of the Great Lakes' region. Also during these years, below-average 850-mb temperatures and 500-mb heights are observed along the northwestern edge of the U.S. and western Canada. These climate patterns are characteristic of an increased precipitation-producing storm-track displacement, where the trough over the western U.S. indicates a southward displacement, and the ridge over northeastern U.S. indicates a northward displacement. Hence, it appears that wintertime cyclones that travel over the U.S. during increasing lake-level years tend to be of more southern origin, allowing more moisture from the Gulf of Mexico to reach the Great Lakes' region. Another way to interpret moisture transport is by examination of the SLP plot (Figure 9c). The anomalous cyclonic rotation of the low SLPs in the southwest (Colorado low), combined with the anomalous anticyclonic rotation of the anomalous high SLPs in the northeast, set up an anomalous northeastward flow (Panhandle hook) directly transporting moisture from the Gulf of Mexico toward the Great Lakes, resulting in greater precipitation amounts there.

We additionally examined potential differences in precipitable water (kg/m^2) associated with storm-track displacements (Figure 10; analogous to composite plots of Figure 9). A statistically significant positive anomaly in precipitable water includes most of the Great Lakes' region and is consistent with that which would be expected from a northward storm track displacement over the eastern U.S, further substantiating our findings that the wintertime lake-level changes are associated with moisture anomalies stemming from variations in storm track.

Figure 9. Composite differences between the 12 years with the greatest increase and 12 years with the greatest decrease of wintertime (NDJ) lake-level, for high-pass filtered seasonal anomalies of (**a**) SSTs; (**b**) 850-mb temperatures; (**c**) SLPs; and (**d**) 500-mb geopotential heights, for 1949–1998. Areas encompassed by white lines indicate significant differences at the 5% level.

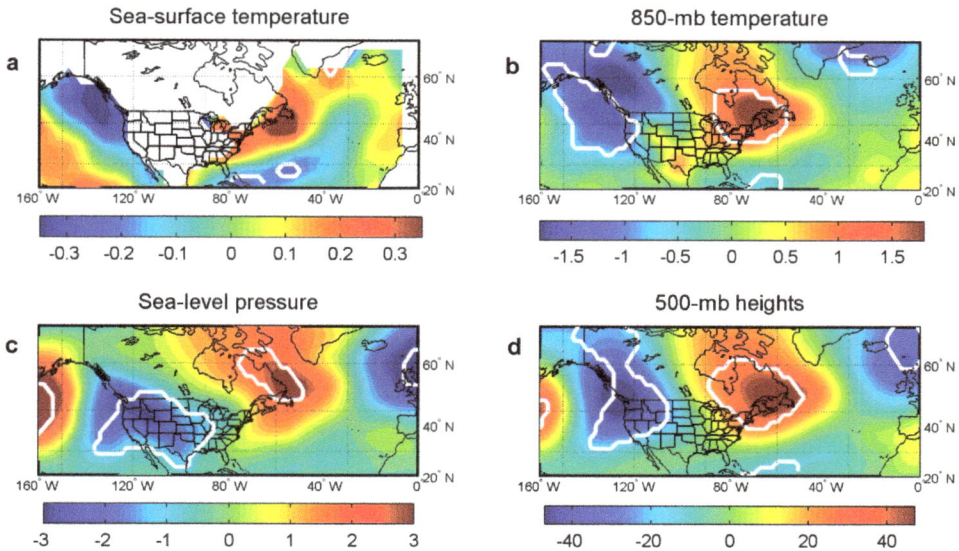

Figure 10. Composite differences between the 12 years with the greatest increase and 12 years with the greatest decrease of wintertime (NDJ) lake-level for high-pass filtered seasonal anomalies of precipitable water (kg/m²). Areas encompassed by white lines indicate significant differences at the $p < 0.05$ level.

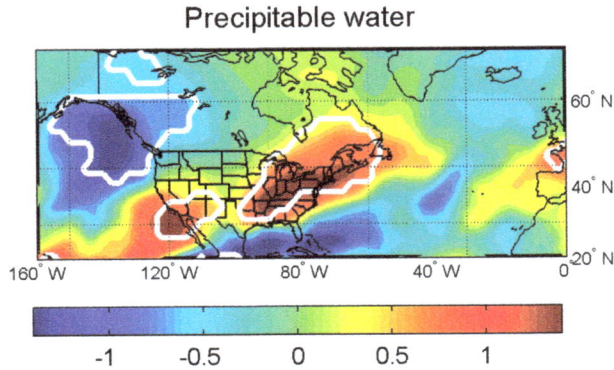

Precipitable water

4.1.2. The Pacific/North American Index

The negative height anomaly located over the western U.S. and positive anomaly over the east-central North Pacific during the years of greatest wintertime lake-level increase (Figure 9d), is a pattern consistent with the negative phase of the Pacific/North American (PNA) index. In contrast, the anomalous high over the northeastern U.S. is positioned further north than that which is typically characterized by the PNA pattern. However, it is the western North American component of this pattern that has previously been linked to increased precipitation amounts during the winter months [24]. The negative PNA phase is characterized by a storm track that is more zonal than its positive counterpart, resulting in a more southerly storm track and enhanced moisture advection into the Great Lakes' region [4]. Indeed, the wintertime (November–January) Climate Prediction Center (CPC) PNA index [25] is well correlated with the lake-level changes ($r = -0.50$). In spite of this, the 8-y reconstructed lake-level time series dL_y^{rc8} is not well correlated with the PNA index ($r = -0.06$), indicating that the PNA pattern is not associated with 8-y lake-level cycle.

Another way to examine the PNA/lake-level connection is by computing the correlations between the 500-mb height anomalies and the wintertime lake-level changes dL_y^w. These correlations are illustrated in Figure 11a, where the white lines indicate areas that are significantly correlated at the 5% confidence level. The negative phase of the PNA pattern is indeed observed with a north-south dipole pattern in the North Pacific and an upper-level trough located in the northwest region of North America.

Because the PNA index is not well correlated with the 8-y cycle, we evaluated the residual lake-level signal, or the lake-level signal that is not associated with the PNA. To achieve this, we used the CPC PNA index as a predictor I and computed the best fit β from the linear model:

$$dL_y^w = I \times \beta + \varepsilon \tag{6}$$

where dL_y^w is the wintertime lake-level change. The residual lake-level change ε is significantly correlated with the 8-y reconstruction dL_y^{rc8} ($r = 0.39$), and the correlations between this residual signal ε and the 500-mb heights are illustrated in Figure 11b. It appears that while part of the wintertime lake-level signal is stemming from the North Pacific region, the 8-y cycle is instead rooted in the North Atlantic region. Statistically significant correlations between ε and SSTs were also identified in the North Atlantic (not shown).

Figure 11. Correlations between 500-mb height anomalies and (**a**) the actual Michigan-Huron wintertime lake-level changes (20-y moving average removed), and (**b**) the residual lake-level changes after accounting for the PNA signal. Statistically significant correlations ($p < 0.05$) are encompassed by the white lines.

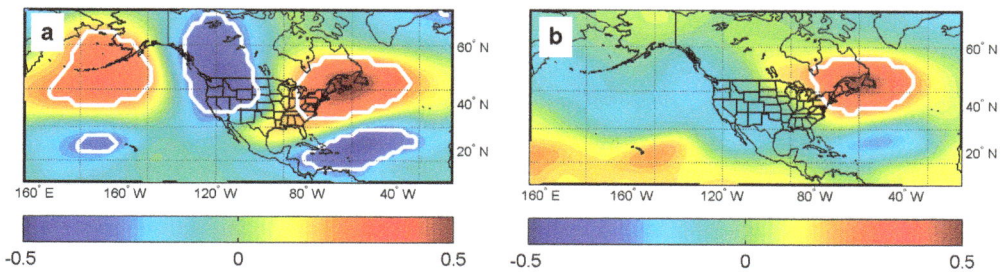

The North Atlantic SST's north-south dipole pattern is similar in structure to spatial patterns identified by Moron *et al.* [26] (Figure 10) and Da Costa and De Verdiere [27] (Figure 2), who analyzed SSTs and SLP fields and found a 7.7-y oscillation possibly rooted in coupled dynamics; the time scale and pattern of this oscillation (not shown) is remarkably similar to the ones associated with the ~8-y oscillation in the Michigan-Huron water levels, thus further substantiating our findings that the 8-y wintertime lake-level cycle is driven by processes in the North Atlantic region as opposed to those associated with the PNA in the North Pacific. We thus conclude that the 8-y cycle is initiated by temperature changes in the North Atlantic which modify long-wave synoptic patterns thus altering moisture transport and precipitation totals in the Great Lakes' region, thereby producing quasi-periodic lake-level changes during the winter months.

4.2. The 12-y Summertime Cycle

We concluded in Section 3 that the second near-decadal cycle, with a period of about 12 years, is associated with changes in precipitation frequency during the summer months of June–August. We thus compared summertime lake-level changes dL_y^s to summertime atmospheric fields and SSTs, analogous to the 8-y wintertime cycle analysis as described in Section 4.1. The resulting composite differences are illustrated in Figure 12, where the white lines again indicate significantly different means at the 5% level, between the years of most positive and most negative anomalous summertime lake-level changes.

Figure 12. Composite differences between the 12 years with the greatest increase and 12 years with the greatest decrease of summertime (JJA) lake-level change, for high-pass filtered seasonal anomalies of (**a**) SSTs; (**b**) 850-mb temperatures; (**c**) SLPs, and (**d**) 500-mb geopotential heights, for 1949–1998. Areas encompassed by white lines indicate significant differences at the 5% level.

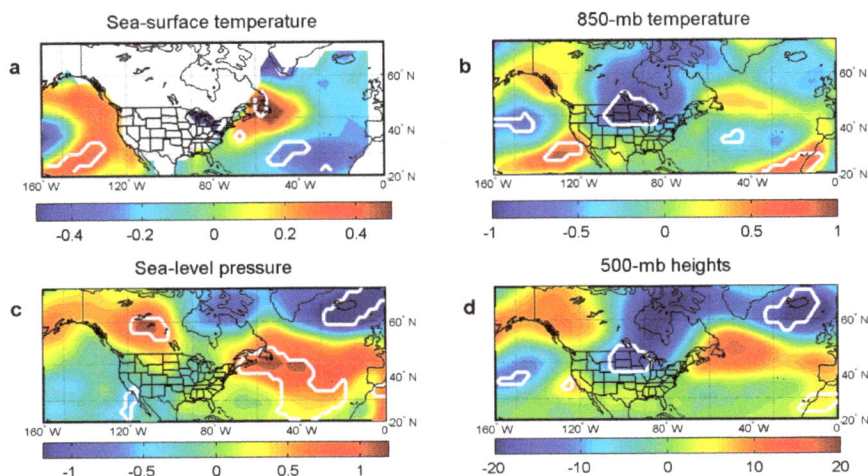

The years of most-increasing summertime lake-levels correspond to a large area of below-average 850-mb temperatures (Figure 12b) which underlie an upper-level trough over the Great Lakes region (Figure 12d). In agreement with Juckes and Smith [28], who examined the relationship between upper-level troughs and convective available potential energy (CAPE), Gold and Nielsen-Gammon [29] determined that the presence of an upper-level potential vorticity anomaly can increase CAPE in the region, thus triggering more frequent convection when conditions are favorable [30]. Thus, the existence of a quasi-periodic upper-level trough over the Great Lakes may alter the frequency of convective precipitation events during the summer months, producing lake-level changes which also exhibit this near-decadal cyclic behavior.

As with the wintertime composite plots, the SST anomalies (Figure 12a) largely appear to extend up to the 850-mb level, which is evident from the cold temperatures encompassed by the warm anomaly over the North Pacific. The warm SST anomalies near the Tropical Pacific region during the greatest summertime lake-level increases also agree with the findings of Wang *et al.* [31] who used El Niño events to skillfully forecast increased summertime precipitation in the Midwest. The North Atlantic SST pattern is similar to that of an identified ~13-y cycle by Moron *et al.* [26] (their Figure 9); the 13-y cycle SST pattern they described exhibits alternating SST signs between the tropical Atlantic off of Africa's western coast, and the eastern coast of the U.S. Hence, as with the 8-y lake-level cycle, it appears that the 12-y cycle can be associated with distinct differences in large-scale atmospheric and oceanic SST patterns.

We determined in Section 2 that the interannual lake-level changes are primarily precipitation driven, therefore indicating that both of the near-decadal cycles are precipitation driven. An evaluation of evaporative losses during different phases of the 12-y cycle, however, reveals that evaporation may also play a small role in the evolution of this periodicity through below-average evaporative losses when precipitation is high and above-average losses when precipitation is low (not shown). In Section 2.1, comparison of the interannual variability associated with both precipitation and evaporation indicated that the historic year-to-year changes in evaporation were too small to impact the lake levels in a significant way. While its historic impact may have been minimal, the existence of this cycle in the summertime evaporation time series substantiates our findings of differing atmospheric phases during increasing and decreasing years of this cycle.

In conclusion, while the 8-y wintertime lake-level cycle is resulting from changes in precipitation amounts related to storm-track alterations linked to processes in the North Atlantic region, the summertime 12-y cycle is driven by changes in precipitation frequency associated with a preferred upper-air trough pattern over the Great Lakes' region. The latter signature is also apparent in the historic time series of summertime evaporation.

5. Summary and Conclusions

The water levels of Lakes Michigan and Huron exhibit considerable variability over a wide range of time scales, and predicting their extremes for socioeconomic planning has proven problematic. Potentially predictable near-decadal lake-level cycles were identified by Hanrahan et al. [6] and were linked to changes in precipitation. However, while this and other studies [2–5] have suggested that regional precipitation changes are the primary interannual lake-level driver, no clear illustrations that connect precipitation variability to the lake-level time series have been made.

In this study, we used daily gridded precipitation data product to explicitly illustrate the impact of precipitation on the lake levels. By weighting the annual precipitation totals according to the historic GLERL component data, we isolated a precipitation-driven fraction of the total lake-level time series which closely resembles the full observed lake-level time series, thus confirming that precipitation has indeed been the primary lake-level driver (Section 2). Furthermore, we determined that regional precipitation effects are translated to the water levels essentially instantaneously, with a sub-monthly time lag.

Hanrahan [22] found that the AMO alters precipitation characteristics, and hence the lake levels, throughout the year. That is, lake levels tend to increase during the negative AMO phase and decrease during the positive AMO phase, during all seasons. In contrast, we concluded in Section 3 that the near-decadal lake-level cycles occur with distinct seasonality. This was verified by comparing the lake-level reconstructions to precipitation indices and reanalysis variables.

Our finding of two dominant, 8-y and 12-y signals in the lake-level and precipitation time series is consistent with previous observational studies based on the analysis of instrumental record [26,27]. Near-decadal spectral peaks were also found to be ubiquitous in tree-ring-based proxy reconstructions of the Great Lakes water levels [32–34], as well as in such reconstructions of the NAO index known to be correlated with temperature and precipitation conditions over the Eastern and Central U.S. [35–37].

We found that the 8-y cycle is occurring during winter months and is linked to variations in precipitation amount (Section 3). During the years of greatest wintertime lake-level changes, anomalously cold SSTs near the Gulf of Alaska coincide with locations of cold 850-mb temperatures, and below-average 500-mb heights, whereas anomalously warm SSTs off of the New England coast correspond to above average 850-mb temperatures and 500-mb heights (Section 4.1). This spatial pattern of an equatorward displaced storm track near the Rockies, and a poleward displaced track near the Great Lakes, is consistent with the timing of increased wintertime precipitation amounts. Rodionov [4] connected increased wintertime (December–February) Great Lakes' regional precipitation to an increased number of cyclones that originated from the south, which was attributed to a weakened 700-mb PNA teleconnection pattern. It was found here that while the wintertime lake levels are well correlated with the PNA index, the 8-y reconstructed cycle is not. After accounting for the PNA influence on the Michigan-Huron lake levels, we attribute variability in the residual lake-level time series, which contains the identified 8-y cycle, to processes in the North Atlantic region, also in agreement with Moron *et al.* [26] and Da Costa and De Verdiere [27]. The details pertaining to this connection are not yet entirely clear and are thus left for future work.

The 12-y cycle is primarily exhibited during the summer months and is driven by alternating frequencies in precipitation events (Section 3). During the years of largest summertime lake-level increases, we identified a 500-mb trough which blankets the Great Lakes region resulting in an increased number of precipitation days (Section 4). We further identified an 850-mb cold temperature anomaly that underlies the upper-level trough associated with this cycle. Our finding of increased precipitation frequency, which is characteristic of increased convective activity during the summer, is further supported by previous studies which have concluded that warm-season precipitation in the U.S. is primarily convectively driven [38,39].

In spite of these findings that largely consider precipitation effects, changes possibly associated with anthropogenic climate change may further complicate the connections between atmospheric patterns, precipitation, and the lake levels. Several studies have indicated that under global warming, the climatology of Northern Hemisphere cyclones will be altered in terms of both average storm track and intensity [40–44]. Furthermore, warmer temperatures may produce higher precipitation totals along storm tracks [45,46]. We hypothesize that changes such as these will ultimately modify the dynamics that have historically contributed to the 8-y lake-level cycle. In addition, increasing air temperatures may alter both rainfall frequency and intensity [47–51], therefore also modifying the behavior of the 12-y cycle. Watras *et al.* [7] already identified a recent shift in the amplitude of a similar cycle in small lake and groundwater levels. Hence, more work needs to be done to assess how the precipitation-driven near-decadal lake-level cycles may be altered in the presence of a changing climate.

While our findings indicate that precipitation has been the single most important variable when evaluating historical interannual lake-level fluctuations, recent research has identified changes in the way that the lake levels are responding to atmospheric processes. Specifically, some authors have found recent significant positive trends in evaporation [52,53], and although variability associated with evaporation has historically been too small to significantly affect the lake levels, a persistent positive decadal-scale evaporation anomaly is now resulting in significant cumulative lake-level changes [8]. As discussed in Section 2.1, the precipitation-driven lake-level components have explained a majority of the historic lake-level behavior; however, their time series are beginning to diverge, stemming from increasing evaporative losses over the past couple of decades.

Thus, not only the precipitation patterns themselves are likely to be altered under climate change, the nature of the lake-level response to climate variability is itself changing. Therefore, while the discussed periodicities may still be useful in statistical predictions schemes for future lake-level variations, one must be cautious in interpreting these predictions (as with any statistical forecast scheme), due to possible non-stationarity of lake–climate connections.

Acknowledgments

The spectral analyses were performed using MTM-SSA toolkit developed by the Theoretical Climate Dynamics group at the University of California-Los Angeles [54]. This research was supported by the Office of Science (BER), U. S. Department of Energy (DOE) grant DE-FG02-07ER64428, NSF grant ATM-0852459, NSF grant ATM-1236620, UWM grant RGI05 (SK), UWM grant RGI05 (PR), and the Russian Department of Education and Science project 14.B25.31.0026 (SK).

Author Contributions

Janel Hanrahan conducted the data analyses and wrote the first draft of the manuscript. Paul Roebber and Sergey Kravtsov provided guidance throughout the research and writing stages. All authors reviewed the final manuscript.

Conflicts of Interest

The authors declare no conflict of interest.

References

1. World Lakes Website. Available online: http://www.worldlakes.org (accessed on 22 March 2014).
2. Changnon, S.A., Jr. Climate fluctuations and record-high levels of Lake Michigan. *Bull. Am. Meteorol. Soc.* **1987**, *68*, 1394–1402.
3. Changnon, S.A. Temporal behavior of levels of the Great Lakes and climate variability. *J. Gt. Lakes Res.* **2004**, *30*, 184–200.
4. Rodionov, S.N. Association between winter precipitation and water level fluctuations in the Great Lakes and atmospheric circulation patterns. *J. Clim.* **1994**, *7*, 1693–1706.
5. Polderman, N.J.; Pryor, S.C. Linking synoptic-scale climate phenomena to lake-level variability in the Lake Michigan-Huron Basin. *J. Gt. Lakes Res.* **2004**, *30*, 419–434.
6. Hanrahan, J.L.; Kravtsov, S.V.; Roebber, P.J. Quasi-periodic decadal cycles in levels of lakes Michigan and Huron. *J. Gt. Lakes Res.* **2009**, *35*, 30–35.
7. Watras, C.J.; Read, J.S.; Holman, K.D.; Liu, Z.; Song, Y.Y.; Watras, A.J.; Morgan, S.; Stanley, E.H. Decadal oscillation of lakes and aquifers in the upper Great Lakes region of North America: Hydroclimatic implications. *Geophys. Res. Lett.* **2014**, in press.

8. Hanrahan, J.L.; Kravtsov, S.V.; Roebber, P.J. Connecting past and present climate variability to the water levels of Lakes Michigan and Huron. *Geophys. Res. Lett.* **2010**, *37*, doi:10.1029/2009GL041707.

9. Delworth, T.L.; Mann, M.E. Observed and simulated multidecadal variability in the Northern Hemisphere. *Clim. Dyn.* **2000**, *16*, 661–676.

10. Gray, S.T.; Graumlich, L.J.; Betancourt, J.L.; Pederson, G.T. A tree-ring based reconstruction of the Atlantic Multidecadal Oscillation since 1567 AD. *Geophys. Res. Lett.* **2004**, *31*, doi:10.1029/2004GL019932

11. Knight, J.R.; Allan, R.J.; Folland, C.K.; Vellinga, M.; Mann, M.E. A signature of persistent natural thermohaline circulation cycles in observed climate. *Geophys. Res. Lett.* **2005**, *32*, doi:10.1029/2005GL024233

12. Enfield, D.B.; Mestas-Nuñez, A.M.; Trimble, P.J. The Atlantic multidecadal oscillation and its relation to rainfall and river flows in the continental US. *Geophys. Res. Lett.* **2001**, *28*, 2077–2080.

13. Rogers, J.C.; Coleman, J.S.M. Interactions between the Atlantic Multidecadal Oscillation, El Nino/La Nina, and the PNA in winter Mississippi valley stream flow. *Geophys. Res. Lett.* **2003**, *30*, doi:10.1029/2003GL017216.

14. McCabe, G.J.; Palecki, M.A.; Betancourt, J.L. Pacific and Atlantic Ocean influences on multidecadal drought frequency in the United States. *Proc. Natl. Acad. Sci. USA* **2004**, *101*, 4136–4141.

15. Sutton, R.T.; Hodson, D.L. Atlantic Ocean forcing of North American and European summer climate. *Science* **2005**, *309*, 115–118.

16. Ghil, M.; Allen, M.R.; Dettinger, M.D.; Ide, K.; Kondrashov, D.; Mann, M.E.; Robertson, A.W.; Saunders, A.; Tian, Y.; Varadi, F.; Yiou, P. Advanced spectral methods for climatic time series. *Rev. Geophys.* **2002**, *40*, 3:1–3:41.

17. Ghil, M.; Yiou, P.; Hallegatte, S.; Malamud, B.D.; Naveau, P.; Soloviev, A.; Friederichs, P.; Keilis-Borok, V.; Kondrashov, D.; Kossobokov, V.; *et al.* Extreme events: dynamics, statistics, and prediction. *Nonlin. Process. Geophys.* **2011**, *18*, 295–350.

18. GLERL data: Available online: http://www.glerl.noaa.gov/data/pgs/lake_levels.html (accessed on 1 June 2010).

19. Clites, A.H.; Quinn, F.H. The history of Lake Superior regulation: Implications for the future. *J. Gt. Lakes Res.* **2003**, *29*, 157–171.

20. International Research Institute/Lamont-Doherty Earth Observatory (IRI/LDEO) Climate Data Library. Available online: http://iridl.ldeo.columbia.edu/SOURCES/.NOAA/.NCEP/.CPC/.REGIONAL/.USA/.daily/.gridded/ (accessed on 1 March 2011).

21. Zolina, O.; Kapala, A.; Simmer, C.; Gulev, S.K. Analysis of extreme precipitation over Europe from different reanalyses: A comparative assessment. *Glob. Planet. Chang.* **2004**, *44*, 129–161.

22. Hanrahan, J.L. Connecting Past and Present Climate Variability to the Water Levels of Lakes Michigan and Huron. Ph.D. Thesis, Department of Mathematical Sciences, University of Wisconsin–Milwaukee, WI, USA, 2010.

23. Kaplan, A.; Cane, M.A.; Kushnir, Y.; Clement, A.C.; Blumenthal, M.B.; Rajagopalan, B. Analyses of global sea surface temperature 1856–1991. *J. Geophys. Res. Ocean.* **1998**, *103*, 18567–18589.

24. Leathers, D.J.; Yarnal, B.; Palecki, M.A. The Pacific/North American teleconnection pattern and United States climate. Part I: Regional temperature and precipitation associations. *J. Clim.* **1991**, *4*, 517–528.

25. National Weather Service, Climate Prediction Center. Available online: http://www.cpc.ncep. noaa.gov (accessed on 28 January 2011).

26. Moron, V.; Vautard, R.; Ghil, M. Trends, interdecadal and interannual oscillations in global sea-surface temperatures. *Clim. Dyn.* **1998**, *14*, 545–569.

27. Da Costa, E.D.; de Verdiere, A.C. The 7.7-year North Atlantic Oscillation. *Q. J. R. Meteorol. Soc.* **2002**, *128*, 797–817.

28. Juckes, M.; Smith, R.K. Convective destabilization by upper-level troughs. *Q. J. R. Meteorol. Soc.* **2000**, *126*, 111–123.

29. Gold, D.A.; Nielsen-Gammon, J.W. Potential vorticity diagnosis of the severe convective regime. Part III: The Hesston tornado outbreak. *Mon. Weather Rev.* **2008**, *136*, 1593–1611.

30. Van Klooster, S.L.; Roebber, P.J. Surface-based convective potential in the contiguous United States in a business-as-usual future climate. *J. Clim.* **2009**, *22*, 3317–3330.

31. Wang, H.; Ting, M.; Ji, M. Prediction of seasonal mean United States precipitation based on El Nino sea surface temperatures. *Geophys. Res. Lett.* **1999**, *26*, 1341–1344.

32. Brinkmann, W.A.R. Water supply to the Great Lakes reconstructed from tree-rings. *J. Clim. Appl. Meteorol.* **1987**, *26*, 530–538.

33. Quinn, F.H.; Sellinger, C.E. A reconstruction of Lake Michigan–Huron water levels derived from tree ring chronologies for the period 1600–1961. *J. Great Lakes Res.* **2006**, *32*, 29–39.

34. Wiles, G.C.; Krawiec, A.C.; D'Arrigo, R.D. A 265-year reconstruction of Lake Erie water levels based on North Pacific tree rings. *Geophys. Res. Lett.* **2009**, *36*, doi:10.1029/2009GL037164.

35. Cullen, H.M.; D'Arrigo, R.D.; Cook, E.R.; Mann, M.E. Multiproxy reconstructions of the North Atlantic Oscillation. *Paleoceanography* **2001**, *16*, 27–39.

36. Fye, F.K.; Stahle, D.W.; Cook, E.R.; Cleaveland, M.K. NAO influence on sub-decadal moisture variability over central North America. *Geophys. Res. Lett.* **2006**, *33*, doi:10.1029/2006GL026656.

37. D'Arrigo, R.D.; Anchukaitis, K.J.; Buckley, B.; Cook, E.; Wilson, R. Regional climatic and North Atlantic Oscillation signatures in West Virginia red cedar over the past millennium. *Glob. Planet. Change* **2012**, *84–85*, 8–13.

38. Fritsch, J.M.; Kane, R.J.; Chelius, C.R. The contribution of mesoscale convective weather systems to the warm-season precipitation in the United States. *J. Clim. Appl. Meteorol.* **1986**, *25*, 1333–1345.

39. Heideman, K.F.; Michael Fritsch, J. Forcing mechanisms and other characteristics of significant summertime precipitation. *Weath. Forecast.* **1988**, *3*, 115–130.

40. Lambert, S.J. The effect of enhanced greenhouse warming on winter cyclone frequencies and strengths. *J. Clim.* **1995**, *8*, 1447–1452.

41. McCabe, G.J.; Clark, M.P.; Serreze, M.C. Trends in Northern Hemisphere surface cyclone frequency and intensity. *J. Clim.* **2001**, *14*, 2763–2768.

42. Zhang, X.; Zwiers, F.W.; Hegerl, G.C.; Lambert, F.H.; Gillett, N.P.; Solomon, S.; Stott, P.A.; Nozawa, T. Detection of human influence on twentieth-century precipitation trends. *Nature* **2007**, *448*, 461–465.

43. Catto, J.L.; Shaffrey, L.C.; Hodges, K.I. Northern Hemisphere Extratropical Cyclones in a Warming Climate in the HiGEM High-Resolution Climate Model. *J. Clim.* **2011**, *24*, 5336–5352.

44. McDonald, R.E. Understanding the impact of climate change on Northern Hemisphere extra-tropical cyclones. *Clim. Dyn.* **2011**, *37*, 1399–1425.

45. Bengtsson, L.; Hodges, K.I.; Keenlyside, N. Will extratropical storms intensify in a warmer climate? *J. Clim.* **2009**, *22*, 2276–2301.

46. Kunkel, K.E.; Karl, T.R.; Easterling, D.R.; Redmond, K.; Young, J.; Yin, X.; Hennon, P. Probable maximum precipitation and climate change. *Geophys. Res. Lett.* **2013**, *40*, 1402–1408.

47. Meehl, G.A.; Arblaster, J.M.; Tebaldi, C. Understanding future patterns of increased precipitation intensity in climate model simulations. *Geophys. Res. Lett.* **2005**, *32*, doi: 10.1029/2005GL023680

48. Sun, Y.; Solomon, S.; Dai, A.; Portmann, R.W. How often will it rain? *J. Clim.* **2007**, *20*, 4801–4818.

49. Allan, R.P; Soden, B.J. Atmospheric warming and the amplification of precipitation extremes. *Science* **2008**, *321*, 1481–1484.

50. Lenderink, G.; van Meijgaard, E. Increase in hourly precipitation extremes beyond expectations from temperature changes. *Nat. Geosci.* **2008**, *1*, 511–514.

51. Collins, M.; Knutti, R.; Arblaster, J.; Dufresne, J.L.; Fichefet, T.; Friedlingstein, P.; Gao, X.; Gutowski, W.J.; Johns, T.; Krinner, G.; *et al.* Long-term Climate Change: Projections, Commitments and Irreversibility. In *Climate Change 2013: The Physical Science Basis. Contribution of Working Group I to the Fifth Assessment Report of the Intergovernmental Panel on Climate Change*; Cambridge University Press: Cambridge, UK & New York, NY, USA, 2013.

52. Assel, R.A.; Quinn, F.H.; Sewnger, C.E. Hydroclimatic factors of the recent record drop in Laurentian Great Lakes water levels. *Bull. Am. Meteorol. Soc.* **2004**, *85*, 1143–1151.

53. Sellinger, C.E.; Stow, C.A.; Lamon, E.C.; Qian, S.S. Recent water level declines in the Lake Michigan–Huron System. *Environ. Sci. Technol.* **2007**, *42*, 367–373.

54. *MTM-SSA Toolkit*, software for spectral analysis; Theoretical Climate Dynamics group at the University of California-Los Angeles: Los Angeles, CA, USA, 2010; Available online: http://www.atmos.ucla.edu/tcd/ssa/ (accessed on 1 June 2010).

Tracking Inflows in Lake Wivenhoe during a Major Flood Using Optical Spectroscopy

Rupak Aryal, Alistair Grinham and Simon Beecham

Abstract: Lake Wivenhoe is the largest water storage reservoir in South-East Queensland and is the primary drinking water supply storage for over 600,000 people. The dam is dual purpose and was also designed to minimize flooding downstream in the city of Brisbane. In early January, 2011, record inflows were experienced, and during this period, a large number of catchment pollutants entered the lake and rapidly changed the water quality, both spatially and vertically. Due to the dendritic nature of the storage, as well as multiple inflow points, it was likely that pollutant loads differed greatly depending on the water depth and location within the storage. The aim of this study was to better understand this variability in catchment loading, as well as water quality changes during the flood event. Water samples were collected at five locations during the flood period at three different depths (surface, mid-depth and bottom), and the samples were analysed using UV and fluorescence spectroscopy. Primary inflows were identified to persist into the mid-storage zone; however, a strong lateral inflow signature was identified from the mid-storage zone, which persisted to the dam wall outflow. These results illustrate the heterogeneity of inflows in water storages of this type, and this paper discusses the implication this has for the modelling and management of such events.

Reprinted from *Water*. Cite as: Aryal, R.; Grinham, A.; Beecham, S. Tracking Inflows in Lake Wivenhoe during a Major Flood Using Optical Spectroscopy. *Water* **2014**, *6*, 2339-2352.

1. Introduction

Lake Wivenhoe, situated 80 km west of Brisbane, is one of the largest dams in Australia. The lake has a capacity to store 1.15 million megalitres (ML) of water and is the major water supply to Brisbane, which is the fourth largest city in Australia. Being situated on the banks of the Brisbane River, the city frequently experiences flooding. Wivenhoe Dam lies on this river, approximately 80 km upstream of the city of Brisbane. It was specifically designed to minimize the flood risk to Brisbane. During flood periods, the lake is capable of holding back a total of 1.45 million ML. After a long decade of drought, Brisbane experienced extreme rainfall between the end of December, 2010, and the first week of January, 2011. The resultant runoff rapidly filled Lake Wivenhoe to 190% of its designed storage capacity. The surrounding catchment is heavily modified with only 40% remnant vegetation, and this, combined with the record inflows, resulted in a significant pollutant loading, including sediment, dissolved organic matter and nutrients, over a very short period of time [1].

Dissolved organic matter (DOM) is of great concern, due to its role in the binding of nutrients, heavy meals and other pollutants from surrounding terrestrial environments. DOM influences the physical and chemical environment in lakes through light attenuation and metal complexion [2–4]. DOM is also important in trophic dynamics [5,6], which promote the growth of heterotrophic microorganisms [7,8]. Furthermore, the incorporation of nitrogen and phosphorus into the DOM pool can influence nutrient cycling in lakes and reservoirs [9–12]. DOM can negatively impact water treatment directly through taste, odour and colour issues and during chlorination through the production of disinfection by-products. Finally, DOM can lead to bacterial proliferation within water distribution systems.

DOM comprises a large number of organic molecules of varied composition, and their characterization can be both complicated and labour intensive. However, monitoring the spatial and vertical variation of DOM is useful for gaining a better understanding of aquatic environmental significance, particularly during periods of major catchment inflows. During major inflows, water quality can change significantly in short periods of time, and simple and sensitive tools are required to rapidly provide qualitative information regarding DOM changes. Optical spectroscopy techniques, such as UV and fluorescence spectroscopy, are both rapid and capable of providing useful characterization of a wide range of DOM.

Optical spectroscopies, such as UV and fluorescence spectroscopy, have been extensively used to characterize organic matter that undergoes changes due to chemical, biological and physical processes in water and wastewater. Optical spectroscopies are popular, because they are reasonably sensitive, simple, rapid and economic. The UV technique can be used with absorption on single- or dual-wavelength procedures and can provide information on individual or representative organic chemical species. The specific wavelength can provide information on numerous chemicals present in the environmental samples [13–17]. The fluorescence spectroscopy method, commonly known as the excitation emission matrix (EEM), is a technique that can be used to obtain an optical fingerprint of dissolved organic matter in water and wastewater, and this can provide information on the nature of microbial, humic and fulvic organics and other pollutants, such as hydrocarbons. Its high sensitivity and its specificity to specific chemicals or groups of chemicals, such as amino acids, aromatic amino acids, mycosporine-like amino acids, humics, proteins and fulvic type substances, have made the application of fluorescence widely popular in the last few years in environmental monitoring [18–20]. Although UV and EEM have been used to monitor water and wastewater in the past, their application in tracking DOM and specific chemical constituents both spatially and vertically has not been reported so far.

The main aim of this paper is to demonstrate how simple optical techniques can be used to track DOM inflows in a lake during flood periods where rapid mixing of water both spatially and vertically takes place.

2. Materials and Methods

Sampling was conducted on 21 January 2011, 10 days after the peak inflows occurred. In order to maximize the spatial representation of the lake, sites were selected from the dam wall, through the main body of the lake and at adjacent major inflow points (Figure 1). Water samples were taken from both the surface (20 cm below the water surface), mid (8 m water depth) and bottom (>15 m, 1 m above the sediment surface) with a vertical, 4.2-L Niskin water sampler (Wildco, Wildlife Supply Company, Yulee, FL, USA). Water depth at each site was recorded from an on-board depth sounder (Lowrance Elite-7 HDI, Navico Inc., Ensenada, MX, USA). Prior to sample collection, the

Niskin and sampling bottles (glass) were cleaned with diluted nitric acid followed by Milli-Q water and twice flushed with water from the same sample depth to minimise contamination. Field personnel took care to not handle the inside of the Niskin or sample containers during sampling, and samples were placed on ice after collection for transport to the laboratory.

Figure 1. Lake Wivenhoe and sampling stations.

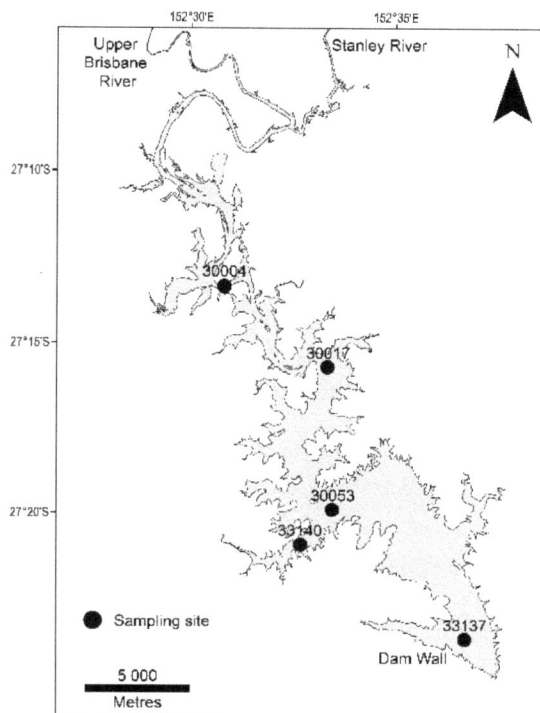

The laboratory samples were filtered through a 1.2-μm filter (Whatman GF/C, GE Healthcare, Little Chalfont, UK) to avoid the influence of turbidity due to suspended solids that cause light scattering, shading and, thus, influence the absorption over the entire spectrum. The filtrate was analysed for dissolved organic carbon (DOC), UV and fluorescence spectra. Details are described below.

2.1. Dissolved Organic Carbon Analysis

Dissolved organic carbon was measured by liquid chromatography with online organic carbon detection (LCOCD, Karlsruhe, Germany) [21]. No replicates were performed in this study.

2.2. UV Analysis

The water samples were analysed using a UV spectrometer (Varian 50 Bio, Victoria, Australia). The instrument was operated at a bandwidth of 1 nm, with a quartz cell of a 10-mm path length, a wavelength of 190 to 400 nm and at a scanning speed of 190 nm/min (slow) at room temperature,

22 ± 2 °C. Milli-Q water was recorded as blank at every set of experiments and subtracted from each sample's UV record.

2.3. Fluorescence Analysis

Three-dimensional fluorescence spectra, also known as excitation emission matrix (EEM) spectra, were obtained using a spectrofluorometer (Perkin Elmer LS 55, Victoria, Australia) with a wavelength range of 200 nm to 500 nm (for excitation); and 280 nm to 500 nm (for emission). The spectra were taken at an incremental wavelength of 5 nm in excitation (Ex); and 2 nm in emission (Em). The EEM value of blank (MQ water) data was subtracted from the analysed samples for blank correction. The fluorescence intensity was corrected by blank subtraction and was expressed in quinine sulphate units (QSU) [22].

A 290-nm emission cut-off filter was used to eliminate the second order Rayleigh light scattering. To eliminate water Raman scatter peaks, Milli-Q water was recorded as the blank and subtracted from each sample. The inner filter effect of EEMs caused by possible higher concentrations of dissolved organic matter (DOM) in the samples was corrected for absorbance by the multiplication of each value in the EEM with a correction factor based on the idea that the average path length of the absorption of the excitation and emission light is 1/2 the cuvette length. For this purpose, the expression was used:

$$F_{corr} = F_{obs} \times 10^{(\lambda_{ex} + \lambda_{em})/2} \qquad (1)$$

where F_{corr} and F_{obs} are the corrected and observed fluorescence intensities and λ_{ex} and λ_{em} are the absorbances at the current excitation and emission wavelengths.

The data obtained from EEM were analysed using an "R" program according to Chen *et al.* (2003) [23] and described below.

$$\emptyset_i = \sum_{ex} \sum_{em} (\lambda_{ex}\lambda_{em})\Delta\lambda_{ex}\lambda_{em} \qquad (2)$$

where $\emptyset_i$ is the volume beneath region i; $(\lambda_{ex}\lambda_{em})$ is the fluorescence intensity at each excitation-emission wavelength pair and $\Delta\lambda_{ex}$ and $\Delta\lambda_{em}$ are the excitation and emission wavelength intervals, respectively.

$$\emptyset_T = \sum \emptyset_i \qquad (3)$$

where $\emptyset_T$ is the cumulative volume.

The EEM spectra was divided into five major regions (shown in Table 1 and Figure 2). The program could calculate area, as well as the contribution percentage of each area.

Table 1. Five major regions in excitation emission matrix (EEM) spectra according to *Chen et al.* (2003) [23]. SMP, soluble microbial by-product; FA, fulvic acid; HA, humic acid. (P1 and P2 = proteins, Ex = excitation, Em = emission, and BOD = biological oxygen demand.)

Region	Chemical composition of organic matter
I (P1): Ex:Em 200–250:280–330	lower molecular weight tyrosine-like aromatic amino acids
II (P2): Ex:Em 200–250:330–380	low molecular weight aromatic proteins and BOD-type substances
III (SMP): Ex:Em 250–340:280–380	large molecular weight peptides and proteins (microorganism related by-products)
IV (FA): Ex:Em 200–250:380–500	fulvic acid type substances
V (HA): Ex:Em 250–500:380–500	humic acid type substances

Figure 2. Five EEM regions selected for this study from the surface water of Site 33137 (regions plotted according to Chen *et al.*, 2003) [23].

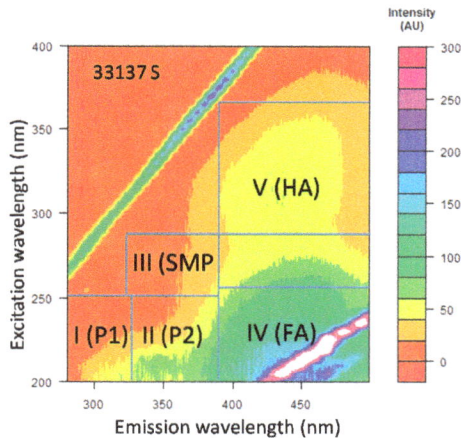

3. Results and Discussion

3.1. Spatial and Vertical Variation in DOM Concentration

Table 2 shows dissolved organic carbon (DOC) across sampling sites and vertically within each site. The DOC concentration was 3–4-times lower than at stratified conditions (9.6–12.8 mg/L recorded almost one year after this study). The lower DOC concentrations indicate dilution by flooding waters. The DOC decreased from inflow sites in the upper lake downstream to the lower lake. Surface waters of upstream sites had relatively elevated DOC levels compared to deeper waters. However, at the middle and downstream sites, the bottom waters had elevated DOC relative to mid-depth and surface waters. Higher DOC concentrations adjacent to a major inflow point (Site 30004) was assumed to be due to catchment inflows having relatively less dilution with lake water compared with sites further into the lake. Decreases in the surface DOC at sites further into the lake were possibly due to settling, microbial/photochemical decomposition and/or subsurface catchment inflows tracking through the lake at deeper depths compared with shallower inflow points [24]. A higher rate of degradation of humic substances at the surface when exposed to UV is reported by Salonen and Vahatalo in a lake in Findland [25].

Table 2. Distribution of dissolved organic carbon $(mg \cdot L^{-1})$ and turbidity (nephelometric turbidity units—NTU) (in brackets) distribution spatially and vertically in Lake Wivenhoe post flood period.

Site (location)	Surface	Mid-depth	Bottom
30004 (upstream)	2.782 (29.8)	2.498 (40.7)	2.622 (191.2)
30017 (upstream)	2.624 (68.9)	2.312 (67.7)	2.248 (212.8)
30053 (middle)	2.129 (77.4)	3.433 (128.2)	2.843 (229)
33140 (middle)	2.244 (90.6)	1.748 (107.8)	2.218 (175.1)
33137 (downstream)	2.373 (111.3)	2.188 (116.5)	2.929 (210.6)

3.2. Optical Analysis

In both UV and fluorescence spectroscopy, incident radiation causes the loosely bound electron present in double or triple bonds and/or in electronegative elements to excite. The absorption of incident radiation is recorded in UV spectroscopy against the wavelength according to the Beer–Lambert Law $(A = \log(I_o/I))$, where I_o is the incident radiation and I is the radiation after passing through the length of solution. In fluorescence spectroscopy, the energy released by excited species to come to the ground state is also recorded. The specific excitation and emission wavelengths are unique for particular species. Two molecules may have similar excitation energies, but different emission energies.

3.2.1. UV Spectra

UV spectroscopy is rapid, simple and requires little sample preparation and small volume samples. Within the absorbance range between 190 and 400 nm, many specific absorbance values are related to a variety of properties, such as aromaticity, hydrophobic content, apparent molecular weight and size and biodegradability [26–28]. Table 3 summarises popular wavelengths widely used to measure chemical species in water and wastewater.

Figure 3 shows a contour diagram of the UV spectral intensity of water recorded at various wavelengths (195, 215, 254 and 330 nm) in Lake Wivenhoe samples collected at the surface (Figure 3a), mid-depth (Figure 3b) and bottom (Figure 3c) during the flood period in January, 2011. Colour patterns in the contour diagram reflect the absorbance intensity at particular wavelengths.

Table 3. UV absorbance recorded at various wavelengths used to measure chemical species in water and wastewater. COD = chemical oxygen demand.

Wavelength (nm)	Property	Reference
195	Proteins	[29]
210	Amino acids	[14,30]
215	Peptides	[30,31]
230	Proteins	[32]
254	Aromaticity	[33]
260	Hydrophobic content/COD	[16,34]
265	Relative abundance of functional group	[35]
272	Aromaticity	[36]
280	Hydrophobic carbon index	[37]
285	Humification index	[27]
300	Characterisation of humic substances	[38]
310–360	Mycosporine-like amino acids	[39–41]
350	Apparent molecular size	[15]
365	Aromaticity, apparent molecular weight	[42]

For wavelengths of 195 nm (proteins), 215 nm (amino acids) and 254 nm (aromaticity), similar colour patterns in the contour diagram indicated that the DOM in the surface and bottom waters showed more homogeneity than in the mid-depth water. In the mid-depth region, UV 215 nm and UV 254 nm showed two distinct colour patterns separating the middle regions from the upstream inflows (surface) and downstream outflows (bottom). The middle region of the lake, where the inflows align, has similar organic characteristics, but at the edge of the lake, different organics are evident. Two possible reasons are proposed: turbid flood runoff being stored in the middle of the reservoir and preferential pathways of stormwater passing through the lake when flowing from upstream to downstream. This is supported by higher DOC and higher turbidity in the middle of the lake at Site 30053. Similar results (higher DOC and turbidity) have been observed by Kim *et al.* (2000) in a deep reservoir at Lake Soyang, Korea [43] when stormwater flooded into the reservoir.

The UV 330 nm wavelength represents mycosporine-like amino acids (MAAs) [39,40]. MAAs are small colourless water soluble compounds composed of cyclohexane or cyclohexenimine chromophore conjugated with nitrogen substituents of amino acids or imino alcohol [44,45] and are very susceptible to photodegradation [41,46]. On the surface, various colour bands were observed in the contour diagram from the upstream inflows to the downstream outflows, but these were not evident in the mid-depth and bottom regions. The results indicate that for the inflows, photodegradation occurred over time in the surface region.

Figure 3. Contour diagram of the UV spectral intensity of flood water recorded at wavelengths 195, 215, 254 and 330 nm collected at the surface (**a**), mid-depth (**b**) and bottom (**c**) of Lake Wivenhoe (colours indicate the absorbance intensity).

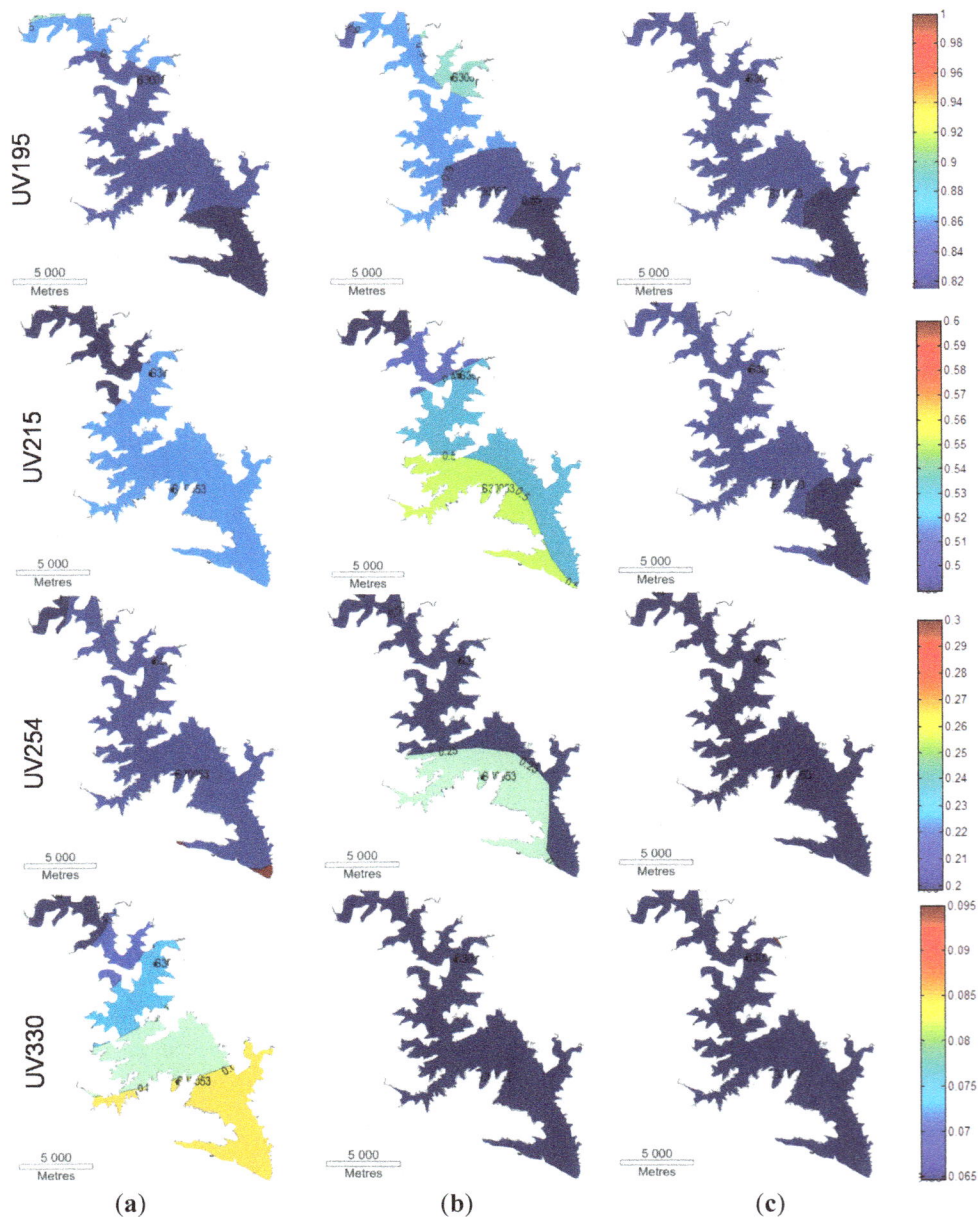

3.2.2. Fluorescence Spectra

Unlike UV, fluorescence spectra provide both excitation and emission information simultaneously, which can distinguish two chemical species from each other that have similar excitation energies.

The fluorescence spectra provide information on different types of organics [47]. There are a number of different methods for interpreting fluorescence data, from selective peak picking [48,49] to complex modelling, such as parallel factor analysis [50,51]. One of the commonly adopted methods is to calculate the area of peak of specific region of the spectrum proposed by (Chen *et al.*, 2003) [23]. In this paper, the EEM spectrum was divided into five regions, and the area of the peak was calculated using "R" software, following the equations described previously [23]. According to Chen *et al.* (2003) [23], Regions I (P1) and II (P2) represent aromatic proteins, Region III represents soluble microbial by-product-like (SMP) substances, Region IV represents fulvic acid-like (FA) substances and Region V represents humic acid-like (HA) substances. Table 1 summarises the representative chemicals in the EEM spectra in the five excitation:emission regions selected for this study.

Figure 4 shows a contour diagram of the five chemical species P1 to HA at three depths. The colour patterns represent fluorescence intensity (or relative concentration) of specific groups of chemicals in the lake, as shown in the legend (far right). Among the five groups of chemicals measured on the surface, P1, P2 and SMP seem to be influenced by the flow pattern, as shown by a preferential flow route for these species. In contrast, the spillover of FA and HA were not affected by the flow. At the mid-depth, we observed a relative decrease in the concentration of P1, P2 and SMP with flow from upstream, but this is not the case with FA and HA. This result indicates the removal of P1, P2 and SMP by sedimentation (binding with particles) and/or by chemical conversion to other organics. We also observed higher fluorescence intensities for P1, P2 and SMP substances in the middle region of the bottom of the lake. The increased microbial activity observed in the middle region of the surface of the water shows possible stagnation of the flow in this region. At the bottom of the lake, sediment particle settling provides increased surface area for microbial colonization, thus allowing increased rates of activity in these waters. According to Kim *et al.* (2000) [43], the carbon and nutrients held in the middle region of turbid water become a good source of food for various bacteria.

4. Conclusions

Lake Wivenhoe is the primary supply water for the city of Brisbane, and it is essential to understand the impact of flood events on water quality in order to fully understand catchment loadings into the system, as well as possible implications for water treatment. The DOC concentration varied spatially and vertically, indicating that the inflow of DOM in the lake varied with space, as well as depth. The UV and fluorescence spectral techniques used in this study showed that organic species were distributed heterogeneously across the lake, both spatially and vertically, and information on specific chemicals or groups of chemicals can be obtained easily. These findings demonstrate the feasibility of optical spectroscopy techniques for understanding the impacts of catchment inflows on DOM species across the lake, and the findings will be highly beneficial for both water treatment and asset management.

252

Figure 4. Contour diagrams of the fluorescence spectral intensity of flood water for the surface, mid-depth and bottom of water across the lake based on the fluorescence area of each region (colours represent fluorescence intensity).

Acknowledgments

The authors wish to thank Seqwater for logistical and financial support and, in particular, Deb Gale and Cameron Veal for their assistance in sample collection.

Author Contributions

The experimental design of the project was undertaken by Rupak Aryal and Alistair Grinham. The majority of field work was conducted by Alistair Grinham. Rupak Aryal, conducted chemical analysis with the help of coauthors Alistair Grinhama and Simon Beecham. All three authors involved equally in interpreting the data and writing this manuscript.

Conflicts of Interest

The authors declare no conflict of interest.

References

1. Grinham, A.; Gibbes, B.; Gale, D.; Watkinson, A.; Bartkow, M. Extreme rainfall and drinking water quality: A regional perspective. *Proc. Water Pollut.* **2012**, *164*, 183–194.
2. Zafiriou, O.C.; Joussot-Dubien, J.; Zepp, R.G.; Zika, R.G. Photochemistry of natural waters. *Environ. Sci. Technol.* **1984**, *18*, 358A–371A.
3. Mostofa, K.M.; Wu, F.; Liu, C.-Q.; Vione, D.; Yoshioka, T.; Sakugawa, H.; Tanoue, E. Photochemical, microbial and metal complexation behavior of fluorescent dissolved organic matter in the aquatic environments. *Geochem. J.* **2011**, *45*, 235–254.
4. Davis, J.A. Complexation of trace metals by adsorbed natural organic matter. *Geochim. Cosmochim. Acta* **1984**, *48*, 679–691.
5. Tranvik, L.; Kokalj, S. Decreased biodegradability of algal DOC due to interactive effects of UV radiation and humic matter. *Aquat. Microb. Ecol.* **1998**, *14*, 301–307.
6. Jansson, M.; Bergström, A.-K.; Blomqvist, P.; Drakare, S. Allochthonous organic carbon and phytoplankton/bacterioplankton production relationships in lakes. *Ecology* **2000**, *81*, 3250–3255.
7. McKnight, D.M.; Smith, R.L.; Harnish, R.A.; Miller, C.L.; Bencala, K.E. Seasonal relationships between planktonic microorganisms and dissolved organic material in an alpine stream. *Biogeochemistry* **1993**, *21*, 39–59.
8. Stone, L.; Berman, T. Positive feedback in aquatic ecosystems: The case of the microbial loop. *Bull. Math. Biol.* **1993**, *55*, 919–936.
9. Schindler, D.; Bayley, S.; Curtis, P.; Parker, B.; Stainton, M.; Kelly, C. Natural and man-caused factors affecting the abundance and cycling of dissolved organic substances in precambrian shield lakes. *Hydrobiologia* **1992**, *229*, 1–21.
10. McKnight, D.; Thurman, E.M.; Wershaw, R.L.; Hemond, H. Biogeochemistry of Aquatic Humic Substances in Thoreau's Bog, Concord, Massachusetts. *Ecology* **1985**, *66*, 1339–1352.

11. Qualls, R.G.; Richardson, C.J. Factors controlling concentration, export, and decomposition of dissolved organic nutrients in the Everglades of Florida. *Biogeochemistry* **2003**, *62*, 197–229.

12. Mladenov, N.; McKnight, D.M.; Wolski, P.; Ramberg, L. Effects of annual flooding on dissolved organic carbon dynamics within a pristine wetland, the Okavango Delta, Botswana. *Wetlands* **2005**, *25*, 622–638.

13. GHOSH, K.; Schnitzer, M. UV and visible absorption spectroscopic investigations in relation to macromolecular characteristics of humic substances. *J. Soil Sci.* **1979**, *30*, 735–745.

14. Aitken, A.; Learmonth, M. *The Protein Protocols Handbook* 1996; Springer: New York, NY, USA, 1996; pp. 3–6.

15. Korshin, G.V.; Li, C.-W.; Benjamin, M.M. Monitoring the properties of natural organic matter through UV spectroscopy: A consistent theory. *Water Res.* **1997**, *31*, 1787–1795.

16. Dilling, J.; Kaiser, K. Estimation of the hydrophobic fraction of dissolved organic matter in water samples using UV photometry. *Water Res.* **2002**, *36*, 5037–5044.

17. Roig, B.; Thomas, O. UV spectrophotometry: A powerful tool for environmental measurement. *Manag. Environ. Qual.* **2003**, *14*, 398–404.

18. Aryal, R.; Kandel, D.; Acharya, D.; Chong, M.N.; Beecham, S. Unusual Sydney dust storm and its mineralogical and organic characteristics. *Environ. Chem.* **2012**, *9*, 537–546.

19. Hong, S.; Aryal, R.; Vigneswaran, S.; Johir, M.A.H.; Kandasamy, J. Influence of hydraulic retention time on the nature of foulant organics in a high rate membrane bioreactor. *Desalination* **2012**, *287*, 116–122.

20. Hussain, S.; van Leeuwen, J.; Chow, C.; Beecham, S.; Kamruzzaman, M.; Wang, D.; Drikas, M.; Aryal, R. Removal of organic contaminants from river and reservoir waters by three different aluminum-based metal salts: Coagulation adsorption and kinetics studies. *Chem. Eng. J.* **2013**, *225*, 394–405.

21. Huber, S.A.; Balz, A.; Abert, M.; Pronk, W. Characterisation of aquatic humic and non-humic matter with size-exclusion chromatography—Organic carbon detection—Organic nitrogen detection (LC-OCD-OND). *Water Res.* **2011**, *45*, 879–885.

22. Aryal, R.K.; Murakami, M.; Furumai, H.; Nakajima, F.; Jinadasa, H.K.P.K. Prolonged deposition of heavy metals in infiltration facilities and its possible threat to groundwater contamination. *Water Sci. Technol.* **2006**, *54*, 205–212.

23. Chen, W.; Westerhoff, P.; Leenheer, J.A.; Booksh, K. Fluorescence excitation-emission matrix regional integration to quantify spectra for dissolved organic matter. *Environ. Sci. Technol.* **2003**, *37*, 5701–5710.

24. Moran, M.A.; Sheldon, W.M., Jr.; Zepp, R.G. Carbon loss and optical property changes during long-term photochemical and biological degradation of estuarine dissolved organic matter. *Limnol. Oceanogr.* **2000**, *45*, 1254–1264.

25. Salonen, K.; Vähätalo, A. Photochemical mineralisation of dissolved organic matter in lake Skjervatjern. *Environ. Int.* **1994**, *20*, 307–312.

26. Ma, H.; Allen, H.E.; Yin, Y. Characterization of isolated fractions of dissolved organic matter from natural waters and a wastewater effluent. *Water Res.* **2001**, *35*, 985–996.

27. Kalbitz, K.; Geyer, S.; Geyer, W. A comparative characterization of dissolved organic matter by means of original aqueous samples and isolated humic substances. *Chemosphere* **2000**, *40*, 1305–1312.

28. Imai, A.; Fukushima, T.; Matsushige, K.; Kim, Y.H. Fractionation and characterization of dissolved organic matter in a shallow eutrophic lake, its inflowing rivers, and other organic matter sources. *Water Res.* **2001**, *35*, 4019–4028.

29. Yabushita, S.; Wada, K.; Inagaki, T.; Arakawa, E. UV and vacuum UV spectra of organic extract from Yamato carbonaceous chondrites. *Mon. Not. R. Astron. Soc.* **1987**, *229*, 45P–48P.

30. Aryal, R.; Vigneswaran, S.; Kandasamy, J. Application of Ultraviolet (UV) spectrophotometry in the assessment of membrane bioreactor performance for monitoring water and wastewater treatment. *Appl. Spectrosc.* **2011**, *65*, 227–232.

31. Kuipers, B.J.; Gruppen, H. Prediction of molar extinction coefficients of proteins and peptides using UV absorption of the constituent amino acids at 214 nm to enable quantitative reverse phase high-performance liquid chromatography-mass spectrometry analysis. *J. Agric. Food Chem.* **2007**, *55*, 5445–5451.

32. Liu, P.-F.; Avramova, L.V.; Park, C. Revisiting absorbance at 230 nm as a protein unfolding probe. *Anal. Biochem.* **2009**, *389*, 165–170.

33. Hur, J.; Schlautman, M.A. Using selected operational descriptors to examine the heterogeneity within a bulk humic substance. *Environ. Sci. Technol.* **2003**, *37*, 880–887.

34. Chevakidagarn, P. Surrogate parameters for rapid monitoring of contaminant removal for activated sludge treatment plants for para rubber and seafood industries in Southern Thailand. *J. Songklanakarin.* **2005**, *27*, 417–424.

35. Chen, J.; Gu, B.; LeBoeuf, E.J.; Pan, H.; Dai, S. Spectroscopic characterization of the structural and functional properties of natural organic matter fractions. *Chemosphere* **2002**, *48*, 59–68.

36. Traina, S.J.; Novak, J.; Smeck, N.E. An ultraviolet absorbance method of estimating the percent aromatic carbon content of humic acids. *J. Environ. Qual.* **1990**, *19*, 151–153.

37. Chin, Y.-P.; Aiken, G.; O'Loughlin, E. Molecular weight, polydispersity, and spectroscopic properties of aquatic humic substances. *Environ. Sci. Technol.* **1994**, *28*, 1853–1858.

38. Artinger, R.; Buckau, G.; Geyer, S.; Fritz, P.; Wolf, M.; Kim, J. Characterization of groundwater humic substances: Influence of sedimentary organic carbon. *Appl. Geochem.* **2000**, *15*, 97–116.

39. Dionisio-Sese, M.L. Aquatic microalgae as potential sources of UV-screening compounds. *Philipp. J. Sci.* **2010**, *139*, 5–16.

40. Winter, A.R.; Fish, T.A.E.; Playle, R.C.; Smith, D.S.; Curtis, P.J. Photodegradation of natural organic matter from diverse freshwater sources. *Aquat. Toxicol.* **2007**, *84*, 215–222.

41. Whitehead, K.; Vernet, M. Influence of mycosporine-like amino acids (MAAs) on UV absorption by particulate and dissolved organic matter in La Jolla Bay. *Limnol. Oceanogr.* **2000**, *45*, 1788–1796.

42. Peuravuori, J.; Pihlaja, K. Molecular size distribution and spectroscopic properties of aquatic humic substances. *Anal. Chim. Acta* **1997**, *337*, 133–149.

43. Kim, B.; Choi, K.; Kim, C.; Lee, U.H.; Kim, Y.-H. Effects of the summer monsoon on the distribution and loading of organic carbon in a deep reservoir, Lake Soyang, Korea. *Water Res.* **2000**, *34*, 3495–3504.

44. Singh, S.P.; Kumari, S.; Rastogi, R.P.; Singh, K.L.; Sinha, R.P. Mycosporine-like amino acids (MAAs): Chemical structure, biosynthesis and significance as UV-absorbing/screening compounds. *Indian J. Exp. Biol.* **2008**, *46*, 7–17.

45. Sinha, R.; Klisch, M.; Gröniger, A.; Häder, D.-P. Ultraviolet-absorbing/screening substances in cyanobacteria, phytoplankton and macroalgae. *J. Photochem. Photobiol. B* **1998**, 47, 83–94.

46. Vincent, W.F.; Roy, S. Solar ultraviolet-B radiation and aquatic primary production: Damage, protection, and recovery. *Environ. Rev.* **1993**, *1*, 1–12.

47. Chong, M.N.; Sidhu, J.; Aryal, R.; Tang, J.; Gernjak, W.; Escher, B.; Toze, S. Urban stormwater harvesting and reuse: A probe into the chemical, toxicology and microbiological contaminants in water quality. *Environ. Monit. Assess.* **2012**, 1–8.

48. Birdwell, J.E.; Engel, A.S. Characterization of dissolved organic matter in cave and spring waters using UV–Vis absorbance and fluorescence spectroscopy. *Org. Geochem.* **2010**, *41*, 270–280.

49. Coble, P.G. Characterization of marine and terrestrial DOM in seawater using excitation-emission matrix spectroscopy. *Mar. Chem.* **1996**, *51*, 325–346.

50. Stedmon, C.A.; Bro, R. Characterizing dissolved organic matter fluorescence with parallel factor analysis: A tutorial. *Limnol. Oceanogr.* **2008**, *6*, 572–579.

51. Stedmon, C.A.; Markager, S.; Bro, R. Tracing dissolved organic matter in aquatic environments using a new approach to fluorescence spectroscopy. *Mar. Chem.* **2003**, *82*, 239–254.

Suitability of a Coupled Hydrodynamic Water Quality Model to Predict Changes in Water Quality from Altered Meteorological Boundary Conditions

Leon van der Linden, Robert I. Daly and Mike D. Burch

Abstract: Downscaled climate scenarios can be used to inform management decisions on investment in infrastructure or alternative water sources within water supply systems. Appropriate models of the system components, such as catchments, rivers, lakes and reservoirs, are required. The climatic sensitivity of the coupled hydrodynamic water quality model ELCOM-CAEDYM was investigated, by incrementally altering boundary conditions, to determine its suitability for evaluating climate change impacts. A series of simulations were run with altered boundary condition inputs for the reservoir. Air and inflowing water temperature (TEMP), wind speed (WIND) and reservoir inflow and outflow volumes (FLOW) were altered to investigate the sensitivity of these key drivers over relevant domains. The simulated water quality variables responded in broadly plausible ways to the altered boundary conditions; sensitivity of the simulated cyanobacteria population to increases in temperature was similar to published values. However the negative response of total chlorophyll-*a* suggested by the model was not supported by an empirical analysis of climatic sensitivity. This study demonstrated that ELCOM-CAEDYM is sensitive to climate drivers and may be suitable for use in climate impact studies. It is recommended that the influence of structural and parameter derived uncertainty on the results be evaluated. Important factors in determining phytoplankton growth were identified and the importance of inflowing water quality was emphasized.

Reprinted from *Water*. Cite as: van der Linden, L.; Daly, R.I.; Burch, M.D. Suitability of a Coupled Hydrodynamic Water Quality Model to Predict Changes in Water Quality from Altered Meteorological Boundary Conditions. *Water* **2015**, *7*, 348-361.

1. Introduction

The Goyder Water Research Institute project C.1.1 was initiated to fill a gap in the current understanding of the potential impacts of climate change on South Australia. The project seeks to understand climate drivers, downscale global circulation (GCM) model projections of future climate and develop a suite of model applications for the evaluation of climate change impacts on society. Current global circulation model (GCM) projections suggest Australian average temperatures will increase by 1.0 to 5.0 degrees by 2070 (compared to 1980–1999), there will be a decrease in average annual rainfall over southern Australia and there will be an increase in the number of hot days and warm nights [1]. Decreases in winter and autumn wind speed and increases in spring and winter downward solar radiation are also projected, but these projections are subject to large uncertainties [2]. Recent efforts to downscale GCM outputs to the catchment scale have identified the potential for reduced catchment yields as the result of reduced precipitation, changes in rainfall seasonality and increased temperatures [3–5]. Besides issues of water quantity, there are potential impacts of climate change on water quality [6,7]. Reservoirs play a major role in determining the water quality within a given water supply system, as they act as both barriers to (e.g., pathogens) and producers of (e.g., cyanobacteria (toxins, tastes and odors), iron and manganese) water quality hazards [8]. Reservoirs integrate the prevailing hydrology, meteorology, biology and biogeochemistry and the resulting quantity and quality of water is a valuable resource that requires sound management to ensure the utility and sustainability of the source water; water quality models are tools to this end.

The potential impacts of climate change on water quality has been evaluated using integrated modeling schemes which include water quality models [9–13]. Such schemes use a combination of catchment and lake/reservoir models that use meteorological boundary conditions as inputs. The meteorological conditions are altered to represent projected future climate and the resulting simulations are taken to represent the potential impacts of those changed climatic conditions. Too few of these studies have been performed to make generalizations about the potential impacts; both positive and negative influences have been identified. Additionally, the differences in model structure and method make it difficult to compare the different studies directly. There are many sources of uncertainty within such a modeling scheme, including the choice of GCM, emissions scenario, downscaling methodology, and the selection of and rigor of application of the hydrological, constituent and lake/reservoir water quality models, including model structure selection and identification of parameters. Each step in the modeling scheme needs to be thoroughly evaluated to ensure the results can be useful.

It is therefore appropriate to adequately test the response of the proposed reservoir water quality model to changes in the environmental variables expected to change in the future. Formalizing our understanding of the way that water quality variables respond to climate related model inputs is fundamentally important to understanding the outputs we generate from models [13]. As these models will be used to project the impacts of downscaled climate scenarios, it is important that the response of the water quality models to the boundary conditions is understood. Water quality models vary in their data input requirements and often contain options for the sub-model structures they contain, making it difficult to assume that they will be equally sensitive in any given application. Responses of chemical and biological processes to the changes in physical state generated by changes in meteorological inputs are dynamic and interactive and therefore difficult to resolve without resolving individual sensitivities in an explicit analysis.

The outputs from any model are dependent on the inputs. It follows that uncertainty in the inputs, either the boundary conditions or the model parameters, contributes to the uncertainty of the model results. Quantification of the influence of the inputs on the model outputs is known as sensitivity analysis and has been extensively described in the literature. Complex models with many parameters, boundary conditions and long runtimes have particular challenges associated with the analysis of their sensitivity and uncertainty. Consequently a great deal of effort has gone towards developing screening methods to identify sensitive parameters and evaluate their influence on model output [14–17]. Less often the influence of boundary conditions or input data is evaluated. Generally, the error associated with these inputs is considered to be less than the uncertainty associated with model parameters as they are quantities that are generally measured at, or proximal to, the lake or reservoir being modeled, using accurate instrumentation. However the range of meteorological boundary conditions are expected to change in the future [18] and given the non-linear and non-monotonic nature of ecosystem models, their behavior in these conditions is uncertain. As suitable observed validation data cannot exist for unobserved future conditions, model behavior under altered boundary conditions can only be validated against qualitative projected responses of ecosystems. These qualitative responses may be derived from space-for-time approaches, robust ecophysiological conceptual models and response data [19] and ensemble model predictions [20].

Therefore, the goal of this work is to answer the question: Does ELCOM-CAEDYM demonstrate appropriate climatic sensitivity to be used as part of a robust integrated modeling scheme? The responsiveness of the ELCOM-CAEDYM model [21,22] to changes in meteorological boundary conditions was analyzed. A previous application of the model to Happy Valley Reservoir (HVR) was used in conjunction with scenarios with altered environmental forcing of incremental changes in flow, air and water temperature, and wind speed. Responses in water quality variables of primary focus were cyanobacteria and soluble metals; further consideration was given to water temperature and water column stratification due to their important role in determining mixing and the rates of biogeochemical reactions. This work does not constitute a model sensitivity analysis, *sensu stricto*, but evaluates the climatic sensitivity or responsiveness of ELCOM-CAEDYM and compares it to other studies and an empirical climate sensitivity analysis of chlorophyll-*a* in Happy Valley Reservoir.

2. Materials and Methods

2.1. Happy Valley Reservoir

Happy Valley Reservoir (35°04'12" S, 138°34'12" E) is situated to the south of Adelaide, the capital of South Australia (Figure 1). It was created by the construction of an earth wall dam between 1892 and 1897. Following a rehabilitation project from 2002 to 2004, it has a capacity of 11,600 ML, a surface area of 178 hectares and average and maximum depths of 6.5 and 18 m, respectively. It is an off stream reservoir and supplies raw water to South Australia's largest water treatment plant, which produces up to 400 ML of filtered water per day, resulting in a hydraulic retention time of 15–30 days. As HVR is isolated from its natural catchment, it is supplied with water from the Onkaparinga River system via an aqueduct from Clarendon Weir, which is in turn supplied from the much larger Mount Bold Reservoir (35°07'12" S, 138°42'00" E). Mount Bold Reservoir collects water from the Mount Lofty Ranges and is supplemented with water pumped from the River Murray, as are most of South Australia's reservoirs. Happy Valley Reservoir has experienced a range of water quality challenges in the past, with blue-green algae (cyanobacteria) causing taste and odor problems in recent decades. The use of artificial destratification (mixing) and algaecides are used for management in the reservoir, while granular activated carbon used in the water treatment process to reduce taste and odor compound concentrations to acceptable levels in the product water. As HVR is supplied with water from an unprotected catchment (*i.e.*, containing various farming activities and human habitation), vigilance against pathogens is required and loads of nutrients are greater than is generally desirable. During the study period, nutrient concentrations were, total phosphorus, 0.05–0.1 mgL^{-1}; total Kjeldahl nitrogen, 0.5–1.0 mgL^{-1}; filterable reactive phosphorus, 0.005–0.03 mgL^{-1}; ammonia, 0.005–0.05 mgL^{-1} and oxidized nitrogen, 0.05–0.5 mgL^{-1}. The seasonal temperature range is generally between 8–10 °C and 25–27 °C, strong persistent stratification is prevented from occurring by the operation of a bubble plume aerator. Due to the importance of Happy Valley Reservoir to Adelaide's water supply, the South Australian Water Corporation has invested heavily in monitoring and research into the processes influencing water quality.

Figure 1. Location of Happy Valley Reservoir. Inset shows 10 m contours of depth and inflow from the aqueduct and the location of the offtake to the water treatment plant (WTP).

2.2. Model Description

   The Estuary and Lake Computer Model (ELCOM) is a hydrodynamic model that simulates the temporal behavior of stratified water bodies with environmental forcing. The model solves the unsteady, viscous Navier-Stokes equations for incompressible flow using the hydrostatic assumption for pressure. The simulated processes include baroclinic and barotropic responses, rotational effects, tidal forcing, wind stresses, surface thermal forcing, inflows, outflows, and transport of salt, heat and passive scalars [21]. When coupled with the Computational Aquatic Ecosystem DYnamics Model [22] water quality model, ELCOM can be used to simulate three-dimensional transport and interactions of flow physics, biology and chemistry. ELCOM uses the Euler-Lagrange method for advection of momentum with a conjugate-gradient solution for the free-surface height. Passive and active scalars (*i.e.*, tracers, salinity and temperature) are advected using a conservative ULTIMATE QUICKEST discretization, see [21] and references within for further details.

The Centre for Water Research was previously engaged to apply ELCOM-CAEDYM to Happy Valley Reservoir [23]. Upon delivery, the model was considered appropriate for the simulation of water movement, contaminant transport, algal growth and biogeochemical cycling [23]. ELCOM was applied at three resolutions (25, 50 and 100 m grid sizes); the finest grid to be used for examining short-circuiting and inflow dilution, and the coarser grids for quicker runtimes and running scenarios relating to stratification, algal growth and soluble metal release from sediments (the 100 m grid was used in this study). The hydrodynamic model was validated against temperature sensor data over two periods, 29 June–6 October 2005 and 23 October 2005–8 February 2006. The parameter set for CAEDYM was derived from applications to other Australian reservoirs and some minor calibration of parameters to suit Happy Valley Reservoir. The manual calibration focused on parameters that could not be derived from literature values and included, the density of particulate organic matter, the maximum rate for microbial decomposition of particulate organic phosphorus (nitrogen), the maximum rate of mineralization of dissolved organic phosphorus (nitrogen), the dissolved oxygen ½ saturation constant for nitrification, the rate of denitrification and the phosphorus ½ saturation constant for algal uptake. Some deficiencies in the calibration of the algal growth components of the model remained.

Two algal groups were included in the model structure, representing chlorophytes (green algae) and cyanophytes (blue-green algae). The phytoplankton growth model was parameterized according to literature values, with only a single parameter being manually calibrated for Happy Valley Reservoir (Table 1). Parameters relating to light, temperature, phosphorus uptake and respiratory losses were different between the two phytoplankton groups. All other parameters were shared and derived from literature values. Notably, buoyancy regulation by cyanobacteria was not invoked in the model structure.

Table 1. Phytoplankton group parameters that differentiate the response to ecophysiological drivers in the ELCOM-CAEDYM model set up.

Parameter	Cyanophyte Value	Chlorophyte Value	Description	Reference
μ_{GTH}	0.8	1.2	Maximum growth rate (d^{-1})	[24]
ϑ_{Ag}	1.09	1.07	Temperature multiplier for growth (-)	[25,26]
μ_{RES}	0.09	0.10	Respiration, mortality and excretion (d^{-1})	[27]
K_P	0.009	0.008	P ½ saturation constant ($mg\ L^{-1}$)	Calibrated
I_K	130	100	Light ½ saturation constant ($\mu E\ m^{-2}\ s^{-1}$)	[28]
T_{STD}	24	20	Standard temperature for algal growth (°C)	[29]
T_{OPT}	30	22	Optimum temperature for algal growth (°C)	[29,30]
T_{MAX}	39	35	Maximum temperature for algal growth (°C)	[29]

For this work, the model was not further calibrated or modified beyond the work of Romero *et al.* [23] and therefore no performance metrics are presented. The lack of extensive calibration to HVR water quality dynamics means the results of the study can be considered to be a general test of the response sensitivity of ELCOM-CAEDYM to climate drivers and not an investigation of the likely effects of climate change on water quality in Happy Valley Reservoir.

262

2.3. Scenarios for Analysis of ELCOM-CAEDYM Climatic Sensitivity

A series of twenty four (24) scenarios were defined, synthetic input data files were generated and ELCOM-CAEDYM simulations were run. As stratification, algal growth and soluble metal concentrations were of key interest, the summer period simulation was used. The 100 m grid version of ELCOM was used to minimize the runtime required, as short-circuiting was not a primary concern of the water quality problems being investigated. The input boundary conditions analyzed were selected to represent the "climate drivers" of precipitation, air temperature and wind speed and are represented by the input files as changes in flow, air and water temperature, and wind speed, respectively (these will be referred to as INFLOW, WIND and TEMP in text). The synthetic input files were generated by applying a linear multiplier, for INFLOW and WIND, and an increment in the case of TEMP (Table 2). Temperature was modified in this fashion to facilitate comparison to potential temperature change magnitudes. For comparison, −5 and +5 degrees correspond to multipliers of 0.8 and 1.25, respectively, at 20 degrees Celsius, similar to the average temperature in the reservoir during the simulations. As ELCOM-CAEDYM will fail if changes to the water budget result in violations in the boundary conditions, changes in the inflow and outflow must be balanced, therefore the outflow (consumption at the offtake) was increased by a corresponding amount. The FLOW scenarios could therefore be considered to represent a change in the consumption of water by the water treatment plant (WTP), rather than changes in precipitation, strictly. This may initially seem artificial; however, as HVR is an offline storage and the inflow to the reservoir is fully regulated by a flume at Clarendon Weir, it can be interpreted as representing changes in demand, especially as a summer period was considered.

Table 2. Boundary condition modifications applied in the sensitivity analysis. A scenario was generated for each change in meteorological variable, resulting in 24 scenarios differing from the base scenario.

Temperature (TEMP) [Increment]	Precipitation (FLOW) [Multiplier]	Wind Speed (WIND) [Multiplier]
−5.0	0.50	0.50
−2.0	0.75	0.75
−1.0	0.90	0.90
−0.5	0.95	0.95
0.5	1.05	1.05
1.0	1.10	1.10
2.0	1.25	1.25
5.0	1.50	1.50

The scenarios were run using the same initial conditions; a "spin-up" period of 1 week was excluded from all summary calculations. As potable water production is the focus of the study, water quality (temperature, suspended solids, chlorophyll, iron and manganese) at the reservoir offtake was analyzed, along with "whole of reservoir" characteristics, such as water temperature and g' (the reduced gravity due to stratification, [21]). Changes in water quality were evaluated as changes in the mean concentration, the maximum concentration and the period of the simulation that the concentration was above a threshold value (green algal and cyanobacterial chlorophyll only, 1 and 10 µg/L, respectively). In order to facilitate the interpretation of the phytoplankton dynamics, summaries of the state variables governing the growth of the two species modeled were calculated as means of the time series values.

2.4. An Empirical Analysis of the Climatic Sensitivity of Chlorophyll-a to Temperature

Historical records of chlorophyll-*a* and water temperature were collated from the primary reservoir surface monitoring location for the period 1998 to 2013. Monthly medians and anomalies were calculated for water temperature and chlorophyll-*a* concentration. The monthly anomalies were normalized to unity, so as to be able to compare directly to modeling results summarized with a similar method. Linear regressions were fitted to the raw anomalies and normalized values, both for the entire year and for the summer months only.

3. Results and Discussion

3.1. Lake Physical Characteristics

The (modeled) physical properties of the lake were altered by the changes in boundary conditions. The degree of stratification, as indicated by average *g'*, was altered in all scenarios; changes in wind speed had a strong negative effect on lake stratification (Table 3). Increasing air and inflowing water temperature resulted in increased reservoir stratification, as did increased flow. Water temperature in the reservoir was not strongly influenced by the INFLOW scenarios, however the WIND and TEMP scenarios had strong effects on the mean of the average, minimum and maximum water temperatures observed over the simulations (Table 3). Only small impacts on reservoir volume and level were observed (not shown).

3.2. Water Quality

An increase in average modeled cyanobacterial chlorophyll was observed with elevated temperature (Figure 2a). The average concentration of reduced soluble iron (FeII) also increased with temperature while soluble manganese was less responsive (Figure 2). Sensitivity responses were close to linear near the origin (±10%), but some became non-linear at the extremes of the scenarios investigated. Exceedance of the threshold selected for cyanobacterial chlorophyll increased approximately linearly with increasing temperature above that of the original scenario, but had little effect below that level (data not shown). The FLOW scenarios had a consistently linear influence on reservoir water quality; increasing average concentrations of chlorophyte and cyanobacterial chlorophyll, MnII and FeII were observed in simulations with reduced flow; only the average concentration of suspended solids (SSOL1) decreased with decreasing flow (Figure 2b). Changes in maximum modeled values behaved similarly as did duration of exceedance for the chlorophyll variables (not shown).

Table 3. Summary of average physical properties for climatic sensitivity analysis of ELCOM-CAEDYM simulations of Happy Valley Reservoir.

Factor	Increment/ Multiplier	g' (/s²)	Temperature Mean (°C)	Temperature Max (°C)	Temperature Min (°C)
Original	-	0.0502	20.5	21.8	16.5
INFLOW	0.50	0.0481	20.9	22.2	16.6
INFLOW	0.75	0.0490	20.8	22.0	16.6
INFLOW	0.90	0.0496	20.6	21.9	16.5
INFLOW	0.95	0.0498	20.6	21.9	16.5
INFLOW	1.05	0.0503	20.5	21.8	16.5
INFLOW	1.10	0.0505	20.5	21.8	16.6
INFLOW	1.25	0.0510	20.3	21.7	16.6
INFLOW	1.50	0.0513	20.2	21.5	16.6
TEMP	−5.0	0.0454	17.0	18.3	13.4
TEMP	−2.0	0.0481	19.1	20.4	15.9
TEMP	−1.0	0.0490	19.8	21.1	16.2
TEMP	−0.5	0.0495	20.2	21.5	16.4
TEMP	+0.5	0.0505	20.9	22.2	16.7
TEMP	+1.0	0.0511	21.3	22.5	17.0
TEMP	+2.0	0.0524	22.0	23.2	17.3
TEMP	+5.0	0.0571	24.1	25.4	17.5
WIND	0.50	0.0984	22.7	25.9	17.0
WIND	0.75	0.0681	21.5	23.4	17.0
WIND	0.90	0.0560	20.9	22.4	16.7
WIND	0.95	0.0528	20.7	22.1	16.6
WIND	1.05	0.0474	20.4	21.6	16.6
WIND	1.10	0.0452	20.2	21.4	17.2
WIND	1.25	0.0397	19.8	20.8	17.4
WIND	1.50	0.0334	19.3	20.1	17.3

The relationship between WIND and algal growth was obviously non-linear with large increases in the average concentrations of both algal groups with decreasing wind speed (Figure 2c). Cyanobacteria were especially favored by low wind speeds. Reduction of wind speed from 90% to 75% of today's averages resulted in a large increase in the duration of exceedance by cyanobacteria (not shown). The simulated phytoplankton production rates were low (~0.1 day^{-1}) compared to what they can potentially be (~0.3–0.5 day^{-1}) and probably are in HVR. This was also noted by Romero *et al.* [23]. The simulated whole lake averages of respiration exceeded that of production in cyanobacteria, indicating that they were limited to growing in a limited volume of the lake where sufficient light was available. Elevated temperatures increased cyanobacterial production rates but these increased production rates were kept in check by elevated respiration. There was very little change in the nutrient (N&P) limitation of phytoplankton, even under the INFLOW scenarios; simulated phytoplankton growth was more limited by light availability (Table 4).

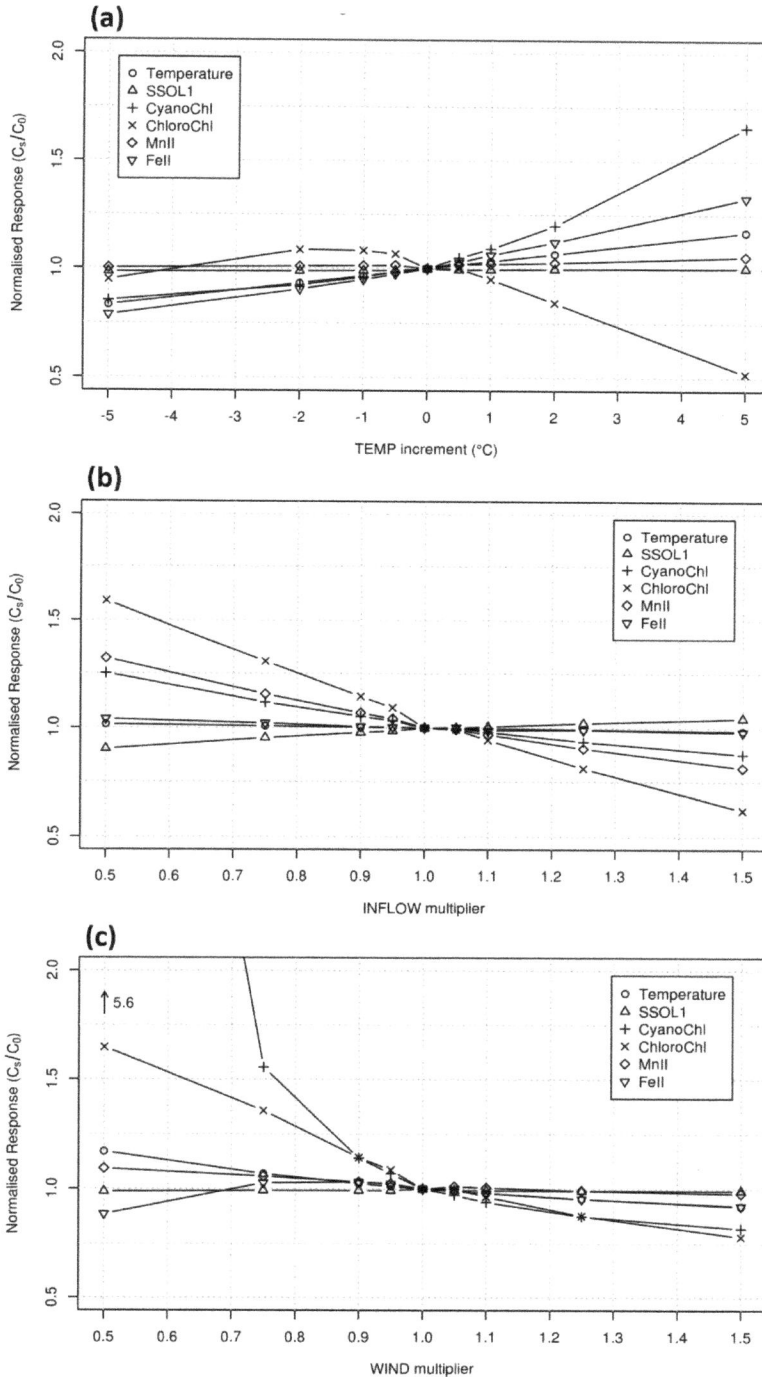

Figure 2. Change in mean modeled water quality values over the summer period in the different sensitivity analysis scenarios where temperature (**a**); rate of inflow and outflow (**b**) or wind speed (**c**) were incrementally changed.

Table 4. Mean cyanobacterial growth characteristics in ELCOM-CAEDYM simulations. The "Limitation by" values indicate the degree of growth limitation by light, phosphorus and nitrogen. It takes a value from 0 to 1; where 1 is unlimited and 0 is completely limited (no growth).

Scenario	Production (day^{-1})	Respiration (day^{-1})	Limitation by		
			Light	Phosphorus	Nitrogen
Original	0.080	0.093	0.099	0.915	0.890
INFLOW by 0.5	0.079	0.096	0.095	0.916	0.883
INFLOW by 1.5	0.081	0.091	0.102	0.916	0.890
TEMP by −5	0.061	0.076	0.101	0.917	0.890
TEMP by +5	0.108	0.115	0.106	0.909	0.884
WIND by 0.5	0.083	0.106	0.086	0.923	0.899
WIND by 1.5	0.075	0.087	0.103	0.917	0.889

3.3. Implied Model Climatic Sensitivity

These scenarios demonstrate that ELCOM-CAEDYM is responsive to changes in environmental drivers that are expected to change under future climate. The model tested was not heavily calibrated and therefore the results are able to be generalized. The observed sensitivities are consistent with qualitative expectations on the basis of contemporary understanding of reservoir processes; for example, that increased temperature and stratification may; increase the prevalence of cyanobacteria; and result in longer periods of decreased dissolved oxygen concentration and higher dissolved metal concentration. Other authors have observed model climatic sensitivities that resulted in increases in the proportion of cyanobacteria by 1%–7.8% per 1 °C increase in temperature (using the model PROTECH [31]). From a review of the literature of the potential impact of climate on phytoplankton communities, Elliott [13] concluded that projected future climate would result in increased relative abundance of cyanobacteria and changes in the phenology of phytoplankton dynamics but not necessarily an increase in the seasonal amount of phytoplankton biomass. These conclusions are consistent with the responses observed in this study.

Important interactions with nutrient availability exist [32] but this was not investigated here. As an independent factor, nutrient addition (*sensu* INFLOW scenarios) did not have a large effect on the phytoplankton dynamics, presumably because of the lack of nutrient limitation (Table 4). The model tested in this study employed a relatively simple representation of phytoplankton community dynamics; only two main functional groups were represented. Furthermore some physiological mechanisms that facilitate cyanobacterial dominance, despite being available in CAEDYM, were not used in the model application of Romero *et al.* [23]. Greater sensitivity and/or more non-linearity may be expected if these mechanisms (e.g., buoyancy regulation) were implemented.

The environmental drivers that were manipulated in the scenarios were not investigated factorially, however they are not completely independent; changes in mean and maximum water temperature occurred in the INFLOW and WIND scenarios (Table 3). This complicates the interpretation of model outputs without extensive comparison of individual simulations; an effort not warranted by the goals of this study. The scenarios were arbitrarily selected to quickly develop a picture of the sensitivity of the model to changed boundary conditions. As such, the important environmental drivers of dilution and nutrient loading are confounded in the multiplication of inflow volumes. Inflow scenarios assumed the same constituent concentrations and therefore the higher flow scenarios had higher nutrient loads. However as chlorophyll concentrations decreased as flow increased; it is apparent that dilution was a more important driver of algal biomass than nutrient load and availability. Despite this, the prediction that phytoplankton growth is rarely limited by nutrient availability may suggest that reducing the external load may be an option for reducing algal growth. The internal load was not investigated as part of this study but given the short water retention time of the reservoir, it is probably of minor importance, compared to the external load. The reduction of nutrient availability represents a potential strategy for adaptation to climate change and the likely negative effects on water quality resulting from increased cyanobacterial growth. Water quality models, such as ELCOM-CAEDYM, have an important role to play in determining the potential benefit of a nutrient reduction program.

3.4. Empirical Reservoir Climatic Sensitivity

Linear regression between water temperature and chlorophyll median monthly anomalies did not resolve slope estimates significantly different from zero (0.105 ± 0.134, $Pr(>|t|) = 0.43$). The weak positive slope estimate combined with a poor predictive relationship ($R^2 = 0.0142$) demonstrates that surface water temperature did not play an important role in determining total chlorophyll in this period (Figure 3b); it also demonstrates that total chlorophyll was not negatively correlated with water temperature, as implied by the water quality model (Figure 3a). This might suggest that deficiencies in definition of model structure or parameter identification have resulted in a non-behavioral model response (one not consistent with our expectations). These deficiencies could, for example, be found in the parameterization of the temperature response functions for growth, or be the product of the over-simplification of the phytoplankton community. This remains speculative, as this simple comparison cannot resolve the differences between the processes structuring algal growth in the model scenarios as compared to those operating over a longer period and in different years, within the reservoir. It must further be noted that the empirical analysis is limited to (monthly) anomalies less than +2 °C and so could not explore the full range of (annual) anomalies as defined by the model scenarios.

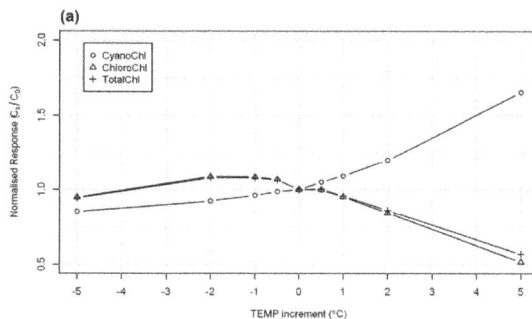

Figure 3. *Cont.*

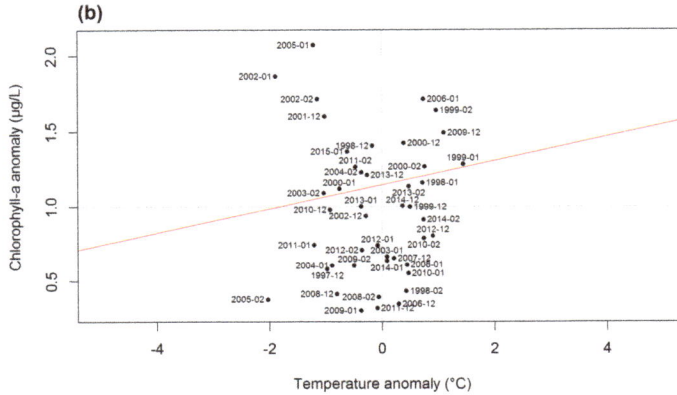

Figure 3. Comparison of (**a**) model derived climate sensitivity to (**b**) empirical reservoir climate sensitivity of chlorophyll-*a* to temperature in summer (December, January, February). In panel (**b**) each point represents the unity normalized anomaly from the monthly median value calculated over the period 1998–2013 and is labeled as yyyy-mm.

4. Conclusions

This study demonstrated that ELCOM-CAEDYM is sensitive to climate drivers and suitable for use in climate impact studies. Rigorous evaluation of the impact of selection of model structures and parameter values on the conclusions drawn from scenarios conducted with altered boundary conditions is advised. This study highlighted factors likely to be important in determining phytoplankton growth in Happy Valley Reservoir. Further it demonstrates that the water quality of the source waters will be of major importance to the reservoir water quality dynamics.

Acknowledgments

This work was funded by the Goyder Water Research Institute Project C1.1. Development of an agreed set of climate projections for South Australia.

Author Contributions

Leon van der Linden conceived and implemented the model climatic sensitivity analysis, the empirical climate sensitivity analysis, analysed the data, interpreted the results and prepared the manuscript; Robert Daly prepared the original input data for ELCOM-CAEDYM and contributed to interpreting the results and preparing the manuscript; Mike Burch conceived the work program, secured funding, managed the research group, and contributed to preparing the manuscript.

Conflicts of Interest

The authors declare no conflict of interest.

References

1. CSIRO. *State of the Climate 2012*; CSIRO and Bureau of Meteorology: Canberra, Australia, 2012; p. 12.
2. CSIRO. *Climate Change in Australia*; CSIRO: Canberra, Australia, 2007.
3. Heneker, T.; Cresswell, D. *Potential Impact on Water Resource Availability in the Mount Lofty Ranges due to Climate Change*; Science, Monitoring and Information Division, Department for Water: Adelaide, Australia, 2010.
4. Charles, S.P.; Heneker, T.; Bates, B.C. Stochastically downscaled rainfall projections and modelled hydrological response for the Mount Lofty Ranges, South Australia. In Proceedings of Water Down Under 2008, Adelaide, Australia, 15–18 April 2008; Lambert, M., Daniell, T., Leonard, M., Eds.; Engineers Australia: Adelaide, Australia, 2008; pp. 428–438.
5. Green, G.; Gibbs, M.; Wood, C. *Impacts of Climate Change on Water Resources, Phase 3: Northern and Yorke Natural Resources Management Region DFW Technical Report 2011/03*; Government of South Australia, Department for Water: Adelaide, Australia, 2011; Volume 1.
6. Whitehead, P.G.; Wilby, R.L.; Battarbee, R.W.; Kernan, M.; Wade, A.J. A review of the potential impacts of climate change on surface water quality. *Hydrol. Sci. J.* **2009**, *54*, 101–123.
7. Delpla, I.; Jung, A.-V.; Baures, E.; Clement, M.; Thomas, O. Impacts of climate change on surface water quality in relation to drinking water production. *Environ. Int.* **2009**, *35*, 1225–1233.
8. Brookes, J.; Burch, M.; Hipsey, M.; Linden, L.; Antenucci, J.; Steffensen, D.; Hobson, P.; Thorne, O.; Lewis, D.; Rinck-Pfeiffer, S.; *et al. A Practical Guide to Reservoir Management*; Cooperative Research Centre for Water Quality and Treatment: Adelaide, Australia, 2008; p. 116.
9. Arheimer, B.; Andréasson, J.; Fogelberg, S.; Johnsson, H.; Pers, C.B.; Persson, K. Climate change impact on water quality: Model results from southern Sweden. *Ambio* **2005**, *34*, 559–566.
10. Mimikou, M.A.; Baltas, E.; Varanou, E.; Pantazis, K. Regional impacts of climate change on water resources quantity and quality indicators. *J. Hydrol.* **2000**, *234*, 95–109.
11. Saloranta, T.; Forsius, M.; Järvinen, M.; Arvola, L. Impacts of projected climate change on the thermodynamics of a shallow and a deep lake in Finland: Model simulations and Bayesian uncertainty analysis. *Hydrol. Res.* **2009**, *40*, 234–248.
12. Thorne, O.M.; Fenner, R.A. Modelling the impacts of climate change on a water treatment plant in South Australia. *Water Sci. Technol. Water Supply* **2008**, *8*, 305–312.
13. Elliott, J.A. Is the future blue-green? A review of the current model predictions of how climate change could affect pelagic freshwater cyanobacteria. *Water Res.* **2012**, *46*, 1364–1371.
14. Arhonditsis, G.B.; Perhar, G.; Zhang, W.; Massos, E.; Shi, M.; Das, A. Addressing equifinality and uncertainty in eutrophication models. *Water Resour. Res.* **2008**, *44*, doi:10.1029/2007WR005862.
15. Makler-Pick, V.; Gal, G.; Gorfine, M.; Hipsey, M.R.; Carmel, Y. Sensitivity analysis for complex ecological models—A new approach. *Environ. Model. Softw.* **2011**, *26*, 124–134.

16. Saltelli, A. Making best use of model evaluations to compute sensitivity indices. *Comput. Phys. Commun.* **2002**, *145*, 280–297.

17. Campolongo, F.; Cariboni, J.; Saltelli, A. An effective screening design for sensitivity analysis of large models. *Environ. Model. Softw.* **2007**, *22*, 1509–1518.

18. Schlabing, D.; Frassl, M.A.; Eder, M.M.; Rinke, K.; Bárdossy, A. Use of a weather generator for simulating climate change effects on ecosystems: A case study on Lake Constance. *Environ. Model. Softw.* **2014**, *61*, 326–338.

19. Paerl, H.W.; Paul, V.J. Climate change: Links to global expansion of harmful cyanobacteria. *Water Res.* **2012**, *46*, 1349–1363.

20. Trolle, D.; Hamilton, D.P.; Pilditch, C.A.; Duggan, I.C.; Jeppesen, E. Predicting the effects of climate change on trophic status of three morphologically varying lakes: Implications for lake restoration and management. *Environ. Model. Softw.* **2011**, *26*, 354–370.

21. Hodges, B.; Dallimore, C. *Estuary, Lake and Coastal Ocean Model: ELCOM v2.2 User Manual*; Centre for Water Research, University of Western Australia: Crawley, Australia, 2007.

22. Hipsey, M.R.; Romero, J.R.; Antenucci, J.P.; Hamilton, D.P. *Computational Aquatic Ecosystem Dynamics Model CAEDYM v2.3 User Manual*; Centre for Water Research, University of Western Australia: Crawley, Australia, 2006.

23. Romero, J.; Antenucci, J.; Okley, P. *Happy Valley Reservoir Modelling Study—Final Report*; Centre for Water Research, University of Western Australia: Crawley, Australia, 2005; p. 41.

24. USCE. *CE-QUAL-R1: A Numerical One-Dimensional Model of Reservoir Water Quality; User's Manual*; Instruction Report E-82-1 (Revised Edition); Department of the Army, U.S. Corps Engineers: Washington, DC, USA, 1995; p. 427.

25. Krüger, G.H.J.; Eloff, J.N. The influence of light intensity on the growth of different Microcystis isolates. *J. Limnol. Soc. South. Afr.* **2010**, *3*, 21–25.

26. Coles, J.F.; Jones, R.C. Effect of temperature on photosynthesis-light response and growth of four phytoplankton species isolated from a tidal freshwater river. *J. Phycol.* **2000**, *36*, 7–16.

27. Schladow, S.; Hamilton, D. Prediction of water quality in lakes and reservoirs: Part II—Model calibration, sensitivity analysis and application. *Ecol. Modell.* **1997**, *96*, 111–123.

28. Hamilton, D.P.; Schladow, S.G. Prediction of water quality in lakes and reservoirs. Part I—Model description. *Ecol. Model.* **1997**, *96*, 91–110.

29. Griffin, S.L.; Herzfeld, M.; Hamilton, D.P. Modelling the impact of zooplankton grazing on phytoplankton biomass during a dinoflagellate bloom in the Swan River Estuary, Western Australia. *Ecol. Eng.* **2001**, *16*, 373–394.

30. Robarts, R.D.; Zohary, T. Temperature effects on photosynthetic capacity, respiration, and growth-rates of bloom-forming cyanobacteria. *N. Z. J. Mar. Freshw. Res.* **1987**, *21*, 391–399.

31. Elliott, J.A.; Jones, I.D.; Thackeray, S.J. Testing the sensitivity of phytoplankton communities to changes in water temperature and nutrient load, in a temperate lake. *Hydrobiologia* **2006**, *559*, 401–411.

32. Mooij, W.M.; Janse, J.H.; Senerpont Domis, L.N.; Hülsmann, S.; Ibelings, B.W. Predicting the effect of climate change on temperate shallow lakes with the ecosystem model PCLake. *Hydrobiologia* **2007**, *584*, 443–454.

Trends in Levels of Allochthonous Dissolved Organic Carbon in Natural Water: A Review of Potential Mechanisms under a Changing Climate

Todd Pagano, Morgan Bida and Jonathan E. Kenny

Abstract: Over the past several decades, dissolved organic carbon (DOC) in inland natural water systems has been a popular research topic to a variety of scientific disciplines. Part of the attention has been due to observed changes in DOC concentrations in many of the water systems of the Northern Hemisphere. Shifts in DOC levels, and changes in its composition, are of concern due to its significance in aquatic ecosystem functioning and its potential and realized negative effects on waters that might be treated for drinking purposes. While it may not be possible to establish sound cause and effect relationships using a limited number of drivers, through long-term DOC monitoring studies and a variety of laboratory/field experiments, several explanations for increasing DOC trends have been proposed, including two key mechanisms: decreased atmospheric acid deposition and the increasing impact of climate change agents. The purpose of this review is three-fold: to outline frequently discussed conceptual mechanisms used to explain DOC increases (especially under a changing climate), to discuss the structure of DOC and the impact of higher levels of DOC on drinking water resources, and to provide renewed/sustained interest in DOC research that can encourage interdisciplinary collaboration. Understanding the cycling of carbon from terrestrial ecosystems into natural waters is necessary in the face of a variable and changing climate, as climate change-related mechanisms may become increasingly responsible for variations in the inputs of allochthonous DOC concentrations in water.

Reprinted from *Water*. Cite as: Pagano, T.; Bida, M.; Kenny, J.E. Trends in Levels of Allochthonous Dissolved Organic Carbon in Natural Water: A Review of Potential Mechanisms under a Changing Climate. *Water* **2014**, *6*, 2862-2897.

1. Introduction

Dissolved organic carbon (DOC) is a complex mixture of aromatic and aliphatic carbon-rich compounds that are important natural components of aquatic ecosystems, modulating many basic biogeochemical and ecological processes [1]. Over the past few decades, several long-term studies have reported an apparent trend of increasing DOC concentrations in inland surface waters over large areas of the Northern Hemisphere. According to a comprehensive review, there remains some uncertainty as to the ubiquity of the trend in increasing DOC levels [2]. However, the review reports that it is clear that a number of studies purport to show an increasing trend in DOC levels over large regions located in the Northern Hemisphere [2], and under a changing climate, it is feasible that DOC changes may become more pronounced—making examination of the potential drivers worthy of attention. At sites where DOC has increased, waters have often become "colored" (brownish tint) or "darker" (more absorbing of radiant light) [3], resulting in changes in light penetration and availability to plants in aquatic ecosystems. DOC concentrations also impact the transport of nutrients as well as pollutants, like heavy metals, in ecosystems. Further, since natural water supplies are sometimes treated and used for drinking water, higher levels of DOC have been shown to contribute to an increased formation of potentially dangerous chemical by-products (the most publicized being the trihalomethanes (THMs)) upon traditional chlorine disinfection treatment. High DOC can also dictate an increased demand for pretreatment steps in the water disinfection process, effectively making water more costly to treat [4,5].

Several factors have been considered to explain the observed increases in surface water DOC, including changes in soil and surface water acidity (e.g., [6,7]) as a result of declining sulfur deposition (e.g., [8,9]), continued nitrogen deposition (e.g., [10]), climate change (e.g., [11,12]), changes in hydrology and precipitation (e.g., [13,14]), changes in land-use patterns (e.g., [15]), as well as combinations of some of these factors. Despite intensive research over a few decades, much remains unknown in regard to our understanding of the terrestrial cycling of carbon. While many studies suggest that changes in DOC are occurring over large regional areas and the potential mechanisms discussed here have the capability to influence these changes to varying degrees, it should be recognized that it may not be possible to establish sound cause and effect relationships using a limited number of drivers. Many drivers, in fact, can work to produce a variety of effects on DOC release from the terrestrial landscape to natural waters and these effects can even work in opposition to each other. Further, many drivers may be covariate and it may, therefore, be difficult to isolate specific effects on DOC, especially in ecosystems where small changes in DOC or in a particular driver are observed. Nonetheless, the drivers outlined in this review are important to consider in areas experiencing changes in allochthonous DOC inputs, but further research should take a critical approach when attempting to ascribe significant statistical relationships between a reduced number of drivers and changes in DOC.

It is also recognized that changes in DOC can vary significantly according to geographic region [16]. For example, arctic and subarctic regions will respond differently to the complex drivers of DOC change when compared to temperate and tropic regions. It is not within the scope of this review to cover in detail how each ecoregion will respond to changes in the carbon cycle and many of the published research articles we reviewed were performed in temperate and subarctic regions. To understand more about carbon cycling in the arctic and changes in arctic carbon stocks, the authors direct readers to a review by McGuire *et al.* [17]. A highly cited review in regard to carbon cycling in tropical forests is also available [18], as well as several articles highlighting organic-matter [19,20] and climate change [21] in the tropics.

Despite the complexity of this issue, much of the attention devoted to explaining the observed long-term changes in DOC has sometimes been reduced to two key hypotheses. The first attributes the increases in DOC in surface waters to a reduction in acidic deposition as a result of tougher air standards internationally [8]. Proponents of this hypothesis argue that the changes in DOC are representative of a return to pre-industrial DOC levels, prior to the addition of sulfate pollution in the atmosphere [8]. There appears to be supporting evidence in the trends of reduced acidity in many surface waters in the U.K. in conjunction with an increase in DOC levels [22], but in some regions not subject to intense acid-deposition, such as northern, high latitude regions, the sulfate reduction hypotheses alone may not be able to completely explain increasing levels of DOC [23].

The second hypothesis offered to explain the DOC increases in water observed in boreal regions and other regions with low acid-deposition, is climate change and all of its inherent consequences (e.g., increasing temperatures, unpredictable weather, increasing atmospheric CO_2, *etc.*) as the main driver for increases in DOC [11]. Increasing temperature and atmospheric CO_2 concentrations can result in greater primary production and the accumulation of degrading biomass to contribute to the ecosystem's DOC pool. Researchers have indicated this observed increase in DOC may be a result of a change in the way in which soils and inland waters store and respire carbon [11,24], potentially representing a systemic environmental change; essentially a response in the carbon cycle to enhanced atmospheric CO_2 concentrations. Other factors of climate change, like varied precipitation and associated droughts and hydrology changes can also affect DOC levels found in water, but their implications are complex and varied in the ways that they may impact DOC trends.

A purpose of this review is to highlight the atmospheric deposition and climate change paradigms used to explain observations of increasing DOC and to briefly discuss other possible drivers discussed in the literature, such as land-use. The topic of DOC increases has garnered much research over the last few decades, perhaps due to the implication of increasing costs for treating water, but also as a means to better understand the processes governing the terrestrial transport of carbon to natural waters [24] and the associated ecological consequences of increasing DOC. Understanding the mechanisms that govern DOC release to natural waters is important, as it can enhance the ability to model future changes in the chemistry and ecosystem functioning of aquatic environments, especially in the face of impending climate change.

2. Characterization of Natural Organic Matter in Water Systems

2.1. Natural Organic Matter Definition and Composition

There is a range of terminology that is used to describe organic matter found in water systems and the terms to discuss its different fractions are sometimes used ambiguously, while accounting for the variety of methods used in different studies to measure organic matter portions (for a quality review of this issue and a comprehensive study of terminology/methodology used in the literature, please see ref. [25]). For the purposes of this review, the following terminology will be used for the various forms of organic matter found in natural waters. The Venn diagram displayed in Figure 1 can help to categorize and visualize the forms of organic matter and related terminology in a simplified way.

Total (or Natural) Organic Matter (TOM) has approximately 50% carbon by weight, and the organic carbon species found in natural water are often referred to as Total Organic Carbon (TOC) [26]. TOC encompasses all molecular organic carbon species found in water (from small molecules like methane to macro-molecular structures such as lignins and proteins) [26]. Dissolved Organic Matter (DOM) is a term used generally when discussing dissolved organic substances, but the term is often used interchangeably with DOC in the literature, despite the fact that DOC makes up a fraction of the DOM profile. Physically, DOC is categorized as organic compounds that can pass through a 0.45 µm filter, though this is merely an operational definition [27]. Any particulates that do not filter through are designated particulate organic carbon (POC) [27].

DOC can be further divided, based on composition, into humic and non-humic fractions. Humic material contains both aromatic and aliphatic components with amide, carboxyl, ketone, and other functional groups [28]. The humic fraction can be further categorized into humic acids, fulvic acids, and humin. The criteria for inclusion into each of these three categories are based on solubility properties at specific pH levels [27]. Humic acids are soluble in water until pH < 2, and contain the highest molecular weight samples of the humics—in the range of 1500–5000 Da in natural waters [29]. Fulvic acids are soluble at all pH levels found in nature, and range in molecular weight between 600 and 1000 Da in natural waters [29], while humin is not soluble in water at any pH [27]. Further, fulvic acid characteristically has more carboxylic groups and oxygen atoms, while humic acid has more phenolic and aromatic groups with longer aliphatic chains [30]. Because humic acids have longer aliphatic chains, they tend to be more nonpolar than fulvic acids, making them less soluble in water [30]. Humics tend to have fairly broad absorption spectra, absorbing most intensely toward the blue region of the visible spectrum [31]. Though models of the molecular makeup of humic compounds are still debated, a representative model structure of a humic acid has been suggested in the literature (see ref. [32] for a representative image). More recent studies have increased the understanding of the composition of DOC beyond the classic categories of humic and fulvic acids using techniques such as nuclear magnetic resonance (NMR) [33], high performance liquid chromatography (HPLC) [34], and Fourier transform ion cyclotron resonance mass spectrometry (FTICRMS) [35]. FTICRMS, coupled with electrospray ionization (ESI), appears to be a promising method for resolving humic and fulvic acids at a molecular level. ESI-FTICRMS has been used, for example, to differentiate between autochthonous and allochthonous dissolved organic matter in marine environments [36], to assess changes in DOM brought about by its interaction with sunlight [37], and to compare and contrast humic and fulvic acids at nearly the molecular level [38,39].

DOC is a complex material containing many molecules with a variety of functional groups—including phenolic compounds (which are of particular interest in this review). The phenolic components of DOC tend to be more recalcitrant to biodegradation [40], especially when compared to aliphatic compounds, and only certain microorganisms, including specialized fungi [41], seem to be able to decompose them [42]. As such, phenolic compounds have the potential to remain in the environment longer and some suggest the breakdown of lignin phenolic compounds may be a rate limiting step in the terrestrial carbon cycle [42,43]. In contrast, lignin phenols can be highly susceptible to photodegradation, and in areas that receive high levels of solar radiation, phenolic compounds have been shown to have faster rates of degradation [40]. Humic material can arise from the microbial breakdown of plants and organisms during the process of humification. The exact molecular mechanism of the humification process, though not fully understood, has been explained by lignin-based and polymerization/condensation-based processes [27]. Humic material can also be completely decomposed over time and ultimately returned to the atmosphere as CO_2.

Figure 1. Simplified Venn representation of the various forms of organic matter found in natural waters. Total Organic Matter (TOM), Total Organic Carbon (TOC), Dissolved Organic Matter (DOM), Dissolved Organic Carbon (DOC), Particulate Organic Carbon (POC), Dissolved Organic Nitrogen (DON), and Dissolved Organic Phosphorus (DOP) are represented. DOC can be further broken down to its humic (humic acid, fulvic acid, and humin) and non-humic material, while new analytical methods continue to reveal more molecular-level detail.

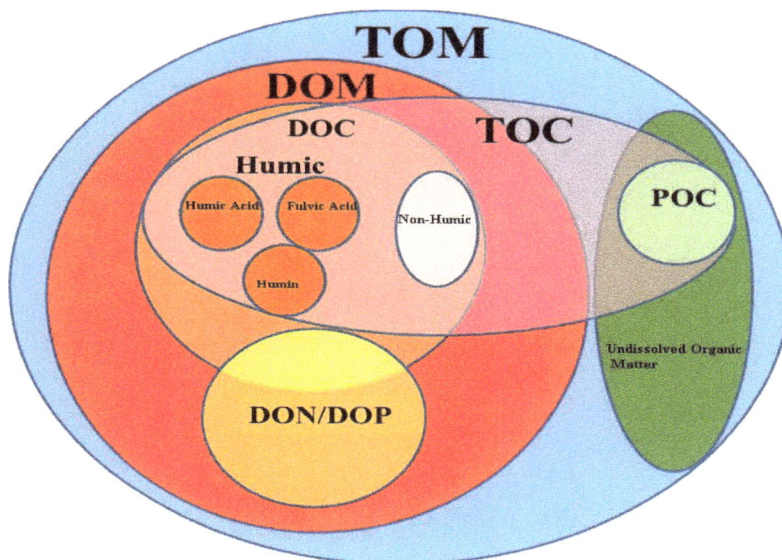

The varied sources of humic material give rise to different spectroscopic properties. As an example, researchers have noted that humics from microbial-derived sources have less aromaticity than humics from terrestrial origins [44–46]. McKnight *et al.*, have proposed that samples with a Fluorescence index (FI) (fluorescence intensity at 450 nm/intensity at 500 nm, both measured at 370 nm excitation) of ~1.9 can be attributed to fulvic acid components of microbial origin [45]. In fact, SanClements *et al.*, showed that increasing DOC in U.S. lakes, which accompanied decreasing sulfate deposition, correlated with a decrease in FI [47]. Similar characteristic spectral responses are the basis for the classic fluorescence spectral classification schemes for DOC proposed by Coble [48,49]. Some studies have further used multidimensional fluorescence spectroscopy coupled with chemometric analyses to categorize modeled components with various types/origins of humic material [50–53].

2.2. DOC Sources and Production

Sobek *et al.*, report that DOC concentrations range from 0.1 to 322 mg/L, with a median concentration of 5.71 mg/L, for 7514 lakes studied in six continents [54]. A main source of DOC in natural waters is the leaching of degraded organic matter from the terrestrial landscape into freshwater systems. The breakdown of aquatic organisms and *in-situ* heterotrophic production can also contribute to DOC concentrations (referred to as autochthonous), but this review focuses on DOC that is exported from terrestrial sources (allochthonous), which is largely influenced by hydrology, temperature, and land-use/cover [55–58]. In a Swedish lake system, it has been estimated that less than 5% of aquatic DOC is autochthonous [59], while in others, particularly systems located in temperate ecoregions and those subject to extensive anthropogenic modification, autochthonous DOC production can exceed allochthonous inputs [55,60,61]. In soils, due to solubility dynamics, an increase of 0.5 pH units can lead to about a 50% corresponding increase in DOC release (though acidity can have an impact on microbial mediated production of DOC as well) [62] and increases in pH have also been shown to enhance the mobilization of DOC as a result of increased enzyme activities [63]. DOC mobilization can be reduced with increased ionic strength of the soil environment, though anions like sulfate, can compete with DOC for soil adsorption sites, thus partially increasing DOC mobility [64] Acidity and ionic strength are important players in the atmospheric chemical deposition mechanism for explaining DOC increases.

With terrestrial origins, allochthonous DOC production is principally described as the result of carbon assimilation via primary production in higher order plants, and is largely a function of the subsequent microbial degradation (or the leaching of plant litter in soils, which may be partially broken down) [64]. Greater amounts of primary production yield a higher accumulation of plant biomass that can become DOC. The plant material can introduce organic matter to its surroundings directly through root exudates, which can be further broken down to become DOC [64]. In addition to the mentioned biotic pathways (enzyme-catalyzed reactions), biomass can also be degraded abiotically (largely through photochemical processes), but often to a lesser extent than biotic processes [65], depending on regional climatic conditions. With plants, a portion of their biomass will be consumed by herbivores [66] whose subsequent death and degradation can be a secondary contribution to the net DOC. Another source of DOC production is through the release of organic matter via the breakdown of the microbial biomass itself [64]. The type of plant species producing the DOC is important in terms of the ultimate quantity that is produced; for example, coniferous dominated ecosystems produce about 50% more DOC than hardwood dominated ecosystems [64]. Likewise, the availability of inorganic nutrients during degradation, microbial environment acidity, and soil moisture (and its related conditions for dictating aerobic or anaerobic microbial degradation) can dictate the amount of DOC produced [67].

It is worth noting the importance of high latitude ecoregions on the bulk production of DOC. The subarctic and boreal regions have been subject to much attention in regard to DOC mobility/runoff into waterways in part because they are responsible for an estimated one-third (390–455 Gt C) of global soil carbon [68], representing 10%–15% of global terrestrial carbon [69], and the arctic delivers 25–36 TgCyr^{-1} [70] to the Arctic Ocean, a quantity that is predicted to increase with a warmer climate. The subarctic is home to large stocks of peat, produced mainly by various species of *Sphagnum*, making it an important region for studies of global carbon sequestration and DOC production. Ultimately, peatlands can add to the global carbon budget in that they are areas where primary production can be greater than decomposition, resulting in a net carbon sink [68] and are therefore of particular significance in the broader context of climate change.

3. Changing DOC Concentrations in Natural Waters

DOC has shown changes in measured concentrations in natural water systems over the past few decades, with a large number of reports coming from North America and northern Europe [22,31,47]. At these sites, over the monitored timeframe (typically one to three decades), the DOC levels have largely been shown to be increasing, and in some cases, may have more than doubled. While the ubiquity of the observed changes in DOC is still up for debate, the vast majority of sites claimed data that show an agreement with the stated trend, though a few geographical areas report no increase and sometimes even a decrease in DOC concentration. Below we will discuss several notable examples of DOC trends in the literature. For a critical review with comprehensive analyses of analytical and data treatment methods of long-term (>10 years) DOC monitoring studies in the literature, readers are encouraged to see ref. [2]. In this review, our intent is to highlight the primary drivers responsible for changes in DOC and to emphasize the complexity of these drivers as they relate to changes in DOC.

A review of the literature indicates that the systematic monitoring of DOC in natural water systems was not pervasive much before the 1980s, though a few studies go back as far as the 1970s—and even the 1960s. DOC in natural waters has been measured in several countries, including parts of Northern Europe, United States, Canada, Korea [71] and Africa [20] and almost all studies report DOC to be increasing [8,9,11,12,14,31,72–95]. Natural waters in the U.K. have been among the most thoroughly studied, in part because of the monitoring conducted by the U.K. Acid Waters Monitoring Network (UKAWMN), which has been measuring the chemical, physical, and biological traits of 11 streams and 11 lakes since 1988 [22,31]. All of the 22 UKAWMN sites showed DOC concentration increases, with an average increase of 91% over a 15 year period [9]. In the U.S., general increases in DOC levels have been reported in the waters of the Adirondack and Catskill Mountain regions of New York State [8,75,81,84,85,96]. Monitored sites in Vermont and Maine showed a few areas of small to moderate rises in DOC levels, while Pennsylvania did not show general increases [8,84]. A study in Canada showed DOC levels increasing 30%–80% in parts of that country during a 20 year period [97]. Areas of Central Europe [98,99] and Siberia [100–102] have also shown marked increases in DOC levels of their natural waters. Models have also predicted DOC level increases in the natural waters of Japan, New Zealand, and portions of China [82].

As one of the most comprehensive intergovernmental studies [84] of DOC, another report of natural waters over a dispersed area (studying 189 water sites in six regions of Europe and six regions of the U.S.) showed some interesting long term trends. All of the European, and all but one of the U.S., regions showed increasing DOC trends in monitored natural water over a twelve year period [84]. The one area that did not show the increasing DOC trend was the Blue Ridge Mountain region of the U.S. [84]. And in another significant study, 68% of 315 varying size natural water sites in the U.K. displayed significant increases in DOC levels over one to four decades [91]. In an even larger study of 522 streams and lakes in North American and Europe, 88% of the surveyed sites showed increases in DOC concentration between the years 1990 and 2004 [8].

4. Mechanisms for Changing DOC Levels

There are several mechanisms that have been used to explain variations in DOC availability and production in freshwaters. Readers are encouraged to see the reviews in refs. [64,103] for discussions of the biogeochemistry and mechanisms potentially responsible for DOC concentration variations in water. As stated, a focus of this review is to explore the widely discussed mechanisms of decreased atmospheric acid deposition and climate change agents, specifically elevated atmospheric CO_2 and temperature, but related precipitation/hydrology and land-use will be introduced as well. Of particular interest here are the mechanisms that are, or will be, factors under a changing climate. Caution should be applied, however, when attempting to explain DOC changes using a reduced number of drivers as changes in DOC are controlled by a variety of complex, interrelated processes, many of which are not fully understood. Fitting schematic diagrams and tabulated information illustrating the complexity of these drivers at work in the terrestrial transport of carbon to inland waters can be found in two quality reviews [104,105].

4.1. Atmospheric Chemical Deposition-Related Increases in DOC

Though the impact of climate change agents are intriguing and are a focus here in subsequent sections of this review, some researchers believe that the impacts of climate change cannot explain the magnitude of reported DOC increases [8] at their study sites. Atmospheric chemical deposition (often referred to as "acid deposition") has been postulated as the main driver responsible for the observed increase in DOC concentrations in natural waters at specific study sites [8,9,31,73–75,83,84,88,94,106–108]. One of the more cited papers on the topic reports that the observed increase in DOC levels is the result of decreases in anthropogenic sulfate, sea salt, and other chemical species in the atmosphere that deposit in the soil [8]. This inverse relationship is believed potentially to be able to account for the magnitude of DOC increase that elevated temperatures and increased atmospheric CO_2 levels alone are seemingly unable to approach, but due to the fact these drivers vary contemporaneously, it is difficult to rule out one driver over another.

As previously highlighted, measurements from 522 lakes and streams in the U.S., Canada, and northern Europe were used to show how increases in DOC concentrations during the time span of 1994–2004 correlate with reductions in atmospheric chemical deposition [8]. In fact, the argument could be made that the observed DOC increases are merely steps toward the DOC levels returning to pre-industrial levels, before anthropogenic sources began adding chemical pollutants to the atmosphere (beginning in about 1750, peaking in the 1970s, and decreasing ever since) [8]. The major chemical in the atmospheric chemical deposition model, sulfate deposition from air pollution, has decreased by about 50% between the years 1986 and 2001 in the U.K. [109] due to notably successful international air regulations. As a consequence, large-scale reductions in sulfate deposition have been observed in many countries.

Simultaneously, increases in nitrogen emissions over the past 150 years have resulted in increases in deposited atmospheric nitrogen (N), which has resulted in an accumulation of N in soils and greater fluxes of N in rivers and streams (in addition to N from agricultural runoff) [110]. It has been hypothesized that increased soil N will stimulate the microbial processing of DOM, resulting in a net decrease in DOC that is available for export to the fluvial network as it is potentially outgassed as CO_2 [10], though experiments designed to test this hypothesis have produced conflicting results [111–114], possibly indicating that N deposition alone is not responsible for changes in DOC [103]. One study points to the significance of nitrogen deposition in bogs, because bogs receive their nutrient supply almost exclusively through atmospheric deposition [114]. The authors of this study show that increased N deposition may result in increased DOC release and an enhancement of phenol oxidase activity in *Sphagnum* litter, promoting peatland carbon release through DOC export and increased outgassing to CO_2. More research into the interplay between C and N cycling is warranted, as much remains not fully understood, particularly the extent to which high N deposition rates will influence plant productivity and litter decomposition. Further, for the purposes of this review we chose to focus primarily on sulfur deposition as it is the more abundant atmospheric deposition chemical referenced in the literature for explaining DOC increases in freshwaters.

It has been observed in the aforementioned studies that sulfate deposition has become significantly reduced [84], while coastal areas in northern Europe recorded reductions in sea salt (a proposed secondary contributor to DOC increases) [115]. And though most of the 522 studied sites showed the inverse relationship between DOC and sulfate concentrations, some coastal areas showed increasing DOC that was not accompanied by significant decreases in sulfate—but these areas showed significant decreases in chlorine concentrations [8]. Models showed that more than 85% of the total anion effect on DOC was a function of anthropogenic sulfate reductions in most of the studied geographic sites [8]. In some areas, the total anion effect was dictated by chlorine to about the same extent as sulfate, and the two anions likely impact DOC concentrations in water by similar mechanisms [8].

The solubility of the bulk of humic matter in soils is largely dependent on acidity [116] and/or ionic strength [117], as increases in either will decrease DOC solubility in laboratory experiments [64]. Acid deposition, specifically by sulfuric acid will increase the acidity as well as increase the total anionic strength (predominantly, increased $[SO_4^{2-}]$) of the soil environment, while sea salt deposition also increases the total anionic strength (increased $[Cl^-]$). A net result of the greater deposition of these chemical species is the decreased solubility of DOC in soils, yielding less mobility to water catchments and resulting in lower DOC levels. Conversely, the observed trends in the reduction of atmospheric chemical deposition should lower soil acidity and anionic strength, allowing for greater solubility of DOC and accounting for the observed increases in DOC concentration in natural waters. Indeed, in a recent study using U.K.-wide upland soil pH datasets, a correlated relationship between DOC and acidity changes was observed [118], lending further support to the chemical deposition paradigm for explaining DOC increases in these systems.

Increasing DOC levels in natural waters can actually be accompanied by increasing organic acidity, as many of the components of DOC are acidic (like humic and fulvic acids). In fact, the presence of these organic acids may be counterbalancing the contributions of declining acidity due to acid deposition—slowing (buffering) recovery from the acidification of the natural waters [8].

Though the trend toward increases in DOC levels, and specifically the link with decreasing sulfate deposition, is compelling, there are a few datasets that show regional areas where DOC is not increasing, despite declining sulfate deposition [76,90,91]. Some argue that DOC concentrations were increasing in the 1960s (or earlier) when sulfate deposition was still increasing and that laboratory studies on the quantitative role of acidification on the mobility of DOC can be ambiguous [119]. In all, it is apparent that a causal relationship between DOC levels and reduced acid deposition may be difficult to establish, especially in areas where the magnitude of DOC changes are small, and that the exact mechanism for changing DOC may include a more complex combination of drivers.

4.2. Climate Change Driven Increases in DOC

Due largely to anthropogenic contributions resulting from the combustion of fossil fuels, the atmospheric concentration of CO_2 has increased by over 60 ppm since 1960 and could rise to about double its current concentration of about 400 ppm by the year 2050 according to IPCC estimates and interpretation of their various models [120,121]. Likewise, an average increase in the global temperature of 0.74 °C has been observed over the past 100 years, and the IPCC estimates that the global temperature could rise by about another 3 °C by the year 2100 (depending on the prediction model chosen) [121]. Many detrimental effects are predicted as a result of climate change, however, the impact of increasing levels of DOC in natural waters has not received as much attention in the IPCC report, perhaps in part because the complex dynamics of terrestrial carbon export are not yet fully understood [121] and available for prediction. Furthermore, net temperature increases are expected to be greatest and most rapid at mid and high latitudes, where the majority of the global peat stock is contained [122].

The various agents associated with climate change have been reported as possible causes for the observed increased DOC levels in several natural water systems [11,12,14,68,69,76,77,79,89,97,98, 100,101,123–129]. As precipitation and its associated hydrology changes are functions of our natural climate, they too have been proposed as dynamics that will change with climate change [121]. Further, hydrologic changes have also been postulated as mechanisms for the observed increases in DOC [92,130,131].

Much of the primary literature related to climate change and its presumed impact on DOC concentration in natural waters deals with increased levels of atmospheric CO_2 and temperature warming. As will be discussed in relation to DOC increases, temperature can play a role in microbial degradation rates and atmospheric CO_2 can increase primary production and root exudates. A secondary outcome of elevated levels of atmospheric CO_2 can be the shift in the population of one plant species to another, where the species taking over the environmental area might assimilate and turnover biomass to DOC more readily. Other drivers are significant in terms of regional observations in DOC, including land-use change and modified hydraulic flow-paths, which will be discussed in subsequent sections.

4.2.1. Increasing Atmospheric CO_2

Increases in plant biomass can accompany elevated atmospheric CO_2 concentrations above ambient levels. This increase is largely due to enhanced levels of primary productivity by plants, and related increases in root exudates [12,69]. In fact, CO_2 "enrichment"/"fertilization" is a technique that has been used by farmers/gardeners in the U.S. to increase their crop yields (strategic biomass) since the mid-1960s [132]. Several controlled laboratory experiments have demonstrated increased biomass accumulation and increased primary productivity under increased atmospheric CO_2. Since 1999, free-air CO_2 enrichment (FACE) technology has been applied to a variety of ecosystem studies; including deciduous, alpine, and evergreen forests; deserts; grasslands; and bogs [133]. FACE experiments have been crucial to our understanding and defining of the ecological processes and responses that occur under elevated atmospheric CO_2 concentrations, especially as it pertains to refining quantitative modeling, but FACE experiments are not always conducive to application to global ecosystem processes [133]. One of the major advantages of FACE experimentation, over laboratory microcosm experimentation, has been the ability to experiment in a larger temporal and spatial context, allowing for the creation of improved models of real-world ecosystems.

Through empirical studies, researchers have predicted an increase of about 20% DOC production when atmospheric CO_2 levels are at twice their recent ambient levels [134]. During a three year experiment of peat monolith samples, the concentration of CO_2 in solardomes was held constant at 372 ppm (matching that of the ambient air at the time) in control samples and at 607 ppm (ambient + 235 ppm) in elevated CO_2 samples [69]. At the end of the three year experiment, pulse isotope labeling studies were conducted on the control and elevated CO_2 samples by exposing them to $^{13}CO_2$ for five hours and monitoring the assimilation of the ^{13}C at various time intervals using collected leachate and plant tissue for analyses [69]. Under conditions of elevated CO_2 levels, measurements of DOC in leachate collections resulted in a 66% increase in concentration over measurements taken from the control samples, while the above and below ground biomass increased by 115% and 96%, respectively [69]. Likewise, $^{13}CO_2$ pulse labeling showed about a ten times increase in "new" (produced within 24 hours of the label exposure) labeled DOC (^{13}DOC, in this case) [69]. The pulse labeling experiments indicate increased exudation (from the roots of increased biomass) as the dominant driver of DOC production under elevated atmospheric CO_2, while microbial degradation still plays a role over the longer term of the isolated ecosystem. Also, the *Sphagnum* moss dominated peat samples shifted greatly toward vascular (lignified tissue) plants, which assimilated 49% more ^{13}C during isotope pulse labeling and led to higher overall DOC production [69].

In terms of allochthonous DOC, some is further broken down and fractions are even returned to the environment as outgassed CO_2, contributing to atmospheric CO_2 concentrations. If further decomposition of organic matter is inhibited, greater quantities of DOC could be released to runoff. In the case of experiments under elevated CO_2, a presence of greater amounts of phenolic compounds (which may further inhibit the breakdown of DOC) was found [89]. The phenolic content of DOC can be an important indicator in the climate change-related mechanisms and its presence is also of concern when water is chlorinated for drinking purposes (discussed in subsequent sections).

One of the more cited articles on the subject of the role of elevated atmospheric CO_2 levels reports increases in DOC concentrations of 14%, 49%, and 61% for bog, fen, and riparian peatlands samples, respectively, at an elevated atmospheric CO_2 concentration of (235 ppm + ambient = 607 ppm) over three years of the controlled study [12]. For clarification, this study [12] was conducted by the same researchers who performed the above experiment [69], and in a parallel timeframe, but on two additional peat types (one of the three peat samples discussed here was the sole sample type reported in the study above). Of course, variations in both estimated and measured DOC levels are expected with the different ecosystems that researchers might be studying (e.g., different forms of peatlands, pine *vs.* deciduous), nutrient availability needed to assist humification, soil moisture, and other variables. This study also showed the similar trend in the shifting of species composition of the isolated environment toward more vascular plants [12]. Fen and riparian peatland soils tend to be rich in the nutrients needed for humification, while bogs tend to be deficient in nutrients, including phosphorus and nitrogen [135]. This nutrient dependence points toward a mechanism of primary production enhancement accompanying elevated atmospheric CO_2 levels for the production of the DOC end product, as the nutrients are needed for primary production of biomass [12]. In soil types where nutrients are less available (like in bogs) [135], the amount of nutrients will be the limiting factor in primary production—showing less sensitivity to environmental stimuli (like increases in atmospheric CO_2 levels) [136].

To support the hypothesis that primary production is responsible for the DOC increases under increased atmospheric CO_2 levels, $^{13}CO_2$ was again used as a marker when it was exposed for five hours to the contained atmosphere of peat monoliths that had been under ambient and elevated (235 ppm + ambient) atmospheric CO_2 levels for about three years [12]. Through photosynthesis and translocation processes, the ^{13}C was traced to the DOC leachate from the monoliths [12]. Not only was more ^{13}C translocated into the sample's biomass (56% more in tissue studies) under elevated atmospheric CO_2 conditions, but the amount of DOC in the soil that was attributed to the "newer" assimilated ^{13}C throughput was an order of magnitude higher [12]. Again, the authors attributed this to the fact that primary production (and associated root exudates) appears to be more sensitive to elevated atmospheric CO_2 levels than microbial degradation [12], which will be discussed further in relation to temperature sensitivity.

Despite the presented data, some believe that elevated atmospheric CO_2 levels cannot account for the magnitude of the observed global trend in DOC concentration increases. The elevated atmospheric CO_2 level (about 607 ppm) used in the two related studies outlined above is much higher than the approximate 20 ppm in atmospheric CO_2 increase that has been measured during the timeframe in which the observed global DOC concentrations were increasing. This poses a challenge to arguments surrounding elevated atmospheric CO_2 levels being solely responsible for the magnitude of observed global DOC concentration increases in freshwaters. Opponents of the elevated atmospheric CO_2 mechanism state that, assuming a linear relationship between DOC increases and elevated atmospheric CO_2 levels, that the group's [12] data for bog, fen, and riparian peat sample would represent only 1.2%, 4.2%, and 5.2% DOC concentration increases, respectively—at the approximate 20 ppm increase in CO_2 concentration over the 20 years of the UKAWMN measurements [9]. It should be noted that bog, fen, and riparian peatlands dominate the U.K. ecosystems of the UKAWMN monitored waters [9]. This is perhaps one reason why the atmospheric chemical deposition theory has been mentioned as an additional or alternative theory to this atmospheric CO_2-controlled increase.

Many FACE experiments have been conducted in a variety of ecosystems. For example, a group at Duke University's Forest FACE experimental station has established an experiment that subjected an actual pine plantation (*Pinus taeda*) in North Carolina to elevated atmospheric CO_2 levels equal to some climate model projections for the year 2050 (about 550 ppm) for twelve years [127]. In a more recent example, a FACE study in the Swiss Alps looked at the effect of elevated atmospheric CO_2 concentrations on a forest near the timberline consisting mainly of a pine species (*Pinus uncinata*) and a deciduous tree species (*Larix decidua*) [137]. More than eight different hypotheses were tested during the most intensive period of FACE experimentation at various sites (1999–2011) and many proved difficult to support [133]. The most well supported hypotheses appear to be in relation to (1) an observed increase in net primary production per unit of leaf-area index (LAI); (2) an apparent closure of leaf stoma in response to elevated CO_2; (3) elevated CO_2 can apparently alter intra- and inter-specific competition for soil resources amongst forest trees, potentially signifying a change in species composition with elevated atmospheric carbon; (4) a weak increase in soil microbial activity under elevated CO_2; and (5) that elevated atmospheric CO_2 does not necessarily imply an increased ecosystem C storage [133], which may have implications for DOC levels within catchments. While FACE experiments have increased our understanding of some ecosystem responses to increased CO_2, many questions remain, but FACE experimentation has demonstrated how complex the interactions of C in an ecosystem are, how they may vary regionally, and how a multidisciplinary approach to DOC research is further warranted to fully understand the dynamics of carbon through ecosystems [133].

4.2.2. Increasing Temperature

Temperature rises are believed to increase DOC levels in natural waters, though the increase has been primarily attributed to improved activity by microorganisms, and specifically by enzyme activity (especially by phenol oxidase) [11]. Phenol oxidase is an enzyme that catalyzes the oxidation of phenolic compounds to quinines [138]—which may occur by extracellular enzyme hydrolysis or by microbial metabolism [12]. Controlled studies on peat samples showed direct increases in DOC production with warmer temperatures [11,129]. In one study, a 10 °C increase in sample environment yielded a 36% increase in phenol oxidase activity, a 33% increase in DOC concentration found in leachate, and a 72% increase in phenolic compounds found in leachate [11]. The presence of higher levels of phenolic compounds is significant, as they are known to inhibit the further breakdown of DOC [139]—making them more available for transport to natural waters [140], including those used as drinking water sources. Additionally, in a study of peat porewater DOC concentrations, warming with infrared lamps (temperature increase of 1.9 °C) produced DOC concentrations about 15% higher than those in controlled, non-heated peatland plots [129]. Further, DOC from the warmed plots degraded faster in lability experiments than the control plots and showed a decreased aromaticity using "specific UV absorbance" (SUVA$_{254}$ = Absorbance @ 254 nm/[DOC]), which suggests an associated increase in microbial activity [129].

Despite the above preliminary findings, the effect on DOC levels using the net global temperature increase estimated by IPCC (0.5–0.7 °C) over the past several decades falls short of explaining the observed DOC trends that have been monitored globally—and a 10 °C increase would likely be a more reasonable temperature rise required to explain the observed global DOC trends [12]. It should be noted that a 10 °C increase in the next century is not projected by any of the current IPCC models. For clarification, the 10 °C rise in temperature chosen in the experimental method of the early study [11] was likely chosen to correlate with the common biological activity coefficient, Q10 (the measure of biological activity over an applied 10 °C increase), not to necessarily match IPCC climate change scenarios. Also, these earlier temperature increase studies were performed on very nutrient rich peat samples that might be expected to be more sensitive to warming (with sufficient nutrient availability to accommodate enhanced microbial activity), but might not be fully representative of all the geographical areas that are producing the observed DOC increases [9]. Nevertheless, other peat soil sample types have been found to be even more sensitive to DOC production due to warming, with even greater increases in aerobic microbial conditions [130]. Though water saturated peat samples are known to produce DOC largely by anaerobic means, there are layers that degrade aerobically due to seasonal drying in the soil [9]. Aerobic degradation and possible future droughts predicted by climate change scenarios could have a synergistic impact on increases in DOC production in soils.

It is estimated that the global 0.5–0.7 °C temperature increase during the time when many DOC levels were actively being measured could potentially be responsible for a 10%–20% increase in DOC concentration in the areas and soil types that were monitored [9]. Still, researchers who make these estimates maintain that they measured a much larger average DOC increase (91%) over about 20 years on a large and geographically diverse data set from the U.K. [9]. Another study cites temperature as potentially being responsible for about 12% of the 78% total DOC increase in a U.K. peat catchment [79]. So, even though rising temperatures appear to be a plausible player in explaining DOC increases, scientists are looking for other mechanisms to account for the larger magnitude of observed increases.

4.2.3. Combined Effect of Increased Atmospheric CO_2 Concentration and Temperature

The combination of elevated CO_2 and temperatures, in conjunction with longer growing seasons, might show additive or synergistic increases in the DOC levels found in natural waters. Experimental investigations on the separate roles of these climate change agents show that they can have at least a partial effect on DOC concentrations in natural waters. Inasmuch as climate change predicted scenarios strongly indicate increases in both atmospheric CO_2 levels and temperature, the interaction of these two factors on natural water DOC concentrations is also of interest. Perhaps, the combined effect can account for both individual models' shortcomings in regard to the magnitude of observed DOC levels that they appear to entail.

In an extension of the studies discussed above (separate/isolated warming and elevated atmospheric CO_2 experiments), the same research group performed a parallel study, again on peat monoliths of riparian peatlands in solardomes and over a three year period [89]. Their data showed a 119% increase in DOC concentration in collected leachate when under a combination of elevated atmospheric CO_2 (again, ambient + 235 ppm CO_2) and temperature (ambient + 3 °C, for this experiment—as opposed to 10 °C discussed for the above experiments) conditions compared to samples kept under ambient (average 372 ppm CO_2 and seasonal average temperatures) conditions [89]. Separate samples kept in elevated atmospheric CO_2 and elevated temperature conditions alone produced 36% and 22% more DOC, respectively, than ambient control samples in collected leachate [89]. It is important to note that the additive effect of the isolated elevated conditions accounts for less than half of the DOC production under the combined condition. The synergistic increase in DOC concentrations for this study is indicative of the potential impact of climate change on DOC flux in natural waters worldwide.

In this same study, both above and below ground DOC concentrations were shown to increase synergistically with the combined treatment conditions [89]; above ground biomass increased by 284% and below ground biomass increased by 407% [89]. Peat Poly-β-Hydroxyalkanoate (PHA) is a microbial nutrient stress indicator that is used to assess competition for inorganic nutrients between the plants and the decomposition microbial agents [128], and was shown to increase 30% under elevated CO_2, 19% under elevated temperature, and 51% under the combined conditions [89]. This was suggestive of an approximate additive relationship in the combined conditions. Phenol oxidase is an enzyme often accredited with the storage and translocation of organic carbon [11]. In this experiment, phenol oxidase decreased by 58% under combined conditions in an apparent synergistic relationship [89]. β-Glucosidase plays a role in the breaking down of general organic substances [89,141] and was shown to decrease by 27% in a slightly synergistic relationship between increased CO_2 and temperature [89]. Similar to the rationale for monitoring PHA, phosphatase can be used to determine whether there has been competition for specific phosphorous nutrients [89]. Phosphatase was shown to increase by 24% under combined conditions, but decrease 9% under elevated temperature and increase 33% under increased CO_2, suggesting neither an additive or synergistic relationship [89]. Phenolic compounds in collected leachate almost doubled in separate elevated atmospheric CO_2 and temperature conditions, and nearly quadrupled in the combined conditions [89]. This finding again indicates that these compounds are more prevalent because of the synergistic reduction of phenol oxidase activity (which breaks down phenolic compounds like those found in lignins, humic acid, and fulvic acid) in the combined conditions [89]. The combined treatment also displayed an additive increase in general microbial nutrient stress, showing microbial competition with the plants, straining the microbes to find available inorganic nutrients—which can be a limiting factor to net DOC production [89]. The results of these chemical measurements led to the development of mechanisms that explain the increased DOC measurements in terms of additive and synergistic effects of elevated CO_2 and elevated temperatures.

The enzyme, phenol oxidase, is an interesting single molecule in the global carbon cycle and has been examined in the literature in detail in regard to its mechanism of carbon assimilation. Phenol oxidase has been labeled the delicate "enzyme latch to the global carbon stock" as it regulates the stability of some 455 Gt C supply in peatlands, and is largely responsible for preventing the release of large amounts of CO_2 back into the environment and could have serious impacts on climate change [142]. Peatlands hold 20%–30% of the world's soil carbon supply, and in anaerobic conditions (true of most water saturated peatlands), the microbial breakdown of phenolic compounds is restrained [142,143]. Under these conditions, phenolic compounds remain in the soil and become available for transport, as part of the net DOC profile, to aquatic systems. Consequently, phenolic compounds are known to inhibit other enzymes from further breaking down organic matter, adding to the net availability of DOC for transport [142]. Some researchers believe that this "enzyme latch" mechanism could be responsible for a portion of observed DOC increases [42,79,142,143]. However, in the presence of droughts (that could accompany climate change), the microbial conditions could become more aerobic, increasing the activity of phenol oxidase, and encouraging the further breakdown of phenolic compounds (and subsequently, other organic compounds) [142,143]. Under these more aerobic conditions, total DOC available for transport may be diminished, but the process could release vast amounts of CO_2 back into the atmosphere [143] with potentially harmful consequences. It is interesting to note that biosensors based on the enzyme interaction of phenol oxidase have been studied for use as phenol sensors in water treatment processes [144].

Along with greater biomass production that is seen with the climate change agents, the suppression of the further breakdown of DOC by phenolic compounds might be a strong accelerator of DOC levels toward current trends. In the study detailed above, there were also changes in plant species composition of the peat samples in the elevated atmospheric CO_2 (slight shift), temperature (moderate shift), and combined (extreme shift) conditions [89]. Some of these potential climate change induced shifts in species composition of an environmental area might also add to the observed global increases of DOC levels in natural waters, as some species exude DOC more readily [89].

Even though it is still a point of debate, while warming and elevated atmospheric CO_2 mechanisms alone appear to fall short in explaining observed DOC level trends, it is conceivable that the synergistic effect of the two conditions combined might approach quantitative explanations of DOC trends. It is also noteworthy that the experimental conditions used in some studies—namely, combined elevated temperature and atmospheric CO_2 levels—are within feasible ranges of IPCC estimates for climate change conditions that could be experienced within the next 50 years or so. Therefore, even opponents of this mechanism for explaining recently observed DOC concentration trends in natural waters should be cognizant of the notion that these could well be the factors driving DOC trends under a changing climate in the near future and beyond.

4.2.4. Changing Hydrology and Its Effect on DOC Concentrations

Many changes to the water cycle are predicted as global temperatures rise, as the water cycle is particularly sensitive to changes in the climate [24]. It is predicted that there will be alterations to precipitation (flooding and drought), discharge, modifications to terrestrial flow-paths of water, and transformations to the hydrologic connectively of the world's water catchments with a warmer climate [121]. Variations in the hydrology of catchments will result in changes to the way DOC is transported, affecting the quantity and composition upon delivery to inland and marine waters. Tranvik and Jasson [145] discussed the significance of hydrology on the transport of DOC from streams to the ocean and stated the ways in which warming might have affected DOC export over a decade ago. Here, we intend to provide several examples of how hydrology affects allochthonous DOC quantity and composition in natural waters. Readers seeking more information, particularly in regard to autochthonous DOC and hydrologic changes, should also consider the comprehensive review of hydrologic changes and DOC associated with climate change provided by Porcal et al. [103].

Precipitation is expected to increase, decrease, and/or become more variable with climate change. Increases in the frequency and severity of droughts have been predicted in some geographic regions, while in others, increases in precipitation are likely, with more severe storm events and floods expected [121]. Changes in the amount, frequency, and seasonal timing of hydrologic events are significant considerations in the prediction of future DOC, as precipitation can increase terrestrial primary production and enhance the export of DOC to natural waters [104,146]. In a study with experimentally controlled drought and rainfall, drought was accompanied by decreases in DOC and its phenolic components, and increased precipitation was accompanied by significant increases in allochthonous DOC and an even more pronounced increase in the phenolic content of allochthonous DOC [42]. Similarly, DOC concentrations decreased in response to a simulated drought that was shown to impact the metabolic activity of biofilms in waters from an experimental wetland [147]. In a study of the effect of drought on litter and peat, drought was shown to be the dominant factor to explain decreases in DOC releases from both sources and the DOC that was released was less hydrophilic in character, suggesting more difficulty in its removal for water treatment purposes [148]. Additionally, in a study in Australia, a severe, decade-long drought was followed by extreme flood events that effected large areas and persisted for several months [149]. During the drought conditions of this study, organic material accumulated on land and in dry fluvial channels, but once this material was again inundated, DOC rapidly leached into the water column of a multitude of streams, rivers, and lakes, resulting in conditions favorable for rapid microbial metabolism and hypoxia [149]. These events, commonly referred to as blackwater events, have also been reported in Brazil [150] and the southern United States [151] and often result in large fish kills. Blackwater events provide insight into the potential water quality challenges under more variable precipitation conditions in the future [149].

Soils that are well-drained, a characteristic of many upland watersheds, tend to have greater proportions of degraded, microbially-derived organic matter than soils located in lowland areas, such as swamps and marshes, where the soil is poorly-drained [152]. This suggests that microorganisms that breakdown DOC and phenolic compounds do so most efficiently in aerobic (water table drawdown) conditions. Under aerobic conditions, phenol oxidase activity increases and organic matter and its constituent phenolic compounds are more readily degraded [128,143]. This hypothesis has been further tested in peatlands, where drought was shown to increase phenol oxidase activity and bacterial growth in a series of in vitro and mesocosm experiments, as well as field observations [143]. Conversely, in high flow or flood conditions (anaerobic), not only might there be higher levels of phenolic compounds [63], but they could be more readily swept away by access to water flow. This was true in peatlands previously experiencing drought, where the effect of high flow conditions resulted in a re-wetting of the peat and carried markedly high concentrations of carbon away, in addition to re-establishing the anoxic pre-drought conditions [143].

The amount of water discharged from fluvial networks is an important consideration, as it can alter both vertical and lateral hydraulic flow-paths in catchments. Multiple studies have indicated that in temperate streams, for example, much of the DOC that is exported occurs during high-flow conditions [153–155] and is largely governed by intra-annual variation in hydrologic regimes [61,156]. Additionally, the composition of DOC exported during high-flow or low-flow conditions has been shown to change and is thought to represent alterations to the flow-path of water due to precipitation. For example, under high-flow conditions, stream waters were shown to contain DOM with large C-to-N ratios, increased aromaticity, and ^{14}C values which indicated that DOM had a less degraded nature, all variables that suggest DOM originating from the upper soil horizons, indicative of a surficial flow-path [152,157]. In contrast, streams under low-flow conditions contained DOM with much older ^{14}C ages, small C-to-N ratios, and a decrease in aromaticity—all indicative of a microbially-processed DOM from lower soil horizons, suggestive of a deeper, ground-water flow-path. Additionally, watersheds with mixed land-use, different land cover types, and anthropogenically modified hydraulic flow-paths can contribute to the DOC quantity and quality in rivers and streams (land-use will be discussed further in a subsequent section).

Jencso *et al.*, highlight the importance of hydrologic connectivity of a watershed in the transport of solutes [158] such as DOC. Hydrologic connectivity is subject to seasonal variation, for example, low connectivity occurs during the winter, when flows over land, to streams, and to rivers are often frozen or otherwise disconnected [158]. In the spring and summer seasons, DOC found in rivers shows signs of more terrestrial inputs (*i.e.*, increased lignin phenols) [159], suggesting a more connected landscape where water carries solutes from all regions of a watershed. Further, the IPCC has predicted that runoff resulting from the thawing of permafrost in the arctic and subarctic areas and subsequent formation of new ponds and lakes will increase the availability of stored carbon stocks [121] through increased hydrologic connectivity [160]. As permafrost thaws, the hydrologic connectivity of arctic and subarctic regions will increase, resulting in greater export of allochthonous DOC to fluvial networks, and ultimately, to the ocean.

Changes in the hydrology experienced by various watersheds will be an important predictor of DOC composition and quantity in the future, especially in areas of high and mid latitudes [161], on both regional and global spatial scales and short and long-term temporal scales. Additionally, increases or decreases in the abundance of ponds, lakes, and man-made reservoirs/impoundments are expected to change the way DOC is exported [104] and processed on land differently in different geographic regions, so additional studies to understand the effect of precipitation, discharge, hydraulic flow-path and the hydrologic connectivity of a catchment are warranted, especially under a changing climate.

4.3. Land-Use and Its Effect on DOC Levels

Analysis of satellite imagery reveals that human land-use activities, such as clearing of forested areas for agriculture and residential land-use, are transforming a large proportion of the earth's surface [162]. Human land-use, including modifications to hydraulic flow-paths, alters biogeochemical cycling, directly affecting water quality, including DOC input to aquatic ecosystems [163–165]. For example, Yallop *et al.*, found that land management may be the most significant driver of humic DOC fluctuations for their study conducted in the U.K. [166] and Gough *et al.*, showed that significant differences in DOC concentration, quality, and THM formation potential are all influenced by land cover type [165]. Further, nutrient enrichment of agricultural land is currently a leading cause of degraded water quality in U.S. coastal waters [167]. Waters with excess nitrogen and phosphorus are impaired by direct effects, like eutrophication, as well as indirect effects, such as changes in temperature, pH, and light attenuation [163,168]. But much less attention has been given to drivers of terrestrial carbon cycling in regard to land-use, especially in terms of land management decisions, where DOC is often overlooked [58]. Further, little is understood about the impact climate change will have on the drivers of DOC in varied land-use catchments.

Researchers have learned about the influence of land-use and land management on stream biochemistry in regard to carbon by analyzing the fluorescence characteristics of DOM [169]. In a recent study, the character of riverine DOM in 34 watersheds in south-central Ontario, Canada was examined in relation to catchment land-use, with findings showing a relationship between agricultural land cover and the composition of DOM in water draining these areas [170]. Values for FI and β/α (a ratio used to estimate the contribution of recently produced DOC, β, to its more degraded form, α) increased as the amount of continuous crop cover increased in the different watersheds [170]. The authors suggest that this indicates a more microbially-derived DOM character [170]. Additionally, they found that there was a strong correlation between total dissolved nitrogen and the DOM composition [170]. It seems that the relationship between DOM composition and the high nitrogen availability associated with agricultural land-use disproportionately increases microbial respiration and changes the composition of DOM, such that it is composed of more microbially-derived moieties. This implies that there may be decreased DOM availability in systems downstream from agriculturally dominated watersheds and that DOM transport distances are decreased [170].

In a similar study, the hypothesis that the optical characteristics of DOM are influenced by microbial activity and land-use, specifically that increased agricultural land-use leads to higher microbial signature in the optical properties of DOM flowing downstream, was tested [55]. They found that the microbial activity increased for DOM in streams with a higher proportion of anthropogenic land-use and that the fluorescence characteristics associated with more labile components in DOM increased with more agricultural land-use [55]. The authors mentioned that their study period was particularly wet and that the increased soil moisture actually made their correlation between more labile, microbially derived DOM stronger [55], pointing, again, to the significance of hydrologic regimes in regional DOC studies.

The main points from these spectroscopic studies are that land-use altered the quantity and quality of DOM exported from human-influenced streams when compared to more natural streams and that DOM from agriculturally dominated watersheds was more labile and supported higher microbial activity when compared to DOM from forested watersheds [55,170]. It is fairly unclear whether land-use will magnify or buffer changes to DOC concentrations and composition in regard to predicted climate change variables, but it is clear that land-use can contribute to changes in DOC levels in natural waters. Therefore, continued studies from varied ecosystem types can contribute to the overall understanding of terrestrial carbon cycling and its prediction in the future.

5. The Impact of Increased DOC on Drinking Water Supply and Treatment

Regardless of the exact driver for the observed increases in water DOC levels, there are several environmental and health concerns associated with the trends, especially in regard to its impact on drinking water. Because increases in aquatic DOC concentration can intensify the "color" of natural waters, higher levels can block the sun's radiation from penetrating to reach deeper ecosystems, even though some DOC is beneficial to blocking solar ultraviolet (UV) radiation from damaging aquatic life. DOC also plays a role in natural nutrient transport [171], which in the case of excessive concentrations of DOC, can also increase the bioavailability of pollutants like mercury in natural waters [172]. As discussed previously, higher levels of DOC, and thus, organic acidity, can also make natural waters more acidic.

Higher concentrations of DOC in water systems that are used for drinking water supplies are a considerable concern for the treatment and disinfection of drinking water [107,173–179]. When treated by the most common chlorination practices, drinking water produced from sources with elevated levels of DOC can produce potentially dangerous by-products; including chloroform, haloacetonitriles, and chloral hydrate—along with the infamous trihalomethane by-products (U.S. EPA maximum drinking water level of THM = 80 ppb [176,180]). Alternatives to traditional chlorination disinfection techniques, like ozonation, and pretreatment strategies, like coagulation/flocculation or filtering, might be necessary for the treatment of natural water that is high in DOC concentration. Since 2006, the U.S. EPA has imposed more rigid regulations on water treatment disinfection by-products through "Stage 2" of their "Disinfection and Disinfection By-product Rule" [181].

In addition to the EPA, disinfection by-products are regulated by the World Health Organization and the European Union [176]. The problem of DOC levels and drinking water treatment have been reported in the U.S. [175,176] Norway [14,173,182], Sweden [178], Australia [173], and the U.K. [31,179], among other locations. In the Central Valley of California, where 23 million people receive their drinking water from the Sacramento-San Joaquin River watershed, DOC levels are a major concern for their drinking water supply [176]. As such, Sacramento's drinking water treatment has been the source of much study [175–177,183].

Studies that are aimed at characterizing the chemical structure of DOC in different input water systems, in order to select the best treatment strategy, are important. In one study, natural water sources of drinking water supplies for Norwegian and Australian systems found differing potentials toward the formation of disinfection by-products upon chlorination related to the chemical characterization of their DOC profiles [173]. In particular, the phenolic/aromatic portion of some DOC profiles are believed to be reactive sites for by-product formation [173]. In fact, a phenolic component of natural DOC, meta-dihydroxybenzene (resorcinol), is thought to be one of the major precursors to the formation of THMs [4]. As well, β-diketones found in DOC are also believed to be THM precursors [184]. Also, methoxyphenol is believed to be one of the most chlorine reactive components of DOC [185]. The phenolic content of DOC may increase under predicted climate change conditions, making DOC levels an even greater concern for water resource managers and treatment facilities.

Using multidimensional fluorescence with parallel factor analysis (PARAFAC) and ^{13}C-NMR studies, quinones (diketones being oxidized-quinones), were shown to represent significant portions of the DOC fluorescence profile [33]. Likewise, phenolic portions of humic acids from the International Humic Substance Society (IHSS) were identified with fluorescence/PARAFAC and reagent-based phenol assays [53] and lignin phenols found in several large artic rivers have been attributed to a PARAFAC-derived component [160]. In kinetic studies of THM formation upon chlorination of natural waters, it was shown that resorcinol components make up 15%–30% of THM precursors in natural water and are likely fast reacting (formed within first hours) THM precursors [4], while other phenolic compounds are likely slow reacting (from hours to weeks) THM precursors [4]. A mechanism has been suggested that involves chlorination by electrophilic aromatic substitutions of phenol that causes ring cleavage, followed by an addition/elimination pathway [186].

Aromatic molecules absorb strongly at 254 nm, thus UV$_{254}$ (observed absorbance at 254 nm) has been suggested as a standard for measuring DOC [20,187]. Specific UV$_{254}$ (SUVA$_{254}$), is the absorbance at 254 nm, measured in inverse meters, divided by an average DOC concentration in milligrams per liter. SUVA$_{254}$ has been proposed as a predicting tool for the evaluation of the suitability of waters containing DOC for the different water treatment methods [188,189]. The variability in "color" of DOC-laden water from different sources is often linked to its aromatic/phenolic content [26]. In the disinfection study mentioned above, phenolic portions of DOC were correlated to THM formation—and the relationship between UV$_{254}$ readings and measured THM concentrations were examined [4]. The relationship between phenolic components of DOC in natural waters and climate change agents has been discussed here in detail, and the inherent importance in monitoring drinking water sources for such components appears to be an area of importance for future studies.

The different sources of natural waters within a system will contain different concentrations and compositions of DOC and inherent phenolic moieties. In effect, two different sampled locations could have the same net quantity of DOC, but because of differences in the composition of the native DOC, they will have different potentials for forming disinfection by-products. Hydrologic factors, including water flow, mixing, and transport, are crucial to understanding the fate of DOC (and its phenolic content) from its source location to its end water supply location. Land-use, anthropogenic sources, and seasonal variations can all impact the composition of DOC and phenolic compounds in water. Additionally, whether a result of climate change or a natural climate function, changes in the water table levels, including severe occurrences of droughts and floods, have an impact on DOC and inherent phenolic compound levels.

In the presence of higher levels of DOC, water treatment facilities may be forced to pre-treat their water with filtration, carbon adsorption, coagulation, or chemical techniques prior to disinfection by chlorination. These costly pretreatment steps often remove much, but not all, of the THM precursors and one study suggests some pretreatment steps may even double the proportion of brominated THMs, a disinfection by-product that is considered to be more carcinogenic than chlorine-derived THMs [179]. Furthermore, less time consuming and more cost effective methods will be required to assist water treatment plants and regulating agencies in monitoring drinking water in regard to potential for forming disinfection by-products. It has been reported that coagulation can reduce THM formation by 40%–50%, depending on the water source [190,191] and different portions of the DOC profile are removed more efficiently than others [50,192]. Pre-oxidation with ozone or chlorine dioxide leads to lower yields of THM upon chlorination [4]. Some scientists suggest that controlling DOC at the geographic source may even be the most economical way to control disinfection by-product formation [176,193]. Sometimes, drinking water supplies, primarily from surface waters, are mixed with groundwater reserves in order to dilute DOC concentrations; however, this practice can be counterproductive if the groundwater also contains DOC (or even a high fraction of phenolic compounds). In groundwater mixing of water supplies, bromine can also be introduced to the system, thereby introducing the potential for formation of brominated disinfection by-products [191]. Readers are also encouraged to see the review by Ritson *et al.*, in regard to changing DOM with climate change and the potential effects on water treatment processes [105].

6. Conclusions

The intent here was to review the issues and research related to observed increases in allochthonous DOC concentrations in several natural water systems, with particular attention to the impact on water resources under a changing climate, in the hope of providing renewed/sustained interest in the study of carbon cycling across the terrestrial landscape and to encourage a multidisciplinary approach toward current research in regard to the drivers of increasing allochthonous DOC levels. Climate change agents such as increases in temperature, variations in precipitation, drought, and related hydrology issues are complex in the way that they impact DOC levels, and may lead to either increases or decreases in aquatic DOC. The experimental evidence that relates elevated CO_2 and temperatures associated with climate change to allochthonous DOC increases in water provides the basis of a plausible mechanism for the increased production, and decreased net degradation, of DOC that is subsequently available for transport to area water catchments. Though it can be inferred that climate change has contributed to the observed increases in the DOC levels of natural water systems over the past few decades, the contribution might be less substantial thus far, as estimates show that elevated temperature and atmospheric CO_2 levels could account for only 10%–20% and 1%–5% of the observed DOC concentration increases, respectively [9]. The experimental controls in the climate change-DOC studies [11,12,69,89,129] often exceed the levels of warming and atmospheric CO_2 coinciding with the current increases in DOC, which makes it difficult to translate the results on a broader scale. In fact, the experimental conditions of most of these experiments exceed the IPCC predicted model for the next 50–100 years. Further, it is possible that climate change-related mechanisms might become increasingly responsible for continued variations in the net DOC concentrations in freshwaters—even before the climatic levels reach those of the experimental conditions.

The atmospheric chemical deposition paradigm predicts that DOC levels might stabilize in the future as sulfate deposition trends reach a lowered steady-state. In short, the current trend of increasing DOC potentially represents a return to DOC levels similar to those of the pre-industrial environment [8,118]. If atmospheric chemical deposition is a key driver responsible for the observed increased DOC levels over recent decades, the prevalence of its role will likely wane with decreasing deposition. Under these conditions, climate change agents may become a dominant mechanism in regulating DOC production and flux. Regardless, the role of DOC levels in selecting appropriate drinking water treatment techniques is an area of paramount importance.

Though not as heavily examined in the primary literature, it is likely that a combination of the major conceptual mechanisms could be occurring. Such a combination has been cited as possible causes for the observed trends in DOC [79,98,130,182,194]. Some studies indirectly state that more than 85% of DOC increases in different geographical areas can be explained by decreases in anthropogenic sulfate deposition, leaving room for other factors (like those of climate change) to contribute [8]. At the core of differentiating the two key conceptual models (or accepting a combination of the two) is whether primary production of DOC and microbial activity (and perhaps the related role of phenolic compounds) or solubility dynamics are driving the observed DOC changes in natural waters. As stated in the introduction, we have simplified our interpretation of current research to a reduced number of drivers of changes in DOC concentration by placing emphasis on two main paradigms presented commonly in the literature. While these drivers may correlate well at specific study sites, the complexity of the processes that govern the release of allochthonous DOC to inland natural waters is immense and may not lend itself to the establishment of distinct cause and effect relationships between a specific driver and specific instances of changing DOC over the spatial and temporal scales included in this review. Future studies should take a critical approach when assigning causal relationships to correlations, especially to explain DOC changes of a lesser magnitude.

It is proposed that future work on this topic should focus, in part, on the analysis of the phenolic composition of DOC. Not only could such a focus potentially assist in deciphering the two main mechanisms for DOC increases, but because of increasing projected climate change conditions (as well as trends of reducing atmospheric chemical deposition), the climate change model (and its inherent projected increases on the phenolic content of DOC) could become the primary driver of DOC levels in the future, especially in areas not subjected to historic acid deposition, areas undergoing land-use change, areas subject to a changing hydrologic regime, and areas situated in mid-high latitudes [118,161]. Studies of the phenolic composition would also be prudent due to concerns related to the role of phenolic compounds in the current and future treatment of drinking water.

Understanding carbon transfer across the terrestrial landscape and into aquatic ecosystems is important, especially considering the unprecedented anthropogenic environmental changes we are experiencing in the biosphere, lithosphere, atmosphere, and hydrosphere and the threat to drinking water sources [152]. To understand the effect of a changing climate on DOC quantity and quality in natural waters and to be better able to predict future DOC changes, it is apparent that work from multiple scientific disciplines is required to unravel the complexity of this issue. The multidisciplinary approach requires a commitment of collaboration from different communities of scientific research that often operate in isolation from each other [152]. Training new scientists to think in a more multidisciplinary way toward the subject of carbon cycling across the terrestrial environment will require exposure to a broad field of subjects in an educational setting. Ultimately, multidisciplinary collaboration will require effective communication of experimental findings and proposed paradigms, as well as an increased exchange between soil and aquatic disciplines [152].

Acknowledgments

Funding to support this work was provided by the National Technical Institute for the Deaf (NTID), through their *Innovation Fund*. The authors would like to thank Susan B. Smith for her helpful comments and technical assistance in preparation of this manuscript. The authors would also like to thank the anonymous reviewers who offered excellent insight and feedback to the manuscript.

Conflicts of Interest

The authors declare no conflict of interest.

References

1. Sinsabaugh, R.L.; Findlay, S. Dissolved organic matter: Out of the black box into the mainstream. In *Aquatic Ecosystems: Interactivity of Dissolved Organic Matter*; Elsevier Science: San Diego, CA, USA, 2003; pp. 479–496.
2. Filella, M.; Rodriguez-Murillo, J.C. Long-Term trends of organic carbon concentrations in freshwaters: Strengths and weaknesses of existing evidence. *Water* **2014**, *6*, 1360–1418.
3. Worrall, F.; Burt, T.P. Has the composition of fluvial DOC changed? Spatiotemporal patterns in the DOC-color relationship. *Glob. Biogeochem. Cycl.* **2010**, *24*, doi:10.1029/2008GB003445.
4. Gallard, H.; von Gunten, U. Chlorination of natural organic matter: Kinetics of chlorination and of THM formation. *Water Res.* **2002**, *36*, 65–74.
5. Sharp, E.L.; Parsons, S.A.; Jefferson, B. Seasonal variations in natural organic matter and its impact on coagulation in water treatment. *Sci. Total Environ.* **2006**, *363*, 183–194.
6. Reynolds, B.; Chamberlain, P.M.; Poskitt, J.; Woods, C.; Scott, W.A.; Rowe, E.C.; Robinson, D.A.; Frogbrook, Z.L.; Keith, A.M.; Henrys, P.A.; *et al.* Countryside survey: National "Soil change" 1978–2007 for topsoils in great britain-acidity, carbon, and total nitrogen status. *Vadose Zone J.* **2013**, *12*, doi:10.2136/vzj2012.0114.
7. Evans, C.D.; Monteith, D.T.; Fowler, D.; Cape, J.N.; Brayshaw, S. Hydrochloric acid: An overlooked driver of environmental change. *Environ. Sci. Technol.* **2011**, *45*, 1887–1894.
8. Monteith, D.T.; Stoddard, J.L.; Evans, C.D.; de Wit, H.A.; Forsius, M.; Hogasen, T.; Wilander, A.; Skjelkvale, B.L.; Jeffries, D.S.; Vuorenmaa, J.; *et al.* Dissolved organic carbon trends resulting from changes in atmospheric deposition chemistry. *Nature* **2007**, *450*, 537–540.
9. Evans, C.D.; Chapman, P.J.; Clark, J.M.; Monteith, D.T.; Cresser, M.S. Alternative explanations for rising dissolved organic carbon export from organic soils. *Glob. Chang. Biol.* **2006**, *12*, 2044–2053.

10. Findlay, S.E.G. Increased carbon transport in the hudson river: Unexpected consequence of nitrogen deposition? *Front. Ecol. Environ.* **2005**, *3*, 133–137.

11. Freeman, C.; Evans, C.D.; Monteith, D.T.; Reynolds, B.; Fenner, N. Export of organic carbon from peat soils. *Nature* **2001**, *412*, 785–785.

12. Freeman, C.; Fenner, N.; Ostle, N.J.; Kang, H.; Dowrick, D.J.; Reynolds, B.; Lock, M.A.; Sleep, D.; Hughes, S.; Hudson, J. Export of dissolved organic carbon from peatlands under elevated carbon dioxide levels. *Nature* **2004**, *430*, 195–198.

13. Wilson, H.F.; Saiers, J.E.; Raymond, P.A.; Sobczak, W.V. Hydrologic drivers and seasonality of dissolved organic carbon concentration, nitrogen content, bioavailability, and export in a forested new england stream. *Ecosystems* **2013**, *16*, 604–616.

14. Hongve, D.; Riise, G.; Kristiansen, J.F. Increased colour and organic acid concentrations in norwegian forest lakes and drinking water—A result of increased precipitation? *Aquat. Sci.* **2004**, *66*, 231–238.

15. Anderson, N.; Dietz, R.; Engstrom, D. Land-Use change, not climate, controls organic carbon burial in lakes. *Proc. R. Soc. B Biol. Sci.* **2013**, *280*, doi:10.1098/rspb.2013.1278.

16. Cool, G.; Lebel, A.; Sadiq, R.; Rodriguez, M.J. Impact of catchment geophysical characteristics and climate on the regional variability of dissolved organic carbon (DOC) in surface water. *Sci. Total Environ.* **2014**, *490*, 947–956.

17. McGuire, A.D.; Anderson, L.G.; Christensen, T.R.; Dallimore, S.; Guo, L.D.; Hayes, D.J.; Heimann, M.; Lorenson, T.D.; Macdonald, R.W.; Roulet, N. Sensitivity of the carbon cycle in the arctic to climate change. *Ecol. Monogr.* **2009**, *79*, 523–555.

18. Detwiler, R.P.; Hall, C.A.S. Tropical forests and the global carbon-cycle. *Science* **1988**, *239*, 42–47.

19. Brown, S.; Lugo, A.E. The storage and production of organic-matter in tropical forests and their role in the global carbon-cycle. *Biotropica* **1982**, *14*, 161–187.

20. Spencer, R.G.M.; Hernes, P.J.; Ruf, R.; Baker, A.; Dyda, R.Y.; Stubbins, A.; Six, J. Temporal controls on dissolved organic matter and lignin biogeochemistry in a pristine tropical river, democratic republic of Congo. *J. Geophys. Res. Biogeosci.* **2010**, *115*, doi:10.1029/2009JG001180.

21. Cramer, W.; Bondeau, A.; Schaphoff, S.; Lucht, W.; Smith, B.; Sitch, S. Tropical forests and the global carbon cycle: Impacts of atmospheric carbon dioxide, climate change and rate of deforestation. *Philos. Trans. R. Soc. Lond. Ser. B Biol. Sci.* **2004**, *359*, 331–343.

22. Monteith, D.; Evans, C.; Henrys, P.; Simpson, G.; Malcolm, I. Trends in the hydrochemistry of acid-sensitive surface waters in the UK 1988–2008. *Ecol. Indic.* **2014**, *37*, 287–303.

23. Evans, C.D.; Monteith, D.T.; Reynolds, B.; Clark, J.M. Buffering of recovery from acidification by organic acids. *Sci. Total Environ.* **2008**, *404*, 316–325.

24. Battin, T.J.; Luyssaert, S.; Kaplan, L.A.; Aufdenkampe, A.K.; Richter, A.; Tranvik, L.J. The boundless carbon cycle. *Nat. Geosci.* **2009**, *2*, 598–600.

25. Filella, M. Freshwaters: Which NOM matters? *Environ. Chem. Lett.* **2009**, *7*, 21–35.

26. Thurman, E.M. *Organic Geochemistry of Natural Waters*; Kluwer Academic: Dordrecht, the Netherlands, 1985.

27. McDonald, S.; Bishop, A.G.; Prenzler, P.D.; Robards, K. Analytical chemistry of freshwater humic substances. *Anal. Chim. Acta* **2004**, *527*, 105–124.

28. Leenheer, J.A.; Croue, J.P. Characterizing aquatic dissolved organic matter. *Environ. Sci. Technol.* **2003**, *37*, 18A–26A.

29. Malcolm, R.L. The uniqueness of humic substances in each of soil, stream and marine environments. *Anal. Chim. Acta* **1990**, *232*, 19–30.

30. Drever, J.I. *The Geochemistry of Natural Waters: Surface and Groundwater Environments*; Prentice Hall: Upper Saddle River, NJ, USA, 1997.

31. Evans, C.D.; Monteith, D.T.; Cooper, D.M. Long-Term increases in surface water dissolved organic carbon: Observations, possible causes and environmental impacts. *Environ. Pollut.* **2005**, *137*, 55–71.

32. Stevenson, I.L.; Schnitzer, M. Transmission electron-microscopy of extracted fulvic and humic acids. *Soil Sci.* **1982**, *133*, 179–185.

33. Cory, R.M.; McKnight, D.M. Fluorescence spectroscopy reveals ubiquitous presence of oxidized and reduced quinones in dissolved organic matter. *Environ. Sci. Technol.* **2005**, *39*, 8142–8149.

34. Parlanti, E.; Morin, B.; Vacher, L. Combined 3D-spectrofluorometry, high performance liquid chromatography and capillary electrophoresis for the characterization of dissolved organic matter in natural waters. *Org. Geochem.* **2002**, *33*, 221–236.

35. Abdulla, H.A.N.; Sleighter, R.L.; Hatcher, P.G. Two dimensional correlation analysis of fourier transform ion cyclotron resonance mass spectra of dissolved organic matter: A new graphical analysis of trends. *Anal. Chem.* **2013**, *85*, 3895–3902.

36. Koch, B.P.; Witt, M.R.; Engbrodt, R.; Dittmar, T.; Kattner, G. Molecular formulae of marine and terrigenous dissolved organic matter detected by electrospray ionization fourier transform ion cyclotron resonance mass spectrometry. *Geochim. Cosmochim. Acta* **2005**, *69*, 3299–3308.

37. Gonsior, M.; Peake, B.M.; Cooper, W.T.; Podgorski, D.; D'Andrilli, J.; Cooper, W.J. Photochemically induced changes in dissolved organic matter identified by ultrahigh resolution fourier transform ion cyclotron resonance mass spectrometry. *Environ. Sci. Technol.* **2009**, *43*, 698–703.

38. Brown, T.L.; Rice, J.A. Effect of experimental parameters on the ESI FT-ICR mass spectrum of fulvic acid. *Anal. Chem.* **2000**, *72*, 384–390.

39. Kujawinski, E.B.; Hatcher, P.G.; Freitas, M.A. High-Resolution fourier transform ion cyclotron resonance mass spectrometry of humic and fulvic acids: Improvements and comparisons. *Anal. Chem.* **2002**, *74*, 413–419.

40. Benner, R.; Kaiser, K. Biological and photochemical transformations of amino acids and lignin phenols in riverine dissolved organic matter. *Biogeochemistry* **2011**, *102*, 209–222.

41. Mutabaruka, R.; Hairiah, K.; Cadisch, G. Microbial degradation of hydrolysable and condensed tannin polyphenol-protein complexes in soils from different land-use histories. *Soil Biol. Biochem.* **2007**, *39*, 1479–1492.

42. Fenner, N.; Freeman, C.; Reynolds, B. Hydrological effects on the diversity of phenolic degrading bacteria in a peatland: Implications for carbon cycling. *Soil Biol. Biochem.* **2005**, *37*, 1277–1287.

43. Orth, A.B.; Denny, M.; Tien, M. Overproduction of lignin-degrading enzymes by and isolate of phanerochaete-chrysosporium. *Appl. Environ. Microbiol.* **1991**, *57*, 2591–2596.

44. McKnight, D.M.; Andrews, E.D.; Spaulding, S.A.; Aiken, G.R. Aquatic fulvic-acids in algal-rich antarctic ponds. *Limnol. Oceanogr.* **1994**, *39*, 1972–1979.

45. McKnight, D.M.; Boyer, E.W.; Westerhoff, P.K.; Doran, P.T.; Kulbe, T.; Andersen, D.T. Spectrofluorometric characterization of dissolved organic matter for indication of precursor organic material and aromaticity. *Limnol. Oceanogr.* **2001**, *46*, 38–48.

46. Jaffe, R.; McKnight, D.; Maie, N.; Cory, R.; McDowell, W.H.; Campbell, J.L. Spatial and temporal variations in DOM composition in ecosystems: The importance of long-term monitoring of optical properties. *J. Geophys. Res. Biogeosci.* **2008**, *113*, doi:10.1029/2008JG000683.

47. SanClements, M.D.; Oelsner, G.P.; McKnight, D.M.; Stoddard, J.L.; Nelson, S.J. New insights into the source of decadal increases of dissolved organic matter in acid-sensitive lakes of the northeastern United States. *Environ. Sci. Technol.* **2012**, *46*, 3212–3219.

48. Coble, P.G.; Green, S.A.; Blough, N.V.; Gagosian, R.B. Characterization of dissolved organic-matter in the black-sea by fluorescence spectroscopy. *Nature* **1990**, *348*, 432–435.

49. Coble, P.G. Characterization of marine and terrestrial DOM in seawater using excitation emission matrix spectroscopy. *Mar. Chem.* **1996**, *51*, 325–346.

50. Gone, D.L.; Seidel, J.; Batiot, C.; Bamory, K.; Ligban, R.; Biemi, J. Using fluoresence spectroscopy eem to evaluate the efficiency of organic matter removal during coagulation-flocculation of a tropical surface water (agbo reservoir). *J. Hazard. Mater.* **2009**, *172*, 693–699.

51. Santin, C.; Yamashita, Y.; Otero, X.L.; Alvarez, M.A.; Jaffe, R. Characterizing humic substances from estuarine soils and sediments by excitation-emission matrix spectroscopy and parallel factor analysis. *Biogeochemistry* **2009**, *96*, 131–147.

52. Walker, S.A.; Amon, R.M.W.; Stedmon, C.; Duan, S.; Louchouarn, P. The use of PARAFAC modeling to trace terrestrial dissolved organic matter and fingerprint water masses in coastal Canadian arctic surface waters. *J. Geophys. Res. Biogeosci.* **2009**, *114*, doi:10.1029/2009JG000990.

53. Pagano, T.; Ross, A.D.; Chiarelli, J.; Kenny, J.E. Multidimensional fluorescence studies of the phenolic content of dissolved organic carbon in humic substances. *J. Environ. Monit.* **2012**, *14*, 937–943.

54. Sobek, S.; Tranvik, L.J.; Prairie, Y.T.; Kortelainen, P.; Cole, J.J. Patterns and regulation of dissolved organic carbon: An analysis of 7,500 widely distributed lakes. *Limnol. Oceanogr.* **2007**, *52*, 1208–1219.

55. Williams, C.J.; Yamashita, Y.; Wilson, H.F.; Jaffe, R.; Xenopoulos, M.A. Unraveling the role of land use and microbial activity in shaping dissolved organic matter characteristics in stream ecosystems. *Limnol. Oceanogr.* **2010**, *55*, 1159–1171.

56. Algesten, G.; Sobek, S.; Bergstrom, A.K.; Jonsson, A.; Tranvik, L.J.; Jansson, M. Contribution of sediment respiration to summer CO_2 emission from low productive boreal and subarctic lakes. *Microb. Ecol.* **2005**, *50*, 529–535.

57. Mattsson, T.; Kortelain, P.; Laubel, A.; Evans, D.; Pujo-Pay, M.; Raike, A.; Conan, P. Export of dissolved organic matter in relation to land use along a european climatic gradient. *Sci. Total Environ.* **2009**, *407*, 1967–1976.

58. Stanley, E.H.; Powers, S.M.; Lottig, N.R.; Buffam, I.; Crawford, J.T. Contemporary changes in dissolved organic carbon (DOC) in human-dominated rivers: Is there a role for DOC management? *Freshw. Biol.* **2012**, *57*, 26–42.

59. Jonsson, A.; Meili, M.; Bergstrom, A.K.; Jansson, M. Whole-Lake mineralization of allochthonous and autochthonous organic carbon in a large humic lake (ortrasket, N. sweden). *Limnol. Oceanogr.* **2001**, *46*, 1691–1700.

60. Royer, T.V.; David, M.B. Export of dissolved organic carbon from agricultural streams in Illinois, USA. *Aquat. Sci.* **2005**, *67*, 465–471.

61. Bida, M.R. Quantity and Composition of Stream Dissolved Organic Matter in the Watershed of Conesus Lake, New York. Master's Thesis, Rochester Institute of Technology, Rochester, NY, USA, 15 July 2013.

62. Tipping, E.; Woof, C. Humic substances in acid organic soils—Modeling their release to the soil solution in terms of humic charge. *J. Soil Sci.* **1990**, *41*, 573–586.

63. Pind, A.; Freeman, C.; Lock, M.A. Enzymatic degradation of phenolic materials in peatlands—Measurement of phenol oxidase activity. *Plant Soil* **1994**, *159*, 227–231.

64. Kalbitz, K.; Solinger, S.; Park, J.H.; Michalzik, B.; Matzner, E. Controls on the dynamics of dissolved organic matter in soils: A review. *Soil Sci.* **2000**, *165*, 277–304.

65. Wershaw, R.L. *Evaluation of Conceptual Models of Natural Organic Matter (humus) from a Consideration of the Chemical and Biochemical Processes of Humification*; U.S. Geological Survey: Reston, VA, USA, 2004.

66. Findlay, S.; Sinsabaugh, R.L. *Aquatic Ecosystems: Interactivity of Dissolved Organic Matter*; Academic Press: San Diego, CA, USA, 2003.

67. Aitkenhead-Peterson, J.A.; McDowell, W.H.; Neff, J.C. Sources, production, and regulation of allochthonous dissolved organic matter inputs to surface waters. In *Aquatic Ecosystems: Interactivity of Dissolved Organic Matter*; Findlay, S.E.G., Sinsabaugh, R.L., Eds.; Elsevier Science: San Diego, CA, USA, 2003; pp. 25–70.

68. Gorham, E. Northern peatlands—Role in the carbon-cycle and probable responses to climatic warming. *Ecol. Appl.* **1991**, *1*, 182–195.

69. Fenner, N.; Ostle, N.J.; McNamara, N.; Sparks, T.; Harmens, H.; Reynolds, B.; Freeman, C. Elevated CO_2 effects on peatland plant community carbon dynamics and DOC production. *Ecosystems* **2007**, *10*, 635–647.

70. Holmes, R.M.; McClelland, J.W.; Peterson, B.J.; Tank, S.E.; Bulygina, E.; Eglinton, T.I.; Gordeev, V.V.; Gurtovaya, T.Y.; Raymond, P.A.; Repeta, D.J.; *et al.* Seasonal and annual fluxes of nutrients and organic matter from large rivers to the arctic ocean and surrounding seas. *Estuaries Coasts* **2012**, *35*, 369–382.

71. Kang, H.; Freeman, C.; Jang, I. Global increases in dissolved organic carbon in rivers and their implications. *Korean J. Limnol.* **2010**, *43*, 453–458.

72. Engelhaupt, E.; Bianchi, T.S. Sources and composition of high-molecular-weight dissolved organic carbon in a Southern Louisiana tidal stream (bayou trepagnier). *Limnol. Oceanogr.* **2001**, *46*, 917–926.

73. Skjelkvale, B.L.; Mannio, J.; Wilander, A.; Andersen, T. Recovery from acidification of lakes in Finland, Norway and Sweden 1990–1999. *Hydrol. Earth Syst. Sci.* **2001**, *5*, 327–337.

74. Dillon, P.J.; Skjelkvale, B.L.; Somers, K.M.; Torseth, K. Coherent responses of sulphate concentration in norwegian lakes: Relationships with sulphur deposition and climate indices. *Hydrol. Earth Syst. Sci.* **2003**, *7*, 596–608.

75. Driscoll, C.T.; Driscoll, K.M.; Roy, K.M.; Mitchell, M.J. Chemical response of lakes in the adirondack region of new york to declines in acidic deposition. *Environ. Sci. Technol.* **2003**, *37*, 2036–2042.

76. Hudson, J.J.; Dillon, P.J.; Somers, K.M. Long-Term patterns in dissolved organic carbon in boreal lakes: The role of incident radiation, precipitation, air temperature, southern oscillation and acid deposition. *Hydrol. Earth Syst. Sci.* **2003**, *7*, 390–398.

77. Pastor, J.; Solin, J.; Bridgham, S.D.; Updegraff, K.; Harth, C.; Weishampel, P.; Dewey, B. Global warming and the export of dissolved organic carbon from boreal peatlands. *Oikos* **2003**, *100*, 380–386.

78. Xenopoulos, M.A.; Lodge, D.M.; Frentress, J.; Kreps, T.A.; Bridgham, S.D.; Grossman, E.; Jackson, C.J. Regional comparisons of watershed determinants of dissolved organic carbon in temperate lakes from the upper great lakes region and selected regions globally. *Limnol. Oceanogr.* **2003**, *48*, 2321–2334.

79. Worrall, F.; Burt, T.; Adamson, J. Can climate change explain increases in DOC flux from upland peat catchments? *Sci. Total Environ.* **2004**, *326*, 95–112.

80. Worrall, F.; Harriman, R.; Evans, C.D.; Watts, C.D.; Adamson, J.; Neal, C.; Tipping, E.; Burt, T.; Grieve, I.; Monteith, D.; *et al.* Trends in dissolved organic carbon in UK rivers and lakes. *Biogeochemistry* **2004**, *70*, 369–402.

81. Fahey, T.J.; Siccama, T.G.; Driscoll, C.T.; Likens, G.E.; Campbell, J.; Johnson, C.E.; Battles, J.J.; Aber, J.D.; Cole, J.J.; Fisk, M.C.; *et al.* The biogeochemistry of carbon at hubbard brook. *Biogeochemistry* **2005**, *75*, 109–176.

82. Harrison, J.A.; Caraco, N.; Seitzinger, S.P. Global patterns and sources of dissolved organic matter export to the coastal zone: Results from a spatially explicit, global model. *Glob. Biogeochem. Cycl.* **2005**, *19*, doi:10.1029/2005GB002480.

83. Skjelkvale, B.L.; Borg, H.; Hindar, A.; Wilander, A. Large scale patterns of chemical recovery in lakes in Norway and Sweden: Importance of seasalt episodes and changes in dissolved organic carbon. In Proceedings of the 7th International Conference on Acid Deposition, Prague, Czech Republic, 12–17 June 2005; pp. 1174–1180.

84. Skjelkvale, B.L.; Stoddard, J.L.; Jeffries, D.S.; Torseth, K.; Hogasen, T.; Bowman, J.; Mannio, J.; Monteith, D.T.; Mosello, R.; Rogora, M.; *et al.* Regional scale evidence for improvements in surface water chemistry 1990–2001. *Environ. Pollut.* **2005**, *137*, 165–176.

85. Burns, D.A.; McHale, M.R.; Driscoll, C.T.; Roy, K.M. Response of surface water chemistry to reduced levels of acid precipitation: Comparison of trends in two regions of New York, USA. *Hydrol. Process.* **2006**, *20*, 1611–1627.

86. Gueguen, C.; Guo, L.D.; Wang, D.; Tanaka, N.; Hung, C.C. Chemical characteristics and origin of dissolved organic matter in the Yukon river. *Biogeochemistry* **2006**, *77*, 139–155.

87. Roulet, N.; Moore, T.R. Environmental chemistry—Browning the waters. *Nature* **2006**, *444*, 283–284.

88. Vuorenmaa, J.; Forsius, M.; Mannio, J. Increasing trends of total organic carbon concentrations in small forest lakes in finland from 1987 to 2003. *Sci. Total Environ.* **2006**, *365*, 47–65.

89. Fenner, N.; Freeman, C.; Lock, M.A.; Harmens, H.; Reynolds, B.; Sparks, T. Interactions between elevated CO_2 and warming could amplify DOC exports from peatland catchments. *Environ. Sci. Technol.* **2007**, *41*, 3146–3152.

90. Worrall, F.; Burt, T.P. Flux of dissolved organic carbon from UK rivers. *Glob. Biogeochem. Cycl.* **2007**, *21*, doi:10.1029/2006GB002709.

91. Worrall, F.; Burt, T.P. Trends in DOC concentration in great britain. *J. Hydrol.* **2007**, *346*, 81–92.

92. Worrall, F.; Burt, T.P. The effect of severe drought on the dissolved organic carbon (DOC) concentration and flux from British rivers. *J. Hydrol.* **2008**, *361*, 262–274.

93. Worrall, F.; Guilbert, T.; Besien, T. The flux of carbon from rivers: The case for flux from england and wales. *Biogeochemistry* **2007**, *86*, 63–75.

94. De Wit, H.A.; Mulder, J.; Hindar, A.; Hole, L. Long-Term increase in dissolved organic carbon in streamwaters in Norway is response to reduced acid deposition. *Environ. Sci. Technol.* **2007**, *41*, 7706–7713.

95. Couture, S.; Houle, D.; Gagnon, C. Increases of dissolved organic carbon in temperate and boreal lakes in Quebec, Canada. *Environ. Sci. Pollut. Res.* **2012**, *19*, 361–371.

96. Lawrence, G.B.; Dukett, J.E.; Houck, N.; Snyder, P.; Capone, S. Increases in dissolved organic carbon accelerate loss of toxic al in Adirondack lakes recovering from acidification. *Environ. Sci. Technol.* **2013**, *47*, 7095–7100.

97. Schindler, D.W.; Curtis, P.J.; Bayley, S.E.; Parker, B.R.; Beaty, K.G.; Stainton, M.P. Climate-Induced changes in the dissolved organic carbon budgets of boreal lakes. *Biogeochemistry* **1997**, *36*, 9–28.

98. Hejzlar, J.; Dubrovsky, M.; Buchtele, J.; Ruzicka, M. The apparent and potential effects of climate change on the inferred concentration of dissolved organic matter in a temperate stream (the Malse River, South Bohemia). In Proceedings of the Detecting Environmental Change—Science and Society Conference, London, England, 16–20 July 2001; pp. 143–152.

99. Oulehle, F.; Chuman, T.; Majer, V.; Hruska, J. Chemical recovery of acidified bohemian lakes between 1984 and 2012: The role of acid deposition and bark beetle induced forest disturbance. *Biogeochemistry* **2013**, *116*, 83–101.

100. Frey, K.E.; Smith, L.C. Amplified carbon release from vast west Siberian Peatlands by 2100. *Geophys. Res. Lett.* **2005**, *32*, doi:10.1029/2004GL022025.

101. Prokushkin, A.S.; Kajimoto, T.; Prokushkin, S.G.; McDowell, W.H.; Abaimov, A.P.; Matsuura, Y. Climatic Factors influencing fluxes of dissolved organic carbon from the forest floor in a continuous-permafrost Siberian Watershed. In Proceedings of the 12th Annual Conference of the International-Boreal-Forest-Research-Association, Fairbanks, AK, USA, 3–6 May 2004; pp. 2130–2140.

102. Gordeev, V.V.; Kravchishina, M.D. River flux of dissolved organic carbon (DOC) and particulate organic carbon (POC) to the Arctic Ocean: What are the consequences of the global changes? In *Influence of Climate Change on the Changing Arctic and Sub-Arctic Conditions*; Nihoul, J.C.J., Kostianoy, A.G., Eds.; Springer: Dordrecht, the Netherlands, 2009; pp. 145–160.

103. Porcal, P.; Koprivnjak, J.F.; Molot, L.A.; Dillon, P.J. Humic substances-part 7: The biogeochemistry of dissolved organic carbon and its interactions with climate change. *Environ. Sci. Pollut. Res.* **2009**, *16*, 714–726.

104. Tranvik, L.J.; Downing, J.A.; Cotner, J.B.; Loiselle, S.A.; Striegl, R.G.; Ballatore, T.J.; Dillon, P.; Finlay, K.; Fortino, K.; Knoll, L.B.; *et al.* Lakes and reservoirs as regulators of carbon cycling and climate. *Limnol. Oceanogr.* **2009**, *54*, 2298–2314.

105. Ritson, J.P.; Graham, N.J.D.; Templeton, M.R.; Clark, J.M.; Gough, R.; Freeman, C. The impact of climate change on the treatability of dissolved organic matter (DOM) in upland water supplies: A UK perspective. *Sci. Total Environ.* **2014**, *473*, 714–730.

106. Stoddard, J.L.; Jeffries, D.S.; Lukewille, A.; Clair, T.A.; Dillon, P.J.; Driscoll, C.T.; Forsius, M.; Johannessen, M.; Kahl, J.S.; Kellogg, J.H.; *et al.* Regional trends in aquatic recovery from acidification in North America and Europe. *Nature* **1999**, *401*, 575–578.

107. Ulrich, K.U.; Paul, L.; Meybohm, A. Response of drinking-water reservoir ecosystems to decreased acidic atmospheric deposition in SE Germany: Trends of chemical reversal. *Environ. Pollut.* **2006**, *141*, 42–53.

108. Rowe, E.C.; Tipping, E.; Posch, M.; Oulehle, F.; Cooper, D.M.; Jones, T.G.; Burden, A.; Hall, J.; Evans, C.D. Predicting nitrogen and acidity effects on long-term dynamics of dissolved organic matter. *Environ. Pollut.* **2014**, *184*, 271–282.

109. Fowler, D.; Smith, R.I.; Muller, J.B.A.; Hayman, G.; Vincent, K.J. Changes in the atmospheric deposition of acidifying compounds in the UK between 1986 and 2001. *Environ. Pollut.* **2005**, *137*, 15–25.

110. Holland, E.A.; Braswell, B.H.; Sulzman, J.; Lamarque, J.F. Nitrogen deposition onto the United States and Western Europe: Synthesis of observations and models. *Ecol. Appl.* **2005**, *15*, 38–57.

111. Pregitzer, K.S.; Zak, D.R.; Burton, A.J.; Ashby, J.A.; MacDonald, N.W. Chronic nitrate additions dramatically increase the export of carbon and nitrogen from northern hardwood ecosystems. *Biogeochemistry* **2004**, *68*, 179–197.

112. Yano, Y.; McDowell, W.H.; Aber, J.D. Biodegradable dissolved organic carbon in forest soil solution and effects of chronic nitrogen deposition. *Soil Biol. Biochem.* **2000**, *32*, 1743–1751.

113. Currie, W.S.; Aber, J.D.; McDowell, W.H.; Boone, R.D.; Magill, A.H. Vertical transport of dissolved organic C and N under long-term N amendments in pine and hardwood forests. *Biogeochemistry* **1996**, *35*, 471–505.

114. Bragazza, L.; Freeman, C.; Jones, T.; Rydin, H.; Limpens, J.; Fenner, N.; Ellis, T.; Gerdol, R.; Hajek, M.; Hajek, T.; *et al.* Atmospheric nitrogen deposition promotes carbon loss from peat bogs. *Proc. Natl. Acad. Sci. USA* **2006**, *103*, 19386–19389.

115. Evans, C.D.; Reynolds, B.; Hinton, C.; Hughes, S.; Norris, D.; Grant, S.; Williams, B. Effects of decreasing acid deposition and climate change on acid extremes in an upland stream. *Hydrol. Earth Syst. Sci.* **2008**, *12*, 337–351.

116. Krug, E.C.; Frink, C.R. Acid-Rain on acid soil—A new perspective. *Science* **1983**, *221*, 520–525.

117. Evans, A.; Zelazny, L.W.; Zipper, C.E. Division s-7—Forest and range soils—Solution parameters influencing dissolved organic-carbon levels in 3 forest soils. *Soil Sci. Soc. Am. J.* **1988**, *52*, 1789–1792.

118. Evans, C.D.; Jones, T.G.; Burden, A.; Ostle, N.; Zielinski, P.; Cooper, M.D.A.; Peacock, M.; Clark, J.M.; Oulehle, F.; Cooper, D.; *et al.* Acidity controls on dissolved organic carbon mobility in organic soils. *Glob. Chang. Biol.* **2012**, *18*, 3317–3331.

119. Worrall, F.; Gibson, H.S.; Burt, T.P. Production *vs.* Solubility in controlling runoff of DOC from peat soils—The use of an event analysis. *J. Hydrol.* **2008**, *358*, 84–95.

120. Solomon, S. *Climate Change 2007: The Physical Science Basis: Part of the Working Group I Contribution to the Fourth Assessment Report of the Intergovernmental Panel on Climate Change*; Cambridge University Press: Cambridge, MA, USA, 2007.

121. Stocker, T.F.; Qin, D.; Plattner, G.-K.; Tignor, M.; Allen, S.K.; Boschung, J.; Nauels, A.; Xia, Y.; Bex, V.; Midgley, P.M. Climate change 2013: The physical science basis. In *Intergovernmental Panel on Climate Change, Working Group I Contribution to the IPCC Fifth Assessment Report (AR5)*; Cambridge University Press: New York, NY, USA, 2013.

122. Houghton, J.T. *Climate Change 2001: The Scientific Basis: Contribution of Working Group I to the Third Assessment Report of the Intergovernmental Panel on Climate Change*; Cambridge University Press: Cambridge, MA, USA, 2001.

123. Harrison, A.F.; Taylor, K.; Scott, A.; Poskitt, J.; Benham, D.; Grace, J.; Chaplow, J.; Rowland, P. Potential effects of climate change on DOC release from three different soil types on the northern pennines UK: Examination using field manipulation experiments. *Glob. Chang. Biol.* **2008**, *14*, 687–702.

124. Pace, M.L.; Cole, J.J. Synchronous variation of dissolved organic carbon and color in lakes. *Limnol. Oceanogr.* **2002**, *47*, 333–342.

125. Tipping, E.; Smith, E.J.; Bryant, C.L.; Adamson, J.K. The organic carbon dynamics of a moorland catchment in NW England. *Biogeochemistry* **2007**, *84*, 171–189.

126. Kominoski, J.S.; Moore, P.A.; Wetzel, R.G.; Tuchman, N.C. Elevated CO_2 alters leaf-litter-derived dissolved organic carbon: Effects on stream periphyton and crayfish feeding preference. *J. North Am. Benthol. Soc.* **2007**, *26*, 663–672.

127. Norby, R.J.; DeLucia, E.H.; Gielen, B.; Calfapietra, C.; Giardina, C.P.; King, J.S.; Ledford, J.; McCarthy, H.R.; Moore, D.J.P.; Ceulemans, R.; *et al.* Forest response to elevated CO_2 is conserved across a broad range of productivity. *Proc. Natl. Acad. Sci. USA* **2005**, *102*, 18052–18056.

128. Freeman, C.; Baxter, R.; Farrar, J.F.; Jones, S.E.; Plum, S.; Ashendon, T.W.; Stirling, C. Could competition between plants and microbes regulate plant nutrition and atmospheric CO_2 concentrations? *Sci. Total Environ.* **1998**, *220*, 181–184.

129. Kane, E.S.; Mazzoleni, L.R.; Kratz, C.J.; Hribljan, J.A.; Johnson, C.P.; Pypker, T.G.; Chimner, R. Peat porewater dissolved organic carbon concentration and lability increase with warming: A field temperature manipulation experiment in a poor-fen. *Biogeochemistry* **2014**, *119*, 161–178.

130. Clark, J.M.; Chapman, P.J.; Adamson, J.K.; Lane, S.N. Influence of drought-induced acidification on the mobility of dissolved organic carbon in peat soils. *Glob. Chang. Biol.* **2005**, *11*, 791–809.

131. Worrall, F.; Burt, T.P.; Adamson, J.K. Linking pulses of atmospheric deposition to DOC release in an upland peat-covered catchment. *Glob. Biogeochem. Cycl.* **2008**, *22*, 15.

132. Wittwer, S.; Robb, W. Carbon dioxide enrichment of greenhouse atmospheres for food crop production. *Econ. Bot.* **1964**, *18*, 34–56.

133. Norby, R.J.; Zak, D.R. Ecological lessons from free-air CO_2 enrichment (FACE) experiments. *Annu. Rev. Ecol. Evol. Syst.* **2011**, *42*, 181–203.

134. Clair, T.A.; Arp, P.; Moore, T.R.; Dalva, M.; Meng, F.R. Gaseous carbon dioxide and methane, as well as dissolved organic carbon losses from a small temperate wetland under a changing climate. In Proceedings of the Advances in Terrestrial Ecosystem: Carbon Inventory Measurements and Monitoring Conference, Raleigh, NC, USA, 3–5 October 2000; pp. S143–S148.

135. Mitsch, W.J.; Gosselink, J.G. *Wetlands*; Van Nostrand Reinhold: New York, NY, USA, 1993.

136. Woodin, S.; Graham, B.; Killick, A.; Skiba, U.; Cresser, M. Nutrient limitation of the long-term response of heather [Calluna-vulgaris (L) Hull] to CO_2 enrichment. *New Phytol.* **1992**, *122*, 635–642.

137. Dawes, M.A.; Hagedorn, F.; Handa, I.T.; Streit, K.; Ekblad, A.; Rixen, C.; Korner, C.; Hattenschwiler, S. An alpine treeline in a carbon dioxide-rich world: Synthesis of a nine-year free-air carbon dioxide enrichment study. *Oecologia* **2013**, *171*, 623–637.

138. Jiang, H.B.; Wang, Y.; Kanost, M.R. Pro-Phenol oxidase activating proteinase from an insect, manduca sexta: A bacteria-inducible protein similar to drosophila Easter. *Proc. Natl. Acad. Sci. USA* **1998**, *95*, 12220–12225.

139. Wetzel, R.G. Gradient-Dominated ecosystems—Sources and regulatory functions of dissolved organic-matter in fresh-water ecosystems. In Proceedings of the Symposium on Dissolved Organic Matter in Lacustrine Ecosystems: Energy Source and System Regulator, Helsinki, Finland, 6–10 August 1990; pp. 181–198.

140. Freeman, C.; Lock, M.A.; Marxsen, J.; Jones, S.E. Inhibitory effects of high-molecular-weight dissolved organic-matter upon metabolic processes in biofilms from contrasting rivers and streams. *Freshw. Biol.* **1990**, *24*, 159–166.

141. Sinsabaugh, R.L.; Antibus, R.K.; Linkins, A.E. An enzymatic approach to the analysis of microbial activity during plant litter decomposition. In Proceedings of the Workshop on Modern Techniques in Soil Ecology, Atlanta, GA, USA, 11–15 September 1989; Elsevier Science Bv: Atlanta, GA, USA; pp. 43–54.

142. Freeman, C.; Ostle, N.; Kang, H. An enzymic "latch" on a global carbon store—A shortage of oxygen locks up carbon in peatlands by restraining a single enzyme. *Nature* **2001**, *409*, 149.

143. Fenner, N.; Freeman, C. Drought-Induced carbon loss in peatlands. *Nat. Geosci.* **2011**, *4*, 895–900.

144. Nistor, C.; Rose, A.; Farre, M.; Stoica, L.; Wollenberger, U.; Ruzgas, T.; Pfeiffer, D.; Barcelo, D.; Gorton, L.; Emneus, J. In-Field monitoring of cleaning efficiency in waste water treatment plants using two phenol-sensitive biosensors. In Proceedings of the 8th Workshop on Biosensors for Environmental Monitoring, Lisbon, Portugal, 11–13 September 2000; pp. 3–17.

145. Tranvik, L.J.; Jansson, M. Climate change—Terrestrial export of organic carbon. *Nature* **2002**, *415*, 861–862.

146. Evans, C.D.; Freeman, C.; Monteith, D.T.; Reynolds, B.; Fenner, N. Climate change—Terrestrial export of organic carbon—Reply. *Nature* **2002**, *415*, 862–862.

147. Freeman, C.; Gresswell, R.; Guasch, H.; Hudson, J.; Lock, M.A.; Reynolds, B.; Sabater, F.; Sabater, S. The role of drought in the impact of climatic-change on the microbiota of peatland streams. *Freshw. Biol.* **1994**, *32*, 223–230.

148. Tang, R.; Clark, J.M.; Bond, T.; Graham, N.; Hughes, D.; Freeman, C. Assessment of potential climate change impacts on peatland dissolved organic carbon release and drinking water treatment from laboratory experiments. *Environ. Pollut.* **2013**, *173*, 270–277.

149. Whitworth, K.L.; Baldwin, D.S.; Kerr, J.L. Drought, floods and water quality: Drivers of a severe hypoxic blackwater event in a major river system (the Southern Murray-Darling Basin, Australia). *J. Hydrol.* **2012**, *450*, 190–198.

150. Hamilton, S.K.; Sippel, S.J.; Calheiros, D.F.; Melack, J.M. An anoxic event and other biogeochemical effects of the pantanal wetland on the Paraguay River. *Limnol. Oceanogr.* **1997**, *42*, 257–272.

151. Fontenot, Q.C.; Rutherford, D.A.; Kelso, W.E. Effects of environmental hypoxia associated with the annual flood pulse on the distribution of larval sunfish and shad in the Atchafalaya River basin, Louisiana. *Trans. Am. Fish. Soc.* **2001**, *130*, 107–116.

152. Marin-Spiotta, E.; Gruley, K.E.; Crawford, J.; Atkinson, E.E.; Miesel, J.R.; Greene, S.; Cardona-Correa, C.; Spencer, R.G.M. Paradigm shifts in soil organic matter research affect interpretations of aquatic carbon cycling: Transcending disciplinary and ecosystem boundaries. *Biogeochemistry* **2014**, *117*, 279–297.

153. Vidon, P.; Wagner, L.E.; Soyeux, E. Changes in the character of DOC in streams during storms in two midwestern watersheds with contrasting land uses. *Biogeochemistry* **2008**, *88*, 257–270.

154. Inamdar, S.; Finger, N.; Singh, S.; Mitchell, M.; Levia, D.; Bais, H.; Scott, D.; McHale, P. Dissolved organic matter (DOM) concentration and quality in a forested mid-Atlantic watershed, USA. *Biogeochemistry* **2012**, *108*, 55–76.

155. Fellman, J.B.; Hood, E.; Edwards, R.T.; D'Amore, D.V. Changes in the concentration, biodegradability, and fluorescent properties of dissolved organic matter during stormflows in coastal temperate watersheds. *J. Geophys. Res. Biogeosci.* **2009**, *114*, doi:10.1029/2008JG000790.

156. Winterdahl, M.; Erlandsson, M.; Futter, M.N.; Weyhenmeyer, G.A.; Bishop, K. Intra-Annual variability of organic carbon concentrations in running waters: Drivers along a climatic gradient. *Glob. Biogeochem. Cycle* **2014**, *28*, 451–464.

157. Sanderman, J.; Lohse, K.A.; Baldock, J.A.; Amundson, R. Linking soils and streams: Sources and chemistry of dissolved organic matter in a small coastal watershed. *Water Resour. Res.* **2009**, *45*, doi:10.1029/2008WR006977.

158. Jencso, K.G.; McGlynn, B.L.; Gooseff, M.N.; Wondzell, S.M.; Bencala, K.E.; Marshall, L.A. Hydrologic connectivity between landscapes and streams: Transferring reach-and plot-scale understanding to the catchment scale. *Water Resour. Res.* **2009**, *45*, doi:10.1029/2008WR007225.

159. Amon, R.M.W.; Rinehart, A.J.; Duan, S.; Louchouarn, P.; Prokushkin, A.; Guggenberger, G.; Bauch, D.; Stedmon, C.; Raymond, P.A.; Holmes, R.M.; *et al.* Dissolved organic matter sources in large arctic rivers. *Geochim. Cosmochim. Acta* **2012**, *94*, 217–237.

160. Walker, S.A.; Amon, R.M.W.; Stedmon, C.A. Variations in high-latitude riverine fluorescent dissolved organic matter: A comparison of large arctic rivers. *J. Geophys. Res. Biogeosci.* **2013**, *118*, 1689–1702.

161. Laudon, H.; Tetzlaff, D.; Soulsby, C.; Carey, S.; Seibert, J.; Buttle, J.; Shanley, J.; McDonnell, J.J.; McGuire, K. Change in winter climate will affect dissolved organic carbon and water fluxes in mid-to-high latitude catchments. *Hydrol. Process.* **2013**, *27*, 700–709.

162. Foley, J.A.; DeFries, R.; Asner, G.P.; Barford, C.; Bonan, G.; Carpenter, S.R.; Chapin, F.S.; Coe, M.T.; Daily, G.C.; Gibbs, H.K.; *et al.* Global consequences of land use. *Science* **2005**, *309*, 570–574.

163. Findlay, S.; Quinn, J.M.; Hickey, C.W.; Burrell, G.; Downes, M. Effects of land use and riparian flowpath on delivery of dissolved organic carbon to streams. *Limnol. Oceanogr.* **2001**, *46*, 345–355.

164. Houghton, R.A. Land-Use change and the carbon-cycle. *Glob. Chang. Biol.* **1995**, *1*, 275–287.

165. Gough, R.; Holliman, P.J.; Willis, N.; Jones, T.G.; Freeman, C. Influence of habitat on the quantity and composition of leachable carbon in the O_2 horizon: Potential implications for potable water treatment. *Lake Reserv. Manag.* **2012**, *28*, 282–292.

166. Yallop, A.R.; Clutterbuck, B.; Thacker, J. Increases in humic dissolved organic carbon export from upland peat catchments: The role of temperature, declining sulphur deposition and changes in land management. *Clim. Res.* **2010**, *45*, 43–56.

167. Bricker, S.B.; Longstaf, B.; Dennison, W.; Jones, A.; Boicourt, K.; Wicks, C.; Woerner, J. Effects of nutrient enrichment in the nation's estuaries: A decade of change. *Harmful Algae* **2008**, *8*, 21–32.

168. Carpenter, S.R.; Caraco, N.F.; Correll, D.L.; Howarth, R.W.; Sharpley, A.N.; Smith, V.H. Nonpoint pollution of surface waters with phosphorus and nitrogen. *Ecol. Appl.* **1998**, *8*, 559–568.

169. Fellman, J.B.; Hood, E.; Spencer, R.G.M. Fluorescence spectroscopy opens new windows into dissolved organic matter dynamics in freshwater ecosystems: A review. *Limnol. Oceanogr.* **2010**, *55*, 2452–2462.

170. Wilson, H.F.; Xenopoulos, M.A. Effects of agricultural land use on the composition of fluvial dissolved organic matter. *Nat. Geosci.* **2009**, *2*, 37–41.

171. Schlesinger, W. *Biogeochemistry: An Analysis of Global Change*; Academic Press: San Diego, CA, USA, 1991.

172. Mierle, G.; Ingram, R. The role of humic substances in the mobilization of mercury from watersheds. *Water Air Soil Pollut.* **1991**, *56*, 349–357.

173. Fabris, R.; Chowa, C.W.K.; Drikas, M.; Eikebrokk, B. Comparison of NOM character in selected Australian and Norwegian drinking waters. *Water Res.* **2008**, *42*, 4188–4196.

174. Chow, A.T.; Guo, F.M.; Gao, S.D.; Breuer, R.S. Trihalomethane reactivity of water- and sodium hydroxide-extractable organic carbon fractions from peat soils. *J. Environ. Qual.* **2006**, *35*, 114–121.

175. Fleck, J.A.; Bossio, D.A.; Fujii, R. Dissolved organic carbon and disinfection by-product precursor release from managed peat soils. *J. Environ. Qual.* **2004**, *33*, 465–475.

176. Diaz, F.J.; Chow, A.T.; O'Geen, A.T.; Dahlgren, R.A.; Wong, P.K. Restored wetlands as a source of disinfection byproduct precursors. *Environ. Sci. Technol.* **2008**, *42*, 5992–5997.

177. Saleh, D.K. *Organic Carbon Trends, Loads, and Yields to the Sacramento-San Joaquin Delta, California, Water Years 1980–2000*, 2nd ed.; Water Resources Investigation Report 03-4070; U.S. Geological Survey: Sacramento, CA, USA, 2007.

178. Ledesma, J.L.; Köhler, S.J.; Futter, M.N. Long-Term dynamics of dissolved organic carbon: Implications for drinking water supply. *Sci. Total Environ.* **2012**, *432*, 1–11.

179. Gough, R.; Holliman, P.J.; Willis, N.; Freeman, C. Dissolved organic carbon and trihalomethane precursor removal at a UK upland water treatment works. *Sci. Total Environ.* **2014**, *468*, 228–239.

180. United States Environmental Protection Agency (US EPA). National primary drinking water regulations: Stage 2 disinfectants and disinfection byproducts rule; National primary and secondary drinking water regulations: Approval of analytical methods for chemical contaminants. *Federal Register.* **2003**, *68*, 49548–49681.

181. United States Environmental Protection Agency (US EPA). National primary drinking water regulations: Stage 2 disinfectants and disinfection byproducts rule. *Federal Register.* **2006**, *71*, 388–493.

182. Haaland, S.; Hongve, D.; Laudon, H.; Riise, G.; Vogt, R.D. Quantifying the drivers of the increasing colored organic matter in boreal surface waters. *Environ. Sci. Technol.* **2010**, *44*, 2975–2980.

183. Eckard, R.S.; Hernes, P.J.; Bergamaschi, B.A.; Stepanauskas, R.; Kendall, C. Landscape scale controls on the vascular plant component of dissolved organic carbon across a freshwater delta. *Geochim. Cosmochim. Acta* **2007**, *71*, 5968–5984.

184. Norwood, D.L.; Johnson, J.D.; Christman, R.F.; Hass, J.R.; Bobenrieth, M.J. Reactions of chlorine with selected aromatic models of aquatic humic material. *Environ. Sci. Technol.* **1980**, *14*, 187–190.

185. Hanna, J.V.; Johnson, W.D.; Quezada, R.A.; Wilson, M.A.; Lu, X.Q. Characterization of aqueous humic substances before and after chlorination. *Environ. Sci. Technol.* **1991**, *25*, 1160–1164.

186. Arnold, W.A.; Bolotin, J.; von Gunten, U.; Hofstetter, T.B. Evaluation of functional groups responsible for chloroform formation during water chlorination using compound specific isotope analysis. *Environ. Sci. Technol.* **2008**, *42*, 7778–7785.

187. Hendricks, D.W. *Water Treatment Unit Processes: Physical and Chemical. Civil and Environmental Engineering*; CRC/Taylor & Francis: Boca Raton, FL, USA, 2006.

188. Archer, A.D.; Singer, P.C. An evaluation of the relationship between SUVA and NOM coagulation using the ICR database. *J. Am. Water Works Assoc.* **2006**, *98*, 110–123.

189. Archer, A.D.; Singer, P.C. Effect of SUVA and enhanced coagulation on removal of TOX precursors. *J. Am. Water Works Assoc.* **2006**, *98*, 97–107.

190. Gang, D.C.; Clevenger, T.E.; Banerji, S.K. Effects of alum coagulation on speciation and distribution of trihalomethanes (THMs) and haloacetic acids (HAAs). *J. Environ. Sci. Health Part A Toxic Hazard. Subst. Environ. Eng.* **2005**, *40*, 521–534.

191. Johnstone, D.W.; Sanchez, N.P.; Miller, C.M. Parallel factor analysis of excitation-emission matrices to assess drinking water disinfection byproduct formation during a peak formation period. *Environ. Eng. Sci.* **2009**, *26*, 1551–1559.

192. Hussain, S.; van Leeuwen, J.; Chow, C.; Beecham, S.; Kamruzzaman, M.; Wang, D.S.; Drikas, M.; Aryal, R. Removal of organic contaminants from river and reservoir waters by three different aluminum-based metal salts: Coagulation adsorption and kinetics studies. *Chem. Eng. J.* **2013**, *225*, 394–405.

193. Reckhow, D.A.; Rees, P.L.S.; Bryan, D. Watershed sources of disinfectant byproduct precursors. *Water Supply.* **2004**, *4*, 61–69.

194. Clark, J.M.; Chapman, P.J.; Heathwaite, A.L.; Adamson, J.K. Suppression of dissolved organic carbon by sulfate induced acidification during simulated droughts. *Environ. Sci. Technol.* **2006**, *40*, 1776–1783.

Assessing the Impacts of Sea Level Rise on Salinity Intrusion and Transport Time Scales in a Tidal Estuary, Taiwan

Wen-Cheng Liu and Hong-Ming Liu

Abstract: Global climate change has resulted in a gradual sea level rise. Sea level rise can cause saline water to migrate upstream in estuaries and rivers, thereby threatening freshwater habitat and drinking-water supplies. In the present study, a three-dimensional hydrodynamic model was established to simulate salinity distributions and transport time scales in the Wu River estuary of central Taiwan. The model was calibrated and verified using tidal amplitudes and phases, time-series water surface elevation and salinity distributions in 2011. The results show that the model simulation and measured data are in good agreement. The validated model was then applied to calculate the salinity distribution, flushing time and residence time in response to a sea level rise of 38.27 cm. We found that the flushing time for high flow under the present condition was lower compared to the sea level rise scenario and that the flushing time for low flow under the present condition was higher compared to the sea level rise scenario. The residence time for the present condition and the sea level rise scenario was between 10.51 and 34.23 h and between 17.11 and 38.92 h, respectively. The simulated results reveal that the residence time of the Wu River estuary will increase when the sea level rises. The distance of salinity intrusion in the Wu River estuary will increase and move further upstream when the sea level rises, resulting in the limited availability of water of suitable quality for municipal and industrial uses.

Reprinted from *Water*. Cite as: Liu, W.-C.; Liu, H.-M. Assessing the Impacts of Sea Level Rise on Salinity Intrusion and Transport Time Scales in a Tidal Estuary, Taiwan. *Water* **2014**, *6*, 324–344.

1. Introduction

Global warming is irrefutably causing sea level to rise. The global mean sea level raised by ~20 cm, along with a rise in the regional mean sea level, as the global air temperature increased by ~0.5–0.6 °C during the 20th century [1,2]. In Taiwan, the surface temperature has raised approximately 1.0–1.4 °C over the last 100 years [3]. Over the past 80 years, the annual precipitation has increased in northern Taiwan and declined in central and southern Taiwan [4]. The changing climate has also caused some impacts on river ecosystems in Taiwan; more-frequent habitat disturbances have caused both a shift in aquatic organism distributions and population decline [5].

Sea level rise can cause saline water to migrate upstream to points where freshwater previously existed [6]. Several studies indicated that sea level rise would increase the salinity in estuaries [7,8], which would result in changes in stratification and estuarine circulation [9]. Salinity migration could cause shifts in salt-sensitive habitats and could thus affect the distribution of flora and fauna.

Salinity intrusion may decrease the water quality in an estuary, so that its water becomes unsuitable for certain uses, such as agricultural, industrial and drinking purposes. Therefore, the determination of the salinity distribution along an estuary is a major interest for water managers in estuaries and coastal regions. The evaluation of transport time scales is highly related to the water quality and ecological health of different aquatic systems [10].

Several numerical modeling studies have shown that increases in sea level have impacts on estuarine salinity. Hull and Tortoriello [11] used a one-dimensional model to estimate the impacts of sea level rise and found that a sea level rise of 0.13 m would result in a salinity increase of 0.4 psu (practical salinity unit) in the upper portion of the Delaware Bay during low-flow periods. Grabemann et al. [12] simulated a 2-km upstream advance of the brackish water zone for a sea level rise of 0.55 m in the Weser Estuary, Germany. Hilton et al. [7] found an average salinity increase of approximately 0.5 with a 0.2 m sea level rise based on model simulations in Chesapeake Bay. Chua et al. [13] found that the intrusion of salt water into San Francisco Bay and the flushing rate both increase as the sea level rises. Bhuiyan and Dutta [8] applied a one-dimensional model to investigate the impact of sea level rise on river salinity in the Gorai River network and found that a sea level rise of 0.59 m increased salinity by 0.9 at a distance of 80 km upstream of the river mouth. Rice et al. [14] concluded that salinity in the James River would intrude about 10 km farther upstream for a sea level rise of 1.0 m using a three-dimensional hydrodynamic model.

Numerous studies have reported the influences of sea level rise on estuarine salinity, stratification, exchange flow, residence time, material transport processes and other relevant processes in estuaries [8,9,14]. However, the reports regarding the impacts of sea level rise on salinity intrusion and transport time scales have not yet been studied in Taiwan's estuaries. The objective of the present study is to examine the salinity intrusion, flushing time and residence time in response to sea level rise in the Wu River estuary of central Taiwan using a three-dimensional hydrodynamic and salinity transport model. The model was validated with observed amplitudes and phases, water levels and salinity to ascertain the model's accuracy and capability. The model was then applied to the Wu River estuary to calculate the salinity distributions and transport time scales based on sea level rise projections. The model results were used to investigate how sea level rise affects salinity intrusion, flushing time and residence time in Taiwan's Wu River estuary.

2. Study Area

The Wu River system is the most important river in central Taiwan (Figure 1a). The mean tidal range at the mouth of the Wu River is 3.8 m above mean sea level. Tidal propagation is the dominant mechanism controlling the water surface elevation. The M_2 (principal lunar semi-diurnal) tide is the primary tidal constituent at the mouth of the Wu River [15]. The main tributaries are the Fazi River, Dali River, Han River and Maoluo River. The downstream reaches of the main Wu River are affected by tides, whereas the tributaries are not subject to tidal effects and are therefore not affected by salt water intrusion. The drainage basin of the Wu River, which is the fourth-largest river basin in Taiwan, covers approximately 2026 km^2. The total channel length is 117 km, and the mean channel slope is 1/92. The morphology of the Wu River displays different features in each segment, molded by natural forces, as well as anthropogenic activities exerted upon the paleo-riverbed built ages ago. The riverbed is composed of silt and sand in the estuary. The mean annual precipitation in this region is 2087 mm. The ample flow season is from May to September, accounting for 70% of the river discharge, and the dry season is from January to February. The daily flow data from 1969 to 2011 at the Dadu Bridge, collected by the Water Resources Bureau of Taiwan, are analyzed in this study. The data analysis indicates that the Q_{75} low flow is 41.2 m³/s. The definition of Q_{75} is the flow that is equaled or exceeded for 75% of the time. The river, which flows into the Taiwan Strait, is located in a temperate area characterized by intense agricultural and industrial activities. The Wu River catchment is also an important water supply source for central Taiwan. Figure 1b shows the topography of the Wu River estuary and its adjacent coastal sea. This figure indicates that the greatest depth in the study area is 70 m (below mean sea level) near the corner of the coastal sea.

Figure 1. (**a**) Map of the Wu River system and (**b**) bathymetry of the Wu River estuary and adjacent coastal sea.

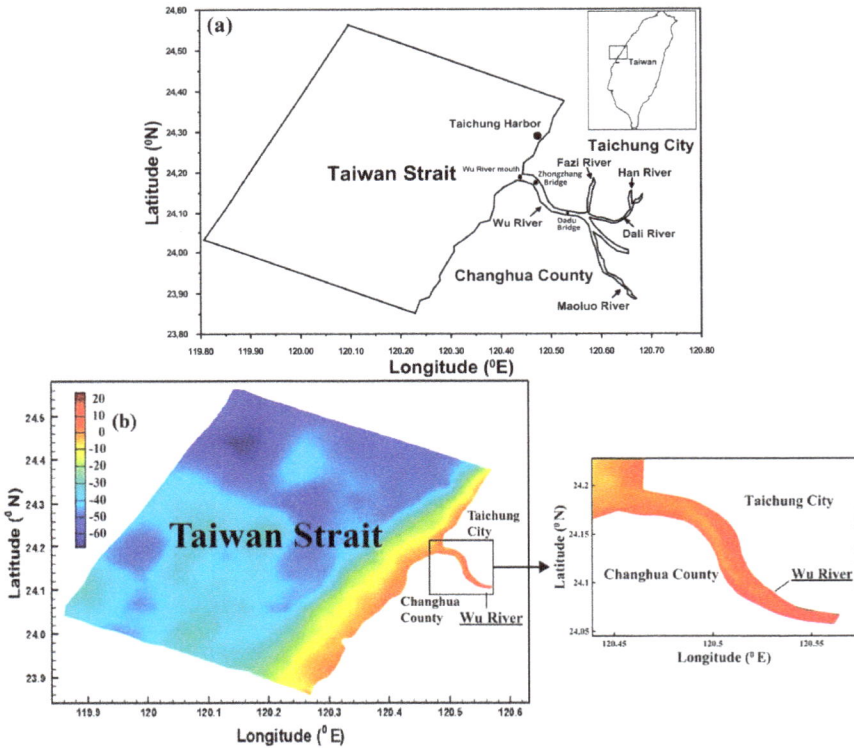

3. Materials and Methods

3.1. Sea Level Rise Projection

The existence of sea level rise is undeniable. Church and White [16] estimated the global mean sea level rise rates from tidal gauges and satellite altimetry as follows: 1.7 ± 0.2 mm year^{-1} for 1900–2009 and 1.9 ± 0.4 mm year^{-1} for 1961–2009, both of which are comparable to the ~1.8 mm year^{-1} rate obtained from GPS-derived crustal velocities and tidal gauges around North

America [17]. Global tide gauge records, satellite data and modeling provide mean historical rates of ~1.7–1.8 mm year^{-1} for the 20th century and ~3.3 mm year^{-1} for the last few decades [18].

The purpose of this study is to identify the response of the Wu River estuary to potential future sea level rise based on the analyzed results of observed sea level. Tseng *et al.* [19] investigated the pattern and trends of sea level rise in the region seas around Taiwan through the analyses of long-term tide-gauge and satellite altimetry data. They found that consistent with the coastal tide-gauge records, satellite altimetry data showed similar increasing rates (+5.3 mm/year) around Taiwan. They did not include wave breaking around the river mouth and coastal seas, resulting in water level rise due to wave set-up. In this study, the wave set-up issue also did not take into account. The linear regression method was used to yield the sea level rise trend according to the monthly average water surface elevation collected from 1971 to 2011 at the Taichung Harbor station, which is shown in Figure 2. The equation of linear regression can be expressed as:

$$Y = 0.43X - 591.8 \tag{1}$$

where X is the time (year) and Y is the sea water level (cm). We found that the rate of sea level rise was 4.3 mm/year at the Taichung Harbor station. Huang *et al.* [20] estimated the sea level rise at the Taichung Harbor station using the data of tidal gauge and satellite altimetry and found that the rate of sea level rise was 3.7 mm/year. Their results are similar to our estimation on the rate of sea level rise. The sea level rise in 2011 was set up zero to project the sea level rise in 2010. The future projected sea level rise of 38.27 cm in 2100 was used in the simulation scenario.

Figure 2. Linear regression for the sea level rise trend at the Taichung Harbor station.

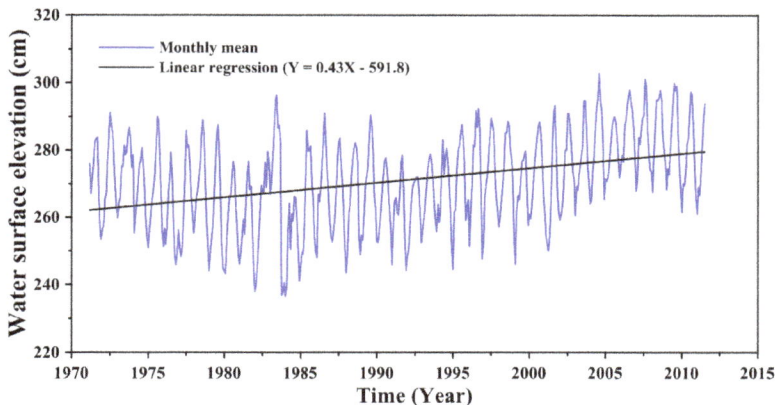

3.2. Three-Dimensional Hydrodynamic Model

A three-dimensional, semi-implicit Euler-Lagrange finite-element model (SELFE) [21] was implemented to simulate the hydrodynamics and salinity transport in the Wu River estuary and its adjacent coastal sea. SELFE solves the Reynolds-stress averaged Navier-Stokes equations, which consist of the conservation laws for mass and momentum and the use of the hydrostatic and Boussinesq approximations, yielding the following free-surface elevation and three-dimensional water velocity equations:

$$\frac{\partial u}{\partial x}+\frac{\partial v}{\partial y}+\frac{\partial w}{\partial z}=0 \tag{2}$$

$$\frac{\partial \eta}{\partial t}+\frac{\partial}{\partial x}\int_{H_R-h}^{H_R+\eta}udz+\frac{\partial}{\partial y}\int_{H_R-h}^{H_R+\eta}vdz=0 \tag{3}$$

$$\frac{\partial u}{\partial t}+u\frac{\partial u}{\partial x}+v\frac{\partial u}{\partial y}+w\frac{\partial u}{\partial z}=fv-\frac{\partial}{\partial x}\left\{g(\eta-\alpha\varphi)+\frac{P_a}{\rho_o}\right\}$$
$$-\frac{g}{\rho_o}\int_z^{H_R+\eta}\frac{\partial\rho}{\partial x}dz+\frac{\partial}{\partial z}(v\frac{\partial u}{\partial z}) \tag{4}$$

$$\frac{\partial v}{\partial t}+u\frac{\partial v}{\partial x}+v\frac{\partial v}{\partial y}+w\frac{\partial v}{\partial z}=-fu-\frac{\partial}{\partial y}\left\{g(\eta-\alpha\varphi)+\frac{P_a}{\rho_o}\right\}$$
$$-\frac{g}{\rho_o}\int_z^{H_R+\eta}\frac{\partial\rho}{\partial y}dz+\frac{\partial}{\partial z}(v\frac{\partial v}{\partial z}) \tag{5}$$

$$\frac{DS}{Dt}=\frac{\partial}{\partial z}(K_v\frac{\partial S}{\partial z})+F_s \tag{6}$$

$$\rho=\rho_0(p,S) \tag{7}$$

where (x,y) are the horizontal Cartesian coordinates; (ϕ,λ) are the latitude and longitude, respectively; z is the vertical coordinate, positive upward; t is time; H_R is the z-coordinate at the reference level (mean sea level); $\eta(x,y,t)$ is the free-surface elevation; $h(x,y)$ is the bathymetric depth; u,v and w are the velocities in the x, y and z directions, respectively; f is the Coriolis force; g is the acceleration of gravity; $\varphi(\phi,\lambda)$ is the tidal potential; α is the effective earth elasticity factor (=0.69); $\rho(x,t)$ is the water density, of which the default reference value; ρ_o, is set to 1,025 kg/m^3; $P_a(x,y,t)$ is the atmospheric pressure at the free surface; p is the pressure; v is the vertical eddy viscosity; S is the salinity; K_v is the vertical eddy diffusivity for salinity and F_s is the horizontal diffusion for the transport equation.

The vertical boundary conditions for the momentum equation, especially the bottom boundary condition, play an important role in the SELFE numerical formulation, as it involves the unknown velocity. In fact, as a crucial step in solving the differential system, SELFE uses the bottom condition to decouple free-surface Equation (3) from momentum Equations (4) and (5). The vertical boundary conditions for the momentum equation are presented as below.

At the water surface, the balance between the internal Reynolds stress and the applied shear stress yields:

$$v\frac{\partial \vec{u}}{\partial z}=\vec{\tau}_w \text{ at } z=\eta \tag{8}$$

where the specific stress, τ_w, can be parameterized using the approach [22].

The boundary condition at the bottom plays an important role in the SELFE formulation, as it involves unknown velocity. Specifically, at the bottom, the no-slip condition ($U=V=0$) is usually replaced by a balance between the internal Reynolds stress and the bottom frictional stress, i.e.,

$$v\frac{\partial \vec{u}}{\partial z}=\tau_b \text{ at } z=-h \tag{9}$$

where the bottom stress is $\vec{\tau}_b=C_D|\vec{u}_b|\vec{u}_b$.

The velocity profile inside the bottom boundary layers obeys the logarithmic law:

$$\overset{r}{u} = \frac{\ln[(z+h)/z_0]}{\ln(\delta_b/z_0)}\overset{\text{\tiny UI}}{u_b}, \quad (z_0 - h \le z \le \delta_b - h) \tag{10}$$

which is subject to be smoothly matched to the exterior flow. In Equation (10), δ_b is the thickness of the bottom computational cell; z_0 is the bottom roughness, which is determined through model calibration and verification; and u_b is the bottom velocity, measured at the top of the bottom computational cell. The Reynolds stress inside the boundary layer is derived from Equation (11) as:

$$\nu \frac{\partial \overset{1}{u}}{\partial z} = \frac{\nu}{(z+h)\ln(\delta_b/z_0)}\overset{\text{\tiny UI}}{u_b} \tag{11}$$

The SELFE model uses the Generic Length Scale (GLS) turbulence closure approach of Umlauf and Burchard [23], which has the advantage of incorporating most of the 2.5-equation closure model. The SELFE model treats advection in the momentum equation using a Euler–Lagrange methodology. A detailed description of the turbulence closure model, the vertical boundary conditions for the momentum equation and the numerical solution methods can be found in Zhang and Baptista [21].

3.3. Computation of Flushing Time

The flushing time can be conveniently determined by the freshwater fraction approach [24–27], which can be determined from salinity distributions. This technique provides an estimation of the time scale over which contaminants and/or other material released in the estuary are removed from the system. Using the freshwater fraction method, the flushing time in an estuary can be expressed as:

$$T_f = \frac{F}{Q} = \frac{\int_{vol} f \cdot d(V)}{Q} \tag{12}$$

where F is the accumulated freshwater volume in the estuary, which can be calculated by integrating the freshwater volume; $d(V)$, in all the sub-divided model grids over a period of time. In estuaries with unsteady river flow and tidal variations; F and Q are the approximate average freshwater volume and average freshwater input, respectively, over several tidal cycles for a period of time, such as a week or a month [20,21]. The term, f, is the freshwater content or the freshwater fraction, which is described by:

$$f = \frac{S_0 - S}{S_0} \tag{13}$$

where S_0 is the salinity in the ocean; and S is the salinity at the study location.

3.4. Computation of Residence Time

The time scales associated with the residence time of water parcels and their associated dissolved and suspended materials in a specific water body due to different transport mechanisms (*i.e.*, advection and dispersion) are fundamental physical characteristics of that water body. Residence time is defined as the time required for a water parcel to leave the region of interest for the first time [28]. Several methodologies for the computation of residence time have been reported in the literature [29–34]. In the present study, the computational method follows the procedures outlined by Takeoka [30]. Consider that a region of interest contains a finite mass of tracer given by $M(0)$ at the initial time $t = t_0$. If we define the remaining mass of tracer at a certain time, t, within the system as $M(t)$, the distribution function of the residence time can be defined as:

$$T_r = -\frac{1}{M(0)} \frac{dM(t)}{dt} \tag{14}$$

where T_r is the distribution function of residence time. The total mass of the tracer will completely leave the system at a given moment when $\lim_{t \to \infty} M(t)$ is equal to zero. The average residence time of the tracer can be computed by:

$$\overline{T_r} = \int_{t_0}^{\infty} t T_r(t) dt = \int_{t_0}^{\infty} \frac{M(t)}{M(0)} dt \tag{15}$$

The fraction of mass $r(t) = M(t) / M(0)$ is known as the remnant function. Note that $M(t)$, the mass of the tracer that remains in the region of interest at a certain time; t, can be computed numerically based on the tracer concentration by:

$$M(t) = \int C(x,t) dV \tag{16}$$

where $C(x,t)$ is the tracer concentration in a differential volume ; dV, at a given time; t, and position, x, within the system. It is expected that a mass of tracers injected close to the boundaries of a given region has a lower residence time than does the residence time of tracers injected at the center of such a region.

3.5. Model Schematization

An accurate representation of the bottom topography in the model grid is critical for successful estuarine, coastal and ocean modeling. In this study, the bottom topography data in the coastal seas and Wu River estuary were obtained from the databank of the National Science Council and the Water Resources Agency in Taiwan, respectively. The modeling domain in the horizontal plane covers an area of 60 km × 45 km at the coastal sea boundary. Because SELFE uses a combination of Eulerian–Lagrangian and implicit time stepping, it does not have to satisfy the usual Courant–Friedrich–Levy (CFL) constraint for numerical stability [21]. However, 120 s was chosen as the time step (Δt). Trial-and-error tests with other time steps demonstrated that the model results did not improve significantly with lower values. The model meshes for the Wu River estuary and the coastal sea consisted of 3541 polygons and 1974 grids, respectively (Figure 3). Because the model domain covers deep bathymetry in the coastal sea and shallow bathymetry near the coastline, ten levels, varying in thickness from 0.2 to 7 m, were adopted for vertical discretization in the SELFE model.

Figure 3. Unstructured grids in the modeling domain.

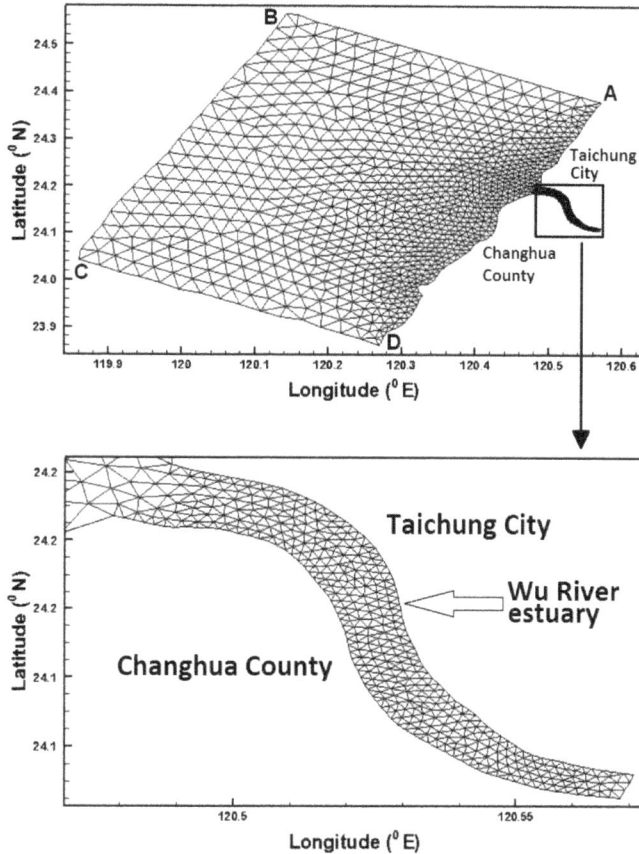

4. Model Calibration and Verification

To ascertain the model accuracy for applications on the assessment of sea level rise on salinity intrusion and transport time scales, a set of observational data collected in 2011 were used to calibrate and verify the model and to validate its capability to predict amplitudes and phases, water surface elevation and salinity distribution.

4.1. Calibration with Amplitudes and Phases

The local bottom roughness height (z_o) is similar to the Manning coefficient, affecting the water level calculations for the coastal sea and estuary. The values of local bottom roughness height were iteratively adjusted by trial and error until the simulated and observed tidal levels were satisfactory [35]. In this study, the bottom roughness was adjusted to calibrate the amplitudes and phases at Taichung Harbor. The model calibration of the amplitudes and phases was conducted using measured data on the daily freshwater discharge at the Dadu Bridge in 2011. A five-constituent tide (*i.e.*, M_2, S_2, N_2, K_1 and O_1) was adopted in the model simulation as a forcing function at the

316

coastal sea boundaries (shown in Table 1). Because the amplitudes of fourth-diurnal, such as M_4 (first overtide of M_2 constituent) and MS_4 (a compound tide of M_2 and S_2), comparing to diurnal and semi-diurnal tides, were relatively small, a five-constituent tide was used to force the open boundaries only. The amplitudes and phases of these five tidal constituents were used to generate time-series water surface elevations along the open boundaries. The freshwater discharge inputs from Dadu Bridge in 2011 are shown in Figure 4. The maximum freshwater discharge reached 690 m³/s during the typhoon event.

The model simulation was run for one year in 2011. Harmonic analysis was performed on the time series of the model simulated water surface elevation at Taichung Harbor. The bottom roughness height was adjusted carefully, and the results are presented in Figure 5. The results show the comparison of the amplitudes and phases of harmonic constants between computed and observed tides. The differences between the computed and observed tidal constituents for amplitude and phase are in the range of 0.01–0.02 m and 0.45°–4.21°, respectively. The differences in amplitude and phase are quite small.

Table 1. The amplitudes and phases used for the model simulation at the coastal sea boundaries.

Constituent	Boundary at Point A		Boundary at Point B		Boundary at Point C		Boundary at Point D	
	Amplitude (m)	Phase (°)	Amplitude (m)	Phase (°)	Amplitude (m)	Phase (°)	Amplitude (m)	Phase (°)
M_2	1.82	266.06	1.88	266.91	1.60	272.39	1.57	267.23
S_2	0.51	14.45	0.53	16.34	0.44	26.29	0.43	20.55
N_2	0.25	28.89	0.26	29.77	0.22	36.02	0.21	31.15
K_1	0.27	161.62	0.29	160.20	0.29	167.21	0.28	167.16
O_1	0.21	279.53	0.22	277.24	0.23	283.55	0.22	284.03

Notes: boundaries at Points A, B, C and D are shown in Figure 3; M_2 is principal lunar semi-diurnal constituent; S_2 is principal solar semi-diurnal constituent; N_2 is larger lunar elliptic semi-diurnal constituent; K_1 is luni-solar declinational diurnal constituent; and O_1 is lunar declinational diurnal constituent.

Figure 4. Freshwater discharge inputs at the Dadu Bridge in 2011.

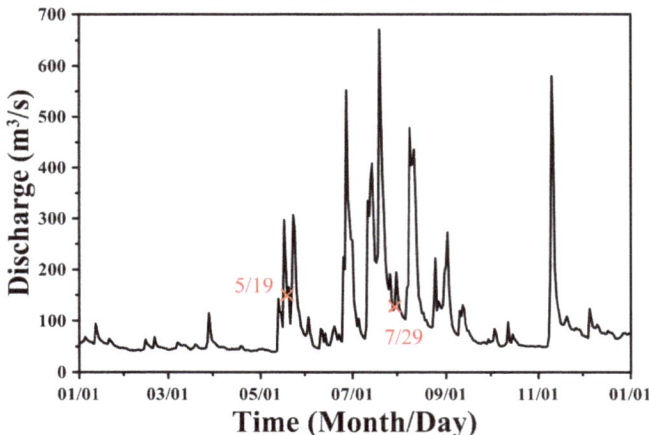

Figure 5. Comparisons of amplitude and phase of five major tidal harmonics computed with a three-dimensional model and obtained from tide measurements (**a**) amplitude; (**b**) phase.

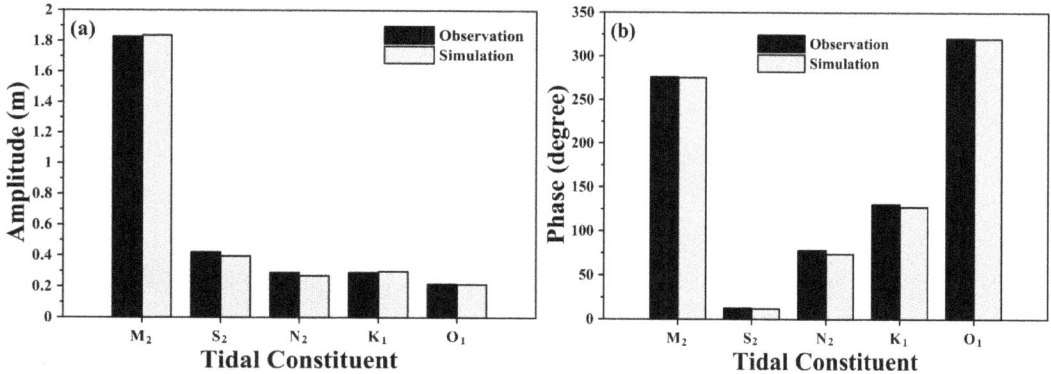

4.2. Verification of Water Surface Elevation

After calibrating the amplitudes and phases, the time-series data of observed water surface elevation were used to verify the model. Figure 6 presents the verified results for water surface elevations at Taichung Harbor station in May and July, 2011. The mean absolute errors of the differences between the measured hourly water levels and the computed water levels for 11–21 May and 22–31 July were 0.147 m and 0.157 m, respectively. The corresponding root-mean-square errors were 0.183 m and 0.193 m, respectively. These results demonstrate that the model can accurately predict the water surface elevation for varying river discharge input and tidal forcing at coastal sea boundaries. A constant bottom roughness height (z_0 = 0.01 cm) was adopted in the model for calibration and verification.

4.3. Calibration and Verification of Salinity Distribution

Salinity distributions reflect the combined results of all processes, including density circulation and mixing processes [36]. In the present study, the salinity distributions were measured *in situ* using conductivity-temperature-depth equipment at six locations in the Wu River estuary during the flood tide surveys. The salinities at four vertical layers of each station in the water column were measured and were then used for model calibration and verification. The salinities of open boundaries in the coastal sea were set to a constant value (*i.e.*, 35 psu). The upstream boundary at the Dadu Bridge was also specified with daily freshwater discharges, and the salinity was set to 0 psu. Figures 7 and 8 present the comparisons of measured and simulated salinity distributions on 19 May and 29 July, 2011, for model calibration and verification purposes, respectively. The freshwater discharges on 19 May and 29 July 2011, were 149.3 m³/s and 127.47 m³/s, respectively (shown in Figure 4). The results show that the model-computed salinity distributions agree well with the field observations. The root-mean-square errors for 19 May and 29 July, 2011, were 0.75 psu and 0.53 psu, respectively.

Figure 6. Comparison of model results and observed water surface elevation during the periods of (**a**) 11–21 May 2011 and (**b**) 22–31 July 2011 at the Taichung Harbor station.

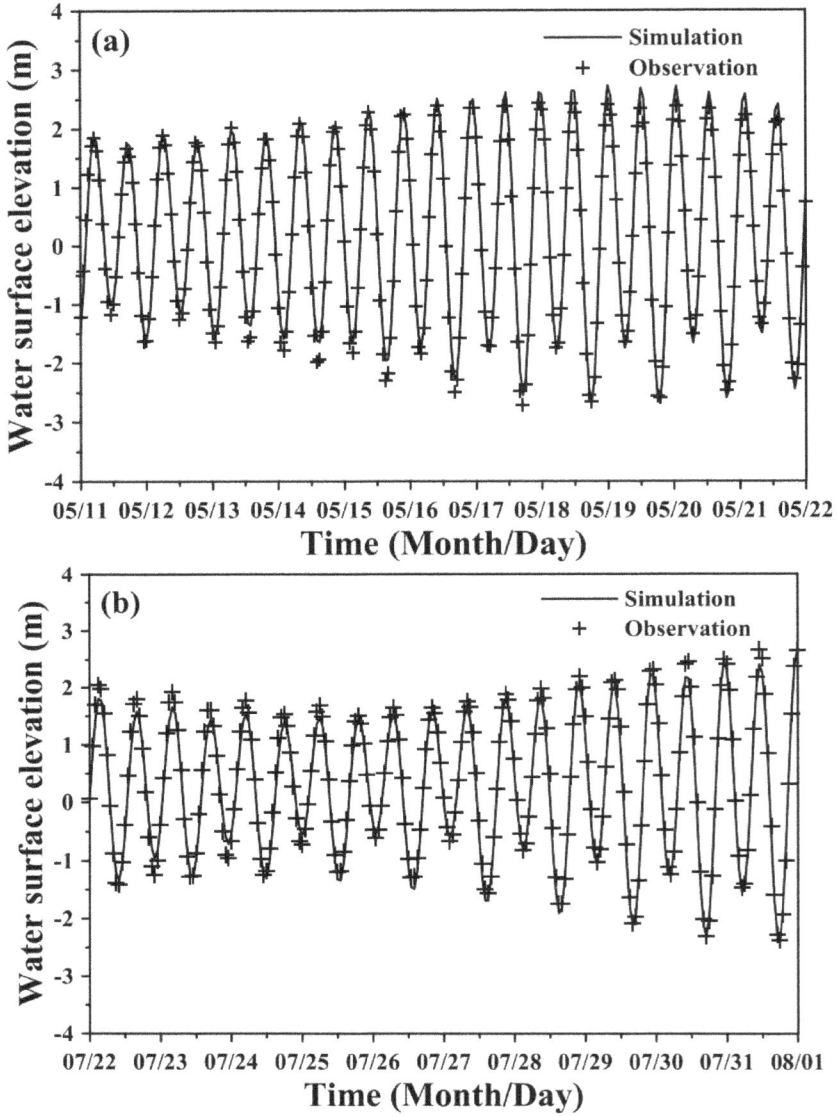

Figure 7. Comparison of salinity distribution along the Wu River estuary. (**a**) measurements and (**b**) model simulation on 19 May 2011, for model calibration.

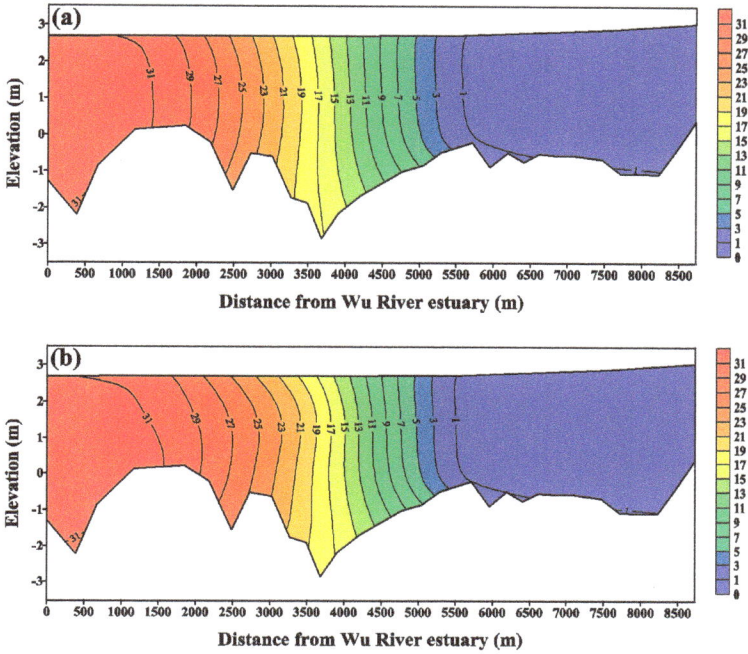

Figure 8. Comparison of salinity distribution along the Wu River estuary. (**a**) measurements and (**b**) model simulation on 29 July 2011 for model verification.

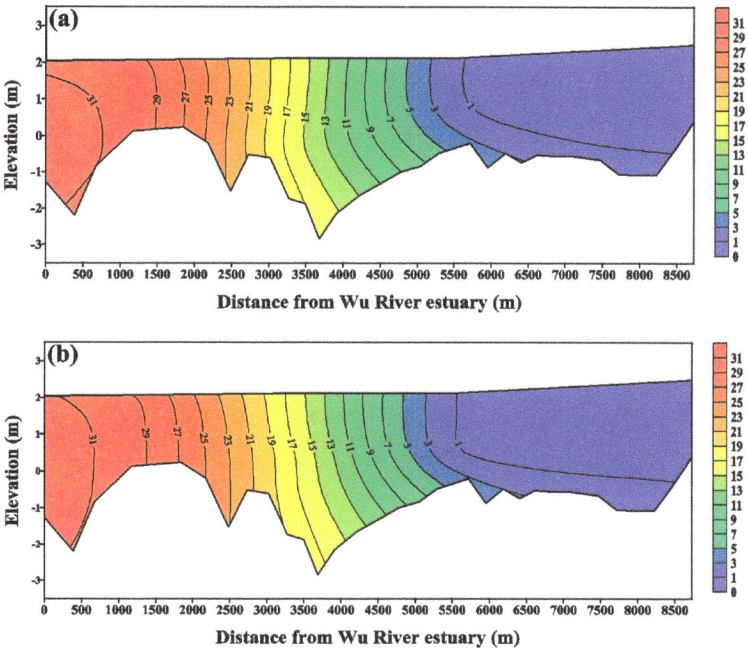

320

5. Results and Discussion

The validated three-dimensional hydrodynamic model was used to calculate the salinity distribution and transport time scale response to different discharges with sea level rise scenarios and without sea level rise (*i.e.*, the present condition) in the Wu River estuary. Figure 9 presents flow duration curves at the Dadu Bridge. The daily flow data from 1969 to 2011, collected by the Central Water Resources Bureau of Taiwan, were analyzed. The freshwater discharges with Q_{10} (the flow that is equaled or exceeded 10% of time) to Q_{90} flow conditions are listed in Table 2. For the cases of Q_{10} and Q_{90} flows, the discharges at the Dadu Bridge are 229.0 and 26.0 m^3/s, respectively. Five tidal constituents (M_2, S_2, N_2, K_1 and O_1) were specified to generate a time-series of water surface elevation as the open boundary conditions at the coastal sea for model simulation. A constant salinity of 35 psu at the open boundaries was used for model simulation. A future sea level rise of 38.27 cm in 2100 was used for the model simulation scenario.

Figure 9. Flow duration curve at Dadu Bridge.

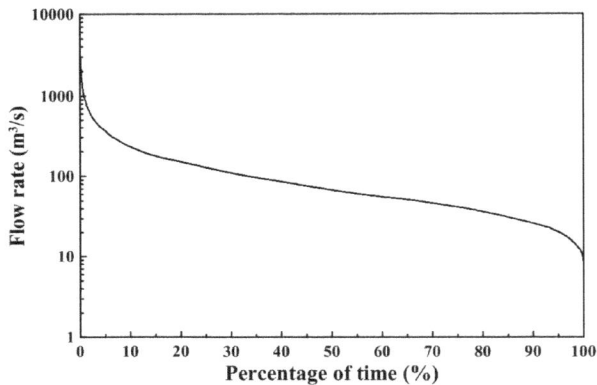

The influence of wind on estuarine circulation has been recognized for many years [37]. In a shallow estuary, the residence time can vary in response to variations in wind-induced flushing [38]. However, in the present study, the wind forcing was excluded in the model simulation for calculating transport time scales with sea level rise scenarios and without sea level rise (*i.e.*, the present condition).

Table 2. Freshwater discharge at upstream boundaries for the model simulation.

Freshwater discharge	Flow rate at Dadu Bridge (m^3/s)
Q_{10}	229.0
Q_{20}	149.0
Q_{30}	108.7
Q_{40}	85.6
Q_{50}	67.5
Q_{60}	55.5
Q_{70}	46.5
Q_{80}	36.5
Q_{90}	26.0

5.1. Sea Level Rise Effects on Salinity Distribution

To quantify the spatial and vertical variations in salinity, the vertical salinity profile along the Wu River estuary shows the detailed changes in the salinity structure with sea level rise. Figures 10 and 11 present the distributions of tidal-averaged salinity along the Wu River estuary under the Q_{10} and Q_{90} flows to represent the high and low flow conditions, respectively, for the present condition and the sea level rise scenario. It is clear that the salinity changes throughout the entire estuary. The limit of salt intrusion is represented by a 1 psu isohaline. The limits of salt intrusion are 3000 m and 6500 m for the present condition and the sea level rise scenario under Q_{10} flow conditions (Figure 10), while they are 5500 m and 8250 m for the present condition and the sea level rise scenario under Q_{90} flow conditions (Figure 11). These two figures indicate that sea level rise pushes the limit of salt intrusion farther upstream in the Wu River estuary. The intensified stratification results in stronger gravitation circulation, which raises the salt content by transporting more saline water into the estuary. However, the sea level rise did not change the tidal amplitude, but the water surface elevation increased in the sea level rise scenario. Moreover, the sea level rise extends to the tidal excursion farther upstream, 500 m and 900 m, respectively, under the Q_{10} and Q_{90} flow conditions (not shown in the figure).

Figure 10. Distribution of the tidal-averaged salinity along the Wu River estuary under the Q_{10} flow condition for (**a**) the present condition and (**b**) the sea level rise scenario. Note that the unit of salinity is psu (practical salinity unit).

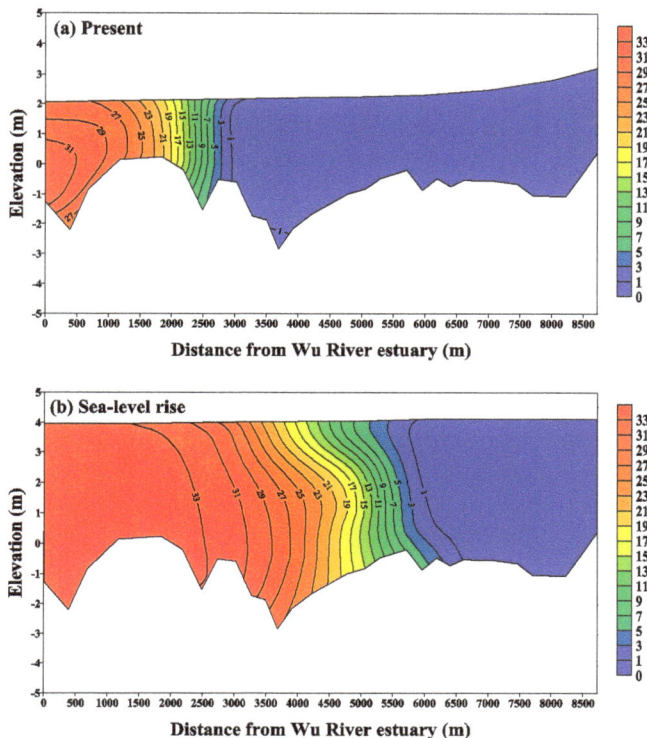

Figure 11. Distribution of the tidal-averaged salinity along the Wu River estuary under the Q_{90} flow condition for (**a**) the present condition and (**b**) the sea level rise scenario. Note that the unit of salinity is psu.

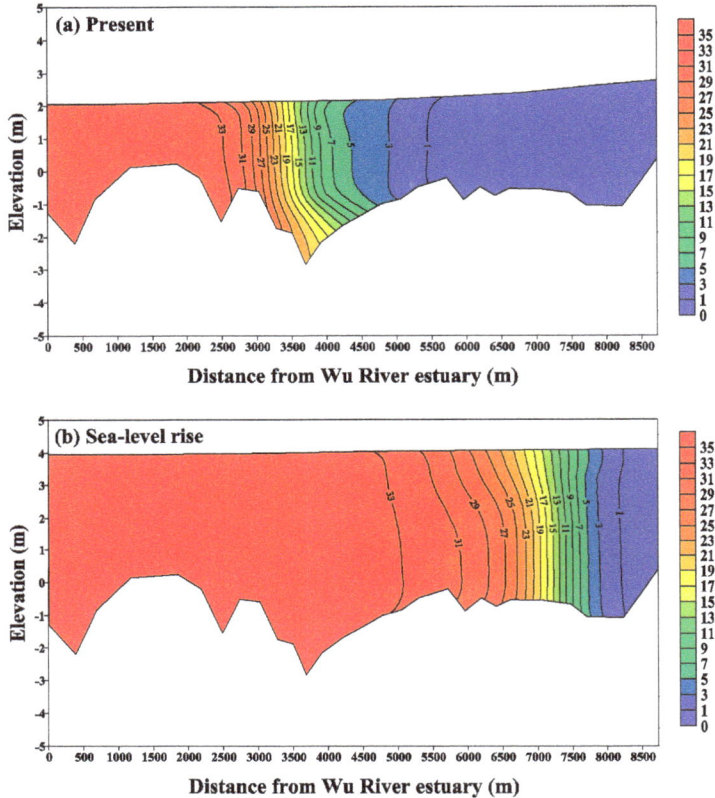

Hong and Shen [9] demonstrated that the mean salinity at the mouth and the water depth within Chesapeake Bay would increase with sea level rise. Bhuiyan and Dutta [8] also described that the sea level rise impact on salinity intrusion would be highly significant. In this study, the limit of salt intrusion will increase 3000 m and 2750 m under high and low flow conditions, respectively, for a 38.27 cm sea level rise. The maximum increased salinity reached 14.2 psu under the Q_{90} low flow. The increased salinity could cause socio-economic problems; the saline water would be unsuitable for drinking and industrial purposes. Salinity intrusion due to sea level rise would constrain the supply of water resources in the river.

5.2. Flushing Time in Response to Sea Level Rise

To calculate the flushing time in the estuary, different freshwater discharges shown in Table 2 were used to serve as the upstream boundary condition for the present condition and for the sea level rise scenario. The model simulated flushing time is plotted against river flow in Figure 12. The increase in river discharge is accompanied by a more rapid exchange of freshwater with the sea.

The volume of fresh water accumulated in the estuary increases to a lesser extent compared to the volume in the discharge. Thus, the flushing time decreases with increasing river discharge. Least squares regression fitting by the power law [39] was conducted to express the empirical function for the present condition and the sea level rise scenario:

$$T_f = 1030.54 \cdot Q^{-0.821}, \ R^2 = 0.99 \ \text{for the present condition} \tag{17}$$

$$T_f = 384.02 \cdot Q^{-0.599}, \ R^2 = 0.98 \ \text{for the sea level rise scenario} \tag{18}$$

where Q is the freshwater discharge. With a higher correlation value (R^2), the power law statistically fits the data better, especially in the low and the high flow ends. The power law reasonably shows physical characteristics between the freshwater fraction and freshwater flow.

Figure 12. Regression between flushing time and freshwater input for the present condition and the sea level rise scenario.

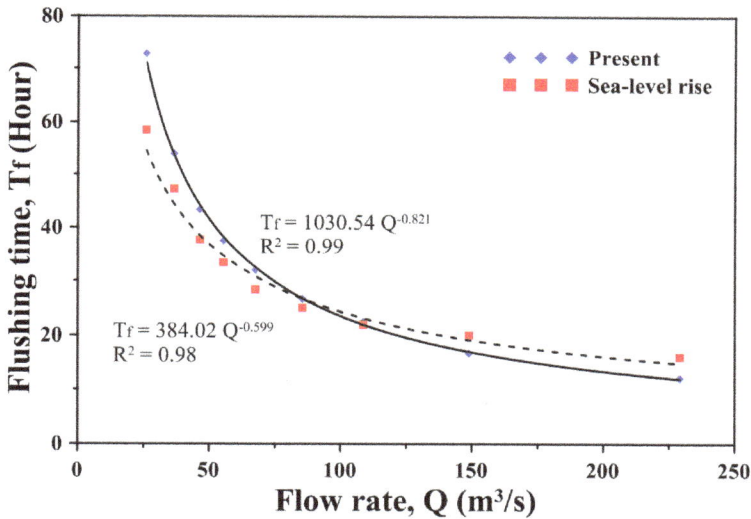

The flushing time is between 12.09 and 72.84 h under the present condition, while it is between 16.04 and 58.57 h under the sea level rise scenario. The results also indicate that the flushing time for high flow under the present condition is lower compared to the sea level rise scenario, while the flushing time for low flow under the present condition is higher compared to the sea level rise scenario. The freshwater volume thus increases under the sea level rise during high flow, and it decreases during low flow. Huang [39] applied a three-dimensional model to estimate the distributions of salinity and the freshwater fractions for flushing time estimation. He found that for the seven-day averaged flow ranging from 10 m³/s to 50 m³/s for a small estuary of North Bay, Florida, corresponding flushing time varies from 3.7 days to 1.8 days. The flushing time in the estuary of North Bay was similar to that in the Wu River estuary.

5.3. Residence Time in Response to Sea Level Rise

Passive tracers are used to simulate the material transport coming from the main river sources at the Dadu Bridge. Changing water levels and the propagation of tidal waves also result in changes in the residence time of water bodies and water constituents within the estuary and in changes in transport time through the estuary towards the sea.

The validated model was applied to explore the impact of sea level rise on residence time in the estuary. We calculated the residence time of the entire Wu River estuary under different freshwater discharge scenarios. After instantly releasing tracers throughout the entire Wu River estuary, the residence time corresponded to the time when the average tracer concentration reached its e-folding value (*i.e.*, e^{-1} value). Model results reveal that the residence time decreases as the freshwater input increases for the present condition and the sea level rise scenario (shown in Figure 13). Finding a general regression relationship between the residence time and the freshwater input would be helpful in understanding the physical and hydrological processes in the estuary. Huang *et al.* [40] conducted regression analyses between estuarine residence time and freshwater input using a power-law function in Little Manatee River, Florida. The authors found that regression by the power law provided a better fit compared to an exponential function. A regression of the residence time (T_r) *versus* freshwater input (Q) was performed and indicated an excellent correlation (R^2) through the power law function:

$$T_r = 207.18 \cdot Q^{-0.541}, \ R^2 = 0.99 \ \text{for the present condition} \tag{19}$$

$$T_r = 143.36 \cdot Q^{-0.385}, \ R^2 = 0.99 \ \text{for the sea level rise scenario} \tag{20}$$

The residence time is between 10.51 and 34.23 h under the present condition, while it is between 17.11 and 38.92 h under the sea level rise scenario. The residence time of the entire Wu River estuary increased 4.7 to 6.6 h based on different freshwater inputs due to sea level rise. The prolonged residence time will result in the deterioration of water quality and induce the limited application of water resources.

Hong and Shen [9] estimated that the residence time could increase five to 20 days in response to different sea level rise scenarios in Chesapeake Bay. The increase of residence time response to sea level rise in the Wu River estuary is smaller than that in Chesapeake Bay, because the entire estuarine system in Chesapeake Bay is much larger than the Wu River estuary.

If the sea level rise rate is changed, the salinity intrusion and transport time scales would be changed. The increase in the sea level rise rate may extend the limit of salt intrusion farther upstream and increase the residence time in the estuary. In a future study, the wind forcing and seasonal freshwater discharge input from the Dadu Bridge can be considered in the model simulations to comprehend how the salinity intrusion and transport time scales respond to these factors.

Figure 13. Regression between residence time and freshwater input for the present condition and the sea level rise scenario.

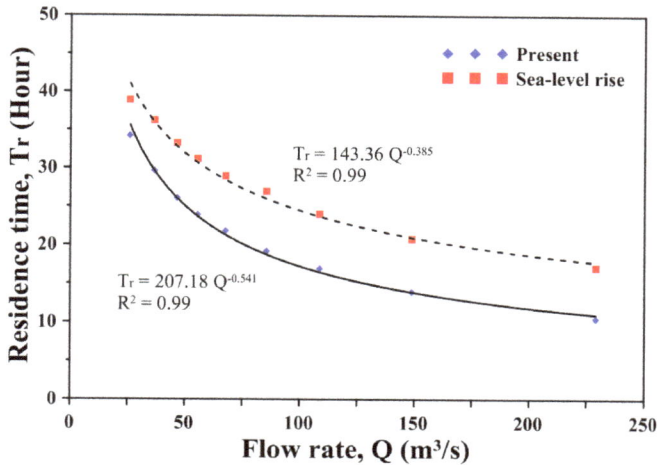

6. Conclusions

A three-dimensional hydrodynamic and salt transport model, SELFE, was established to simulate the hydrodynamics and salinity distributions in the Wu River estuary and adjacent coastal sea in northern Taiwan. The model was calibrated and verified using observational amplitudes and phases, water surface elevations and salinity distributions in 2011. The model simulation results agree well with the field observations.

The validated model was used to perform a series of numerical experiments to identify the potential impacts of future sea level rise on salinity intrusion and transport time scales, including flushing time and residence time, in the Wu River estuary of central Taiwan. The model results indicate that salinity intrusion moves farther upstream by 2750 m and 3500 m under Q_{90} and Q_{10} flow conditions, respectively, due to sea level rise. The flushing time is between 12.09 and 72.84 h for the present condition, and it is between 16.04 and 58.57 h for the sea level rise scenario. We found that the flushing time for high flow under the present condition is lower compared to the sea level rise scenario, while the flushing time for low flow under the present condition is higher compared to the sea level rise scenario. The residence time of the entire Wu River estuary increased by 23.7 h and 21.8 h for high and low flows, respectively, during the sea level rise scenario. We found that the climate change (*i.e.*, sea level rise) scenario implies not only a change in salt intrusion, but also an increase in the residence time. Sea level rise would alter the location of the river estuary, thereby causing a greater change in fish habitat and breeding ground location. Fishes breed in estuarine systems and develop in brackish waters, which is where fresh water and salt water mix. Sea level rise would move this interface backward, changing the habitat of fishing communities in the estuarine system. The increases in transport time scales (*i.e.*, residence time) due to sea level rise would prolong the transport of dissolved substances in the estuary, resulting in the deterioration of water quality.

Acknowledgments

The project under which this study was conducted is supported by the National Science Council, Taiwan, under grant no. NSC 101-2625-M-239-001. The authors would like to express their thanks to the Taiwan Water Resources Agency for providing the observational data. Appreciation and thanks are also given to the anonymous reviewers for their constructive comments and suggestions to improve this paper.

Conflicts of Interest

The authors declare no conflict of interest.

References

1. Yu, Y.F.; Yu, Y.X.; Zuo, J.C.; Wan, Z.W.; Chen, Z.Y. Effect of sea level variation on tidal characteristic values for the East China Sea. *China Ocean Eng.* **2003**, *17*, 369–382.
2. IPCC. Climate Change 2007: The Physical Science Basis. In *Contribution of Working Group 1 to the Fourth Assessment Report of the Intergovernmental Panel on Climate Change*; Solomon, S., Qin, S., Manning, M., Chen, Z., Marquis, M., Averyt, K.B., Tignor, M., Miller, H.L., Eds.; Cambridge University Press: Cambridge, UK; New York, NY, USA, 2007.
3. Hsu, H.H.; Chen, C.T. Observed and projected climate change in Taiwan. *Meteorol. Atmos. Phys.* **2002**, *79*, 87–104.
4. Yu, P.S.; Yang, T.C.; Kuo, C.C. Evaluating long-term trends in annual and seasonal precipitation in Taiwan. *Water Resour. Manag.* **2006**, *20*, 1007–1023.
5. Chiu, M.C. Relationship of Stream Insects with Flooding and Dippers in Wuling Area (in Chinese). Master Thesis, National Chung Hsing University, Taichung, Taiwan, 2009.
6. Poff, N.L.; Brinson, M.M.; Day, J.W., Jr. *Aquatic Ecosystems and Global Climate Change*; Pew Center on Global Change: Arlington, VA, USA, 2002; p. 45.
7. Hilton, T.W.; Najjar, R.G.; Zhong, L.; Li, M. Is there a signal of sea level rise in Chesapeake Bay salinity? *J. Geophys. Res.* **2008**, *113*, doi:10.1029/2007JC004247.
8. Bhuiyan, M.J.A.N.; Dutta, D. Assessing impacts of sea level rise on river salinity in the Gorai river network, Bangladesh. *Estuar. Coast. Shelf Sci.* **2012**, *96*, 219–227.
9. Hong, B.; Shen, J. Responses of estuarine salinity and transport processes to potential future sea level rise in the Chesapeake Bay. *Estuar. Coast. Shelf Sci.* **2012**, *104–105*, 33–45.
10. Lucas, L.V. Implications of Estuarine Transport for Water Quality. In *Contemporary Issues in Estuarine Physics*; Valle-Levinson, A., Ed.; Cambridge University Press: Cambridge, UK, 2010; pp. 273–306.
11. Hull, C.H.J.; Tortoriello, R. *Sea Level Trend and Salinity in the Delaware Estuary*; Staff Report; Delaware Basin Commission: West Trenton, NJ, USA, 1979.
12. Grabemann, H.; Grabemann, I.; Herbers, D.; Muller, A. Effects of a specific climate scenario on the hydrograph and transport of conservative substances in the Weser estuary, Germany: A case study. *Clim. Res.* **2001**, *18*, 77–87.

13. Chua, V.P.; Fringer, O.B.; Monismith, S.G. Influence of sea level rise on salinity in San Francisco Bay. 2011, unpublished work.

14. Rice, K.C.; Hong, B.; Shen, J. Assessment of salinity intrusion in the James and Chickahominy Rivers as a result of simulated sea level rise in Chesapeake Bay, East Coast, USA. *J. Environ. Manag.* **2012**, *111*, 61–69.

15. Chen, W.B.; Liu, W.C.; Wu, C.Y. Coupling of a one-dimensional river routing model and a three-dimensional ocean model to predict overbank flows in a complex river-ocean system. *Appl. Math. Model.* **2013**, *37*, 6163–6176.

16. Church, J.A.; White, N.J. Sea level rise from the late 19th to the early 21st century. *Surv. Geophys.* **2011**, *32*, 585–602.

17. Snay, R.; Cline, M.; Dillinger, W.; Foote, R.; Hilla, S.; Kass, W.; Ray, J.; Rohde, J.; Sella, G.; Soler, T. Using global positions system-derived crustal velocities to estimate rates of absolute sea level change from North American tidal gauge records. *J. Geophys. Res.* **2007**, *112*, doi:10.1029/2006JB004606.

18. Nicholls, R.J.; Cazenave, A. Sea level rise and its impact on coastal zone. *Science* **2010**, *328*, 1517–1520.

19. Tseng, Y.H.; Breaker, L.C.; Cheng, T.Y. Sea level variations in the regional seas around Taiwan. *J. Oceanogr.* **2010**, *66*, 27–39.

20. Huang, C.J.; Hsu, T.W.; Wu, L.C. *Application of Tide-Gauge and Satellite Altimetry Data to Estimate Sea Level Rise*; Report to Water Resources Agency: Taipei, Taiwan, 2009.

21. Zhang, Y.L.; Baptista, A.M. SELFE: A semi-implicit Eulerian-Lagrangian finite-element model for cross-scale ocean circulation. *Ocean Model.* **2008**, *21*, 71–96.

22. Zeng, Z.; Zhao, M.; Dickinson, R.E. Intercomparison of bulk aerodynamic algorithms for the computation of sea surface fluxes using TOGA COARE and TAO data. *J. Clim.* **1998**, *11*, 2628–2644.

23. Umlauf, L.; Buchard, H. A. generic length-scale equation for geophysical turbulence models. *J. Mar. Res.* **2003**, *61*, 235–265.

24. Lauff, G.E. Lyered Sediments of Tidal Flats, Beaches, and Shelf Bottoms of the North Sea. In *American Association for the Advancement of Science Publication No. 83*; American Association for the Advancement of Science: Washington, DC, USA, 1967.

25. Dyer, K.R. *Estuaries: A Physical Introduction*, 2nd ed.; John Wiley: London, UK, 1977; p. 195.

26. Liu, W.C.; Hsu, M.H.; Kuo, A.Y.; Kuo, J.T. The influence of river discharge on salinity intrusion in the Tanshui Estuary, Taiwan. *J. Coast. Res.* **2001**, *17*, 544–552.

27. Huang, W.; Spaulding, M. Modelling residence-time response to freshwater input in Apalachicola Bay, Florida, USA. *Hydrol. Process.* **2002**, *16*, 3051–3064.

28. De Brye, B.; de Brauwere, A.; Gourge, O.; Delhez, E.J.M.; Deleersnijder, E. Water renewal timescales in the Scheldt Estuary. *J. Mar. Syst.* **2012**, *94*, 74–86.

29. Zimmerman, J.T.F. Mixing and flushing of tidal embayment in the western Dutch Wadden Sea. Part I: Description of salinity and calculation of mixing time scales. *Neth. J. Sea Res.* **1976**, *10*, 149–191.

30. Takeoka, H. Fundamental concepts of exchange and transport time scales in a coastal sea. *Cont. Shelf Res.* **1984**, *3*, 311–326.

31. Liu, W.C.; Chen, W.B.; Kuo, J.T.; Wu, C. Numerical determination of residual time and age in a partially mixed estuary using three-dimensional hydrodynamic model. *Cont. Shelf Res.* **2008**, *28*, 1068–1088.

32. Zhang, W.G.; Wilkin, J.L.; Schofield, O.M.E. Simulation of water age and residence time in the New York Bight. *J. Phys. Oceanogr.* **2010**, *40*, 965–982.

33. De Brauwere, A.; de Brye, B.; Blaise, S.; Deleersnijder, E. Residence time, exposure time and connectivity in the Scheldt Estuary. *J. Mar. Syst.* **2011**, *84*, 85–95.

34. Kenov, I.A.; Garcia, A.C.; Neves, R. Residence time of water in the Mondego estuary (Portugal). *Estuar. Coast. Shelf Sci.* **2012**, *106*, 13–22.

35. Shi, J.; Li, G.; Wang, P. Anthropogenic influences on the tidal prism and water exchange in Jiaozhou Bay, Qingdao, China. *J. Coast. Res.* **2011**, *27*, 57–72.

36. Hsu, M.H.; Kuo, A.Y.; Kuo, J.T.; Liu, W.C. Procedure to calibrate and verify numerical models of estuarine hydrodynamics. *J. Hydraul. Eng. ASCE* **1999**, *125*, 166–182.

37. Officer, C.B. *Physical Oceanography of Estuaries (and Associated Coastal Waters)*; Wiley: New York, NY, USA, 1976.

38. Geyer, W.R. Influence of wind on dynamics and flushing of shallow estuaries. *Estuar. Coast. Shelf Sci.* **1997**, *44*, 713–722.

39. Huang, W. Hydrodynamic modeling of flushing time in small estuary of North Bay, Florida, USA. *Estuar. Coast. Shelf Sci.* **2007**, *74*, 722–731.

40. Huang, W.; Liu, X.; Chen, X.; Flannery, M.S. Critical flow for water management in a shallow tidal river based on estuarine residence time. *Water Resour. Manag.* **2011**, *25*, 2367–2385.

MDPI AG
Klybeckstrasse 64
4057 Basel, Switzerland
Tel. +41 61 683 77 34
Fax +41 61 302 89 18
http://www.mdpi.com/

Water Editorial Office
E-mail: water@mdpi.com
http://www.mdpi.com/journal/water

www.ingramcontent.com/pod-product-compliance
Lightning Source LLC
Chambersburg PA
CBHW051924190326
41458CB00026B/6402